AF361794

Regulation of Central Carbon and Amino Acid Metabolism in Plants

Regulation of Central Carbon and Amino Acid Metabolism in Plants

Editors

Stefan Timm
Stéphanie Arrivault

MDPI • Basel • Beijing • Wuhan • Barcelona • Belgrade • Manchester • Tokyo • Cluj • Tianjin

Editors
Stefan Timm
Plant Physiology Department
Germany

Stéphanie Arrivault
Max Planck Institute of Molecular Plant Physiology
Germany

Editorial Office
MDPI
St. Alban-Anlage 66
4052 Basel, Switzerland

This is a reprint of articles from the Special Issue published online in the open access journal *Plants* (ISSN 2223-7747) (available at: https://www.mdpi.com/journal/plants/special_issues/carbon_amino_acid).

For citation purposes, cite each article independently as indicated on the article page online and as indicated below:

LastName, A.A.; LastName, B.B.; LastName, C.C. Article Title. *Journal Name* **Year**, *Volume Number*, Page Range.

ISBN 978-3-0365-1104-7 (Hbk)
ISBN 978-3-0365-1105-4 (PDF)

Contents

About the Editors

Stefan Timm has conducted seminal research on the identification of unknown components of plant photorespiration, including the genetic establishment of a cytosolic bypass, and how photorespiration is imbedded within the central plant carbon metabolism. During recent years, his research has mainly focused on the regulatory impact of photorespiratory metabolites on photosynthetic carbon assimilation, particularly the Calvin Benson cycle. Moreover, he was involved in pioneering the concept that increases the activities of specific photorespiratory enzymes, retrieving the potential to enhance the flux through photorespiration, thereby stimulating photosynthesis and growth. The effects of altered metabolite levels on transcriptional and post-translational regulation are also included in his interests.

Stéphanie Arrivault's research is mainly focused on the elucidation of carbon fluxes in C_3 and C_4 photosynthesis by the use of dynamic $^{13}CO_2$ labelling coupled with LC-MS/MS quantification.

Preface to "Regulation of Central Carbon and Amino Acid Metabolism in Plants"

Over the past few decades, a considerable amount of effort has been dedicated to understanding plant primary metabolism. While the biochemistry and the underlying genetics of central carbon and nitrogen metabolism have been thoroughly studied, there is still a lack of knowledge on how these metabolic branches are regulated, in addition to how they regulate and interact with each other. Improving our current understanding of such regulatory loops is of particular interest given that all oxygenic phototrophs are frequently exposed to environmental changes, including periods of unfavorable conditions that distinctly lower plant growth and yield. Understanding how adjustments of metabolism towards a fluctuating environment are achieved in both short- and long-term timescales will also facilitate genetic engineering approaches. One major goal of such attempts is to produce more robust plant varieties that are able to sustain high photosynthetic efficiencies and yields during persistent phases of abiotic stresses.

This Special Issue of plants aims to highlight the metabolic acclimation and signaling mechanisms of plant central carbon and nitrogen metabolism towards environmental changes, particularly involving alterations in CO_2 and O_2 concentrations, light availability and intensity, as well as fluctuations in temperature and water supply during different stages of plant development. Thus, the major focus will be on the acclimation and the regulatory interplay that—among others—involve the operation and interaction of photosynthesis, photorespiration and respiration.

Stefan Timm, Stéphanie Arrivault
Editors

Editorial

Regulation of Central Carbon and Amino Acid Metabolism in Plants

Stefan Timm [1,*] and Stéphanie Arrivault [2]

[1] Plant Physiology Department, University of Rostock, Albert-Einstein-Straße 3, D-18051 Rostock, Germany
[2] Max Planck Institute of Molecular Plant Physiology, Wissenschaftspark Golm, Am Mühlenberg 1, D-14476 Potsdam-Golm, Germany; Arrivault@mpimp-golm.mpg.de
* Correspondence: stefan.timm@uni-rostock.de; Tel.: +49-(0)381-498-6115; Fax: +49-(0)381-498-6112

Citation: Timm, S.; Arrivault, S. Regulation of Central Carbon and Amino Acid Metabolism in Plants. *Plants* **2021**, *10*, 430. https://doi.org/10.3390/plants10030430

Received: 15 February 2021
Accepted: 22 February 2021
Published: 24 February 2021

1. Introduction

Fluctuations in the prevailing environmental conditions, including light availability and intensity, CO_2/O_2 ratio, temperature, and nutrient or water supply, require rapid metabolic switches to maintain proper metabolism. To achieve this, a multitude of regulatory mechanisms are needed to communicate between the various metabolic branches to adjust fluxes through all routes related to central carbon and nitrogen metabolism. Despite considerable effort in understanding the genetics and biochemistry of plant primary metabolism, insights into the underlying mechanisms governing such acclimations and the involved regulatory circuits are still fragmentary. A better knowledge on those aspects would certainly help to engineer crop plants to maintain high yields under fluctuating environmental conditions.

Over the past decades, tremendous progress has been made to unravel the extraordinary complexity of plant primary metabolism from a biochemical and genetic point of view. Special attention has been paid to carbon and nitrogen assimilation and their subsequent metabolism [1–3]. Biochemical reactions of canonical pathways, such as the Calvin–Benson cycle, the oxidative pentose phosphate pathway, glycolysis, the tricarboxylic acid (TCA) cycle, and photorespiration have been thoroughly investigated [4–8]. However, given that many of those pathways run simultaneously, there are gaps regarding how these pathways are coordinated with each other, intertwined, and regulated. In this special issue, we invited authors to contribute to new studies encompassing the field of regulation of pathways that drive plant primary metabolism in oxygenic phototrophs, particularly focusing on photorespiration, photosynthesis, and glycolysis, as well as metabolite regulation, signaling and transport. In addition, natural acclimation strategies to a changing environment are included.

2. Photorespiration—A Key Driver for Adaptation to Molecular Oxygen and Abiotic Stresses

Photorespiration has attracted major interest in plant research over the past decades for different reasons. First, photorespiration is essential in all oxygenic phototrophs, as it degrades and recycles 2-phosphoglycolate (2-PG), which is formed in high amounts in the presence of oxygen during illumination through oxygenation of ribulose-1,5-bisphosphate via Rubisco [9]. Second, photorespiration has been identified as a key target to increase crop yield since the pathway itself releases CO_2 during 2-PG recycling. Several strategies, such as synthetic bypasses or upregulation of enzyme activities, were successfully developed to manipulate photorespiratory flux, with promising outcomes in terms of stimulated photosynthetic carbon fixation [10,11]. Third, photorespiration is orchestrated in four subcellular compartments, including chloroplasts, peroxisomes, mitochondria, and the cytoplasm. Therefore, this pathway is an interesting example to study metabolite transport across membranes and interactions of the different subcellular compartments [12]. Fourth, the photorespiratory pathway also represents an example to study evolutionary aspects of photosynthetic metabolism [13]. Fifth, several pathway intermediates were shown to display a regulatory impact on the pathway itself or on other metabolic branches [11,14]. Hence, manipulation of the photorespiratory flux can be used for applied purposes and for basic research to shed more light on the communication between different metabolic branches under a changing environment [11].

In this issue, three studies tackle different aspects of research on photorespiration. Given the strong similarities of the cyanobacterial and plant photorespiratory pathways, it is assumed that photorespiration co-evolved with oxygenic photosynthesis in cyanobacteria and was endosymbiotically conveyed in eukaryotic algal lineages up to higher land plants [13]. This hypothesis was further confirmed by in silico analysis, but these data also revealed a dual origin of photorespiratory enzymes [15]. In particular, a strong controversy exists regarding the evolutionary origin of glycolate oxidase (GOX), predicted to originate from different sources [16]. However, in this issue, Kern and colleagues re-analyzed GOX evolution via a combination of phylogenetic and biochemical analyses using broad taxon sampling. These analyses strongly support the conclusion that GOX in higher plants evolved from a cyanobacterial ancestor protein [17]. A second study focused on the potential regulation of GOX activity though protein phosphorylation. Using site-directed mutagenesis and enzymatic measurements, Jossier et al. provided evidence that phosphorylation of GOX contributes to the regulation of enzymatic activities [18]. Finally, Timm et al. analyzed the potential for upregulation of photorespiratory enzymes to contribute to abiotic stress tolerance. These authors showed that faster degradation of 2-PG via increased phosphoglycolate phosphatase (PGLP) alleviated negative feedback of 2-PG on carbon-metabolizing reactions. In turn, photosynthesis was shown to be less O_2-inhibited, indicating that it can operate more efficiently under unfavorable environmental conditions [19].

3. Maintenance of Growth and Photosynthesis under Fluctuating Conditions

In addition to evolutionary adaptations, the photosynthetic process also needs to react rapidly to various changes in the prevailing environmental conditions. In this issue, two research papers and one perspective paper dealing with these topics are presented. A key adaptation to declining CO_2 concentrations during evolution was the establishment of inorganic carbon-concentrating mechanisms, including crassulacean acid metabolism (CAM) and C_4 photosynthesis [20]. Since both CAM and C_4 photosynthesis are present in *Portulaca grandiflora* leaves, and are simultaneously active, Guralnick et al. tested the hypothesis that both photosynthetic types are already developed and active in cotyledons [21]. Indeed, the authors provided evidence that both pathways are present and possibly run already at the early stages of leaf development. However, they also hypothesized that the CAM pathway is considerably slower compared to the C_4 cycle [21]. The effects of short-term acclimation to changes in environmental conditions were analyzed using a collection of 36 randomly chosen Arabidopsis accessions. The study of Kaiser et al. demonstrated a large trait variation in growth and photosynthesis within this collection, especially under fluctuating conditions [22]. It seems likely to conclude that using such natural variation represents a valuable starting point to breed crop plants that are more robust under harsh environmental fluctuations. Finally, Walker et al. provided an interesting perspective on the flexibility of photosynthesis, in particular how the process can adapt and maintain high efficiency in a fluctuating environment [23].

4. Metabolite Regulation, Signaling, and Transport

One research paper and two review papers discuss different regulatory aspects in central carbon metabolism. Glycolysis is one of the major catabolic pathways for the breakdown of carbohydrates down to the TCA cycle to supply energy for different processes within the cell. A key enzyme of this pathway is pyruvate kinase, which transfers phosphate from phosphoenolpyruvate to ADP for ATP synthesis. In the study of Wulfert et al., five cytosolic pyruvate kinase isoforms were thoroughly studied. The authors provided strong evidence that these enzymes undergo several levels of regulation in order to adjust carbon flux through the glycolytic pathway. This includes differential transcription, allosteric metabolic regulation, and formation of subcomplexes among several isoforms [24]. With regard to metabolite regulation, Rosado-Souza et al. provided a comprehensive overview on the potential of ascorbate (vitamin C) and thiamine (vitamin B1) as metabolite signals. The authors summarized recent knowledge in the field and discussed that both molecules are of major importance for the communication between different metabolic branches in cellular organelles during acclimation processes [25]. Similarly, Toleco and colleagues focused on metabolite transport though membranes in cells of higher plants. In more detail, mitochondrial carriers were highlighted as gatekeepers, controlling carbon influx and efflux to regulate central carbon metabolism. A particular focus was on the interconnection of the cytoplasm with mitochondria. These compartments need to communicate and exchange metabolites to support the different flux modes of the TCA cycle, in particular through exchange of organic acids, or drive oxidative phosphorylation [26].

5. Conclusions and Outlook

The simultaneous operation and regulation of the different pathways involved in central carbon and nitrogen metabolism are highly complex. This special issue aims to provide a useful extension of the existing knowledge in the field and, hopefully, inspire new research to further develop this field including new strategies and projects. Such work would not only ultimately help to further increase our current understanding on how metabolism works and is regulated in different subcellular organelles, but will also contribute to the overall aim to breed better crops, showing high productivity in a changing environment.

Author Contributions: Writing—original draft preparation, S.T.; review and editing, S.A. All authors read and agreed to the published version of the manuscript.

Funding: This research received no external funding.

Acknowledgments: We gratefully acknowledge the excellent support from the *Plants* editorial office, especially received from Sylvia Guo. We also thank all researchers who contributed papers to the special issue.

Conflicts of Interest: The authors declare no conflict of interest.

References

1. Stitt, M.; Lunn, J.; Usadel, B. Arabidopsis and primary photosynthetic metabolism—More than the icing on the cake. *Plant J.* **2010**, *61*, 1067–1091. [CrossRef]
2. Lawlor, D.W. Carbon and nitrogen assimilation in relation to yield: Mechanisms are the key to understanding production systems. *J. Exp. Botony* **2002**, *53*, 773–787. [CrossRef]
3. Nunes-Nesi, A.; Fernie, A.R.; Stitt, M. Metabolic and Signaling Aspects Underpinning the Regulation of Plant Carbon Nitrogen Interactions. *Mol. Plant* **2010**, *3*, 973–996. [CrossRef]
4. Raines, C.A. The Calvin cycle revisited. *Photosynth. Res.* **2003**, *75*, 1–10. [CrossRef] [PubMed]
5. Nunes-Nesi, A.; Araújo, W.L.; Obata, T.; Fernie, A.R. Regulation of the mitochondrial tricarboxylic acid cycle. *Curr. Opin. Plant Biol.* **2013**, *16*, 335–343. [CrossRef]
6. Plaxton, W.C. The organization and regulation of plant glycolysis. *Annu. Rev. Plant Biol.* **1996**, *47*, 185–214. [CrossRef] [PubMed]
7. Nicholas, J.K.; von Schaewen, A. The oxidative pentose phosphate pathway: Structure and organization. *Curr. Opin. Plant Biol.* **2003**, *6*, 236–246.
8. Bauwe, H.; Hagemann, M.; Fernie, A.R. Photorespiration: Players, partners and origin. *Trends Plant Sci.* **2010**, *15*, 330–336. [CrossRef] [PubMed]
9. Bauwe, H.; Hagemann, M.; Kern, R.; Timm, S. Photorespiration has a dual origin and manifold links to central metabolism. *Curr. Opin. Plant Biol.* **2012**, *15*, 269–275. [CrossRef] [PubMed]
10. South, P.F.; Cavanagh, A.P.; Lopez-Calcagno, P.E.; Raines, C.A.; Ort, D.R. Optimizing photorespiration for improved crop productivity. *J. Integr. Plant Biol.* **2018**, *60*, 1217–1230. [CrossRef]
11. Timm, S.; Hagemann, M. Photorespiration—how is it regulated and how does it regulate overall plant metabolism? *J. Exp. Bot.* **2020**, *71*, 3955–3965. [CrossRef]

12. Eisenhut, M.; Pick, T.R.; Bordych, C.; Weber, A.P.M. Towards closing the remaining gaps in photorespiration—The essential but unexplored role of transport proteins. *Plant Biol.* **2012**, *15*, 676–685. [CrossRef] [PubMed]
13. Eisenhut, M.; Ruth, W.; Haimovich, M.; Bauwe, H.; Kaplan, A.; Hagemann, M. The photorespiratory glycolate metabolism is essential for cyanobacteria and might have been conveyed endosymbiontically to plants. *Proc. Natl. Acad. Sci. USA* **2008**, *105*, 17199–17204. [CrossRef]
14. Flügel, F.; Timm, S.; Arrivault, S.; Florian, A.; Stitt, M.; Fernie, A.R.; Bauwe, H. The Photorespiratory Metabolite 2-Phosphoglycolate Regulates Photosynthesis and Starch Accumulation in Arabidopsis. *Plant Cell* **2017**, *29*, 2537–2551. [CrossRef] [PubMed]
15. Kern, R.; Bauwe, H.; Hagemann, M. Evolution of enzymes involved in the photorespiratory 2-phosphoglycolate cycle from cyanobacteria via algae toward plants. *Photosynth. Res.* **2011**, *109*, 103–114. [CrossRef]
16. Esser, C.; Kuhn, A.; Groth, G.; Lercher, M.J.; Maurino, V.G. Plant and Animal Glycolate Oxidases Have a Common Eukaryotic Ancestor and Convergently Duplicated to Evolve Long-Chain 2-Hydroxy Acid Oxidases. *Mol. Biol. Evol.* **2014**, *31*, 1089–1101. [CrossRef]
17. Kern, R.; Facchinelli, F.; Delwiche, C.; Weber, A.P.M.; Bauwe, H.; Hagemann, M. Evolution of Photorespiratory Glycolate Oxidase among Archaeplastida. *Plants* **2020**, *9*, 106. [CrossRef]
18. Jossier, M.; Liu, Y.; Massot, S.; Hodges, M. Enzymatic Properties of Recombinant Phospho-Mimetic Photorespiratory Glycolate Oxidases from Arabidopsis thaliana and Zea mays. *Plants* **2019**, *9*, 27. [CrossRef]
19. Timm, S.; Woitschach, F.; Heise, C.; Hagemann, M.; Bauwe, H. Faster Removal of 2-Phosphoglycolate through Photorespiration Improves Abiotic Stress Tolerance of Arabidopsis. *Plants* **2019**, *8*, 563. [CrossRef] [PubMed]
20. Sage, R.F. The evolution of C4photosynthesis. *New Phytol.* **2004**, *161*, 341–370. [CrossRef]
21. Guralnick, L.J.; Gilbert, K.E.; Denio, D.; Antico, N. The Development of Crassulacean Acid Metabolism (CAM) Photosynthesis in Cotyledons of the C4 Species, Portulaca grandiflora (Portulacaceae). *Plants* **2020**, *9*, 55. [CrossRef] [PubMed]
22. Kaiser, E.; Walther, D.; Armbruster, U. Growth under Fluctuating Light Reveals Large Trait Variation in a Panel of Arabidopsis Accessions. *Plants* **2020**, *9*, 316. [CrossRef] [PubMed]
23. Walker, B.J.; Kramer, D.M.; Fisher, N.; Fu, X. Flexibility in the Energy Balancing Network of Photosynthesis Enables Safe Operation under Changing Environmental Conditions. *Plants* **2020**, *9*, 301. [CrossRef]
24. Wulfert, S.; Schilasky, S.; Krueger, S. Transcriptional and Biochemical Characterization of Cytosolic Pyruvate Kinases in Arabidopsis thaliana. *Plants* **2020**, *9*, 353. [CrossRef] [PubMed]
25. Rosado-Souza, L.; Fernie, A.R.; Aarabi, F. Ascorbate and Thiamin: Metabolic Modulators in Plant Acclimation Responses. *Plants* **2020**, *9*, 101. [CrossRef]
26. Toleco, M.R.; Naake, T.; Zhang, Y.; Heazlewood, J.L.; Fernie, A.R. Plant Mitochondrial Carriers: Molecular Gatekeepers That Help to Regulate Plant Central Carbon Metabolism. *Plants* **2020**, *9*, 117. [CrossRef] [PubMed]

Article

Transcriptional and Biochemical Characterization of Cytosolic Pyruvate Kinases in *Arabidopsis thaliana*

Sabine Wulfert [†], Sören Schilasky [†] and Stephan Krueger *

Institute for Plant Sciences, University of Cologne, Zülpicherstraße 47b, 50674 Cologne, Germany;
s.wulfert@mail.de (S.W.); soeren.schilasky@googlemail.com (S.S.)
* Correspondence: stephan.krueger@uni-koeln.de
† These authors are equally contributed.

Received: 19 February 2020; Accepted: 9 March 2020; Published: 11 March 2020

Abstract: Glycolysis is a central catabolic pathway in every living organism with an essential role in carbohydrate breakdown and ATP synthesis, thereby providing pyruvate to the tricarboxylic acid cycle (TCA cycle). The cytosolic pyruvate kinase (cPK) represents a key glycolytic enzyme by catalyzing phosphate transfer from phosphoenolpyruvate (PEP) to ADP for the synthesis of ATP. Besides its important functions in cellular energy homeostasis, the activity of cytosolic pyruvate kinase underlies tight regulation, for instance by allosteric effectors, that impact stability of its quaternary structure. We determined five cytosol-localized pyruvate kinases, out of the fourteen putative pyruvate kinase genes encoded by the *Arabidopsis thaliana* genome, by investigation of phylogeny and localization of yellow fluorescent protein (YFP) fusion proteins. Analysis of promoter β-glucuronidase (GUS) reporter lines revealed an isoform-specific expression pattern for the five enzymes, subject to plant tissue and developmental stage. Investigation of the heterologously expressed and purified cytosolic pyruvate kinases revealed that these enzymes are differentially regulated by metabolites, such as citrate, fructose-1,6-bisphosphate (FBP) and ATP. In addition, measured in vitro enzyme activities suggest that pyruvate kinase subunit complexes consisting of cPK2/3 and cPK4/5 isoforms, respectively, bear regulatory properties. In summary, our study indicates that the five identified cytosolic pyruvate kinase isoforms adjust the carbohydrate flux through the glycolytic pathway in *Arabidopsis thaliana*, by distinct regulatory qualities, such as individual expression pattern as well as dissimilar responsiveness to allosteric effectors and enzyme subgroup association.

Keywords: pyruvate kinase; glycolysis; respiratory metabolism

1. Introduction

During plant development and adaptation to environmental changes, the glycolytic network provides an enormous metabolic flexibility. Thereby, flux regulation is achieved by the fine control of key regulatory enzymes, including pyruvate kinase (PK). PK-mediated synthesis of pyruvate represents a bottleneck for acetyl-CoA entering the TCA cycle. On the other hand, reduced pyruvate kinase activity will lead to a backlog of PEP and other glycolytic intermediates, thereby increasing the flux rate of carbon skeletons into branching biosynthetic pathways. The *Arabidopsis thaliana* genome encodes several putative cytosolic and plastidial PKs, and glycolytic metabolites can be exchanged between the cytosol and plastids [1] since both compartments are connected through diverse transporters located in the inner plastid envelope membrane [2–4]. Despite the assumable key regulatory function of pyruvate kinase so far, only plastidial isoforms have been described [5]. Possible reasons for this are numerous. The high number of isoenzymes with potential redundant physiological roles, as well as the compartmentalized system with glycolytic intermediates equilibrating through plastid membrane transporters, may hamper their investigation.

Several ways of regulation for PKs have been verified, including binding of co-substrates and allosteric effectors. In *Saccharomyces cerevisiae* the glycolytic intermediate fructose-1,6-bisphosphate (FBP) increases the affinity to the bivalent cations, Mg^{2+} or Mn^{2+}, which are essential for PK activity [6]. Studies on cytosolic PK from castor bean identified glutamate as the most effective inhibitor, whereas aspartate functioned as an activator [7]. Furthermore, the TCA cycle intermediates citrate, 2-oxoglutarate, fumarate and malate have the potential to decrease activity of some plant PKs, which indicates a role as feedback regulators [8–10]. A further regulatory aspect may arise from pH-dependent alterations in the PK enzyme's affinity to metabolite inhibitors. This was proposed in a study on PK enzymes that were isolated from cotyledons of *Ricinus communis* [8]. An enhanced PK activity was accompanied by a reduced cytosolic pH, which was caused by H^+-symport that affected the uptake of endosperm-derived sugars and amino acids. Protein degradation is another way of controlling PK activity as shown recently for cotton cytosolic pyruvate kinase 6 (GhPK6) [11]. Here phosphorylation-mediated ubiquitination of GhPK6 appears to modulate the cotton fiber elongation process. Subunit association/dissociation was shown to be an additional mechanism to adjust PK activity. Accordingly, human pyruvate kinase muscle isoenzyme 2 (PKM2) assembled to a dimer has remarkably lower affinity to PEP as the respective tetramer [12]. This regulatory mechanism has been proposed for plant PKs as well, as in vitro studies show that specific subgroup combinations are more active than others [5,7]. Plastidial PKs in *Arabidopsis thaliana* have been shown to be essential for seed oil production, whereby enzyme isoforms form higher-order subunit complexes composed of 4α and 4β-subunits [5].

Spatial distribution of glycolytic enzymes within the cell constitutes a further point of regulation, since enzymes may localize at sites of demand for glycolytic intermediates. Proteomic analyses of highly purified mitochondrial fractions revealed the presence of glycolytic key enzymes, including PK isoforms on the outside of the mitochondrion considered to ensure direct import of pyruvate [13]. Furthermore, degradation by the proteasome determines PK-catalyzed pyruvate synthesis, since C-terminal proteolytic processing of cytosolic PK was shown for isoforms derived from soybean [14]. Finally, the multilayered character of PK activity control generates a complex picture on its role as a flux regulator. On the other hand, the complexity of the involved factors underlines the enormous sensitivity by which fine control is attained.

The *Arabidopsis thaliana* genome encodes 14 putative pyruvate kinases, which are likely to be isoforms catalyzing the ADP-dependent conversion of PEP to pyruvate, thereby releasing ATP. These isoenzymes show a broad diversity concerning gene expression rate and tissue specificity [15] and additionally segregate into plastidial and cytosolic subclades according to consensus predictions of their subcellular localization [16].

We identified five cytosolic PK gene candidates that show a significant expression and are likely to be localized to the cytosol. After having confirmed the cytosolic localization of PK2, PK4 and PK5 by heterologous expression of yellow fluorescent protein (YFP) fusion constructs in *Nicotiana benthamiana*, we aimed to identify the different roles of cytosolic PK enzymes in dependence of changing developmental and environmental conditions. By histochemical analysis of plant lines expressing promoter-β-glucuronidase (GUS) fusion constructs, we observed tissue-specific localization of PK expression in diverse developmental stages. Furthermore, biochemical characterization of purified cPK isoenzymes heterologously expressed in *Escherichia coli* showed that PK activity is controlled by the presence of metabolite effectors or binding of enzyme subunits.

In summary, our findings show that regulation of cPK enzyme activity is controlled by distinct gene expression patterns, different sensitivity to allosteric effectors and enzyme subgroup formation.

2. Results

2.1. Selection of Pyruvate Kinase Candidates to be Involved in Cytosolic Glycolysis

The *Arabidopsis thaliana* genome encodes for 14 putative PK isoforms. A phylogenetic tree based on PK amino acid sequence alignment (Appendix A Figures A1 and A2) is shown in Figure 1A. Four

PK isoforms, namely At1g32440, At5g52920, At3g22960 and At3g49160, are predicted to contain a chloroplast transit peptide according to the Aramemnon database [16], and their localization to the chloroplast was confirmed in vitro by Andre and colleagues (2007) [5]. For the remaining isoforms, the consensus predictions give no clear indication for targeting to a certain organelle (Figure 1B). An alignment of protein sequences of the *Arabidopsis* PK candidates with bona fide PKs from other organisms revealed two PK subclades to exist in *Arabidopsis* [5]. In addition to isoforms localized to the plastids, another subclade consisting of enzymes that target the cytosol was hypothesized.

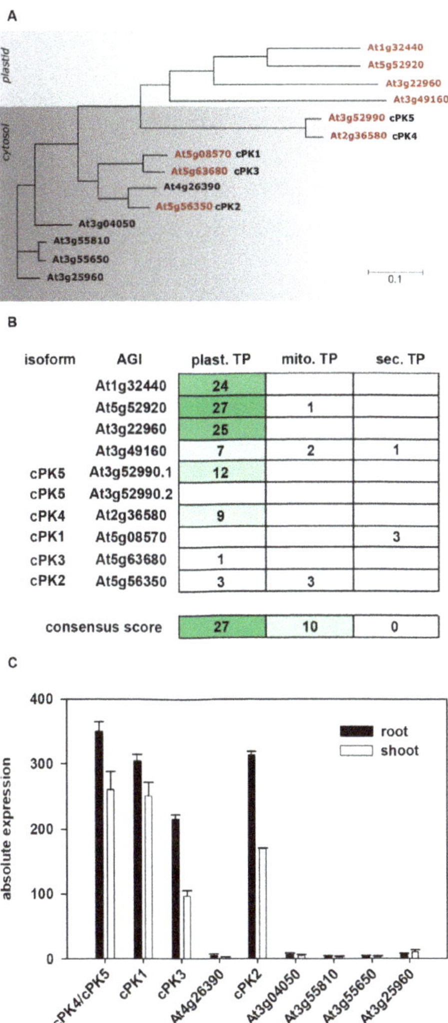

Figure 1. (**A**) Dendrogram of the *Arabidopsis thaliana* pyruvate kinase family based on protein sequences obtained from the Aramemnon database [16]. The alignment was performed using clustalW, and the dendrogram was created using Dendrocsope 3 [17]. The scale bar represents the number of substitutions per site. Significantly expressed genes are highlighted in red (according to the *Arabidopsis* efp browser). (**B**) Consensus prediction of the subcellular location. Consensus scores for PK proteins obtained from the Aramemnon database [16]. (**C**) Expression of putative cytosolic *PK* genes in roots and leaves. Data obtained from published microarray-data, *Arabidopsis* efp browser [15].

According to the *Arabidopsis* efp browser expression database [15], only five out of ten candidate genes coding for putative cytosolic PKs are expressed up to a reasonable level in order to be considered for our analysis (Figure 1C). Relative expression of these putative cytosolic isoforms is generally higher in roots than in leaves as indicated by published microarray data (Figure 1C). Expressed putative cytosolic pyruvate kinase (cPK) candidates were selected for further analysis and are subject of the current study. For clear determination, the genes were named as follows: At5g08570, *cPK1*; At5g56350, *cPK2*; At5g56350, *cPK3*; At2g36580, *cPK4*; and At3g52990, *cPK5*. PK4 and PK5 expression cannot be distinguished from each other, as both genes are identified by the same probe target. Based on an alignment with bona fide PKs from other organisms, the five expressed PK candidates fall into a subclade of cytosol localized isoforms [5]. As for cPK2, cPK4 and cPK5, the localization prediction was unclear, and YFP fusion proteins were constructed for these enzymes and transiently co-expressed with free mCherry fluorescence protein in leaf epidermal cells of *Nicotiana benthamiana*. Confocal laser scanning microscopy analysis of transformed leaf sections revealed that, in contrast to the triose-phosphate/phosphate translocator (TPT) green fluorescent protein (GFP) fusion, all cPK:YFP fusion proteins were co-localized with free mCherry protein in the cytosol and were absent from the chloroplast (Figure 2).

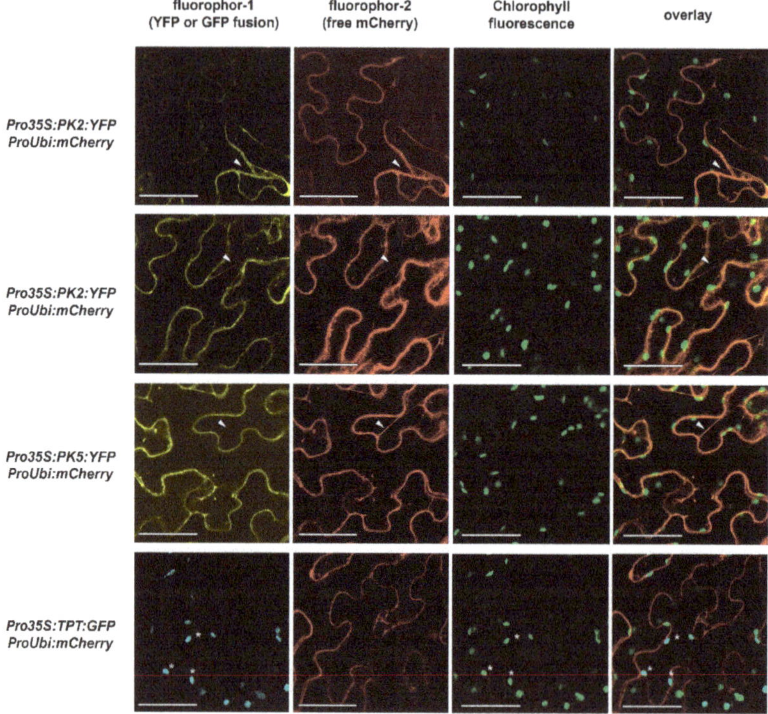

Figure 2. Subcellular localization of pyruvate kinase isoforms. cPK2, cPK4, cPK5-YFP fusion proteins (yellow) and TPT-GFP fusion protein (cyan) were expressed under the control of the cauliflower mosaic virus promoter (*Pro35S*), while free mCherry fluorescence protein was expressed under the control of the ubiquitin promoter (*ProUbi*). All PK-YFP fusion proteins were co-localized with free mCherry in cytosolic plasma strands (white arrows), whereby TPT-GFP fusion protein was co-localized with chlorophyll A fluorescence (green) as indicated by the white asterisks. Constructs were transiently expressed in *Nicotiana benthamiana* leaves. Scale bars = 50 μm.

2.2. Pyruvate Kinase Genes Show an Isoform-Specific Expression Pattern

The high number of expressed cytosolic PK isoforms allows each gene candidate to potentially fulfill distinct tasks during plant glycolysis. To assess tissue-specific expression of respective cPK genes in dependence on the developmental status of the plant, a β-glucuronidase (GUS) reporter approach was taken. Promoters of the five *cPK* candidates were fused to GUS and expressed in *Arabidopsis thaliana*. Subsequently, the plant organs were stained and analyzed in detail. The GUS-staining results suggest that the cytosolic isogenes *cPK1, 2* and *3* are expressed in vegetative and reproductive tissues, as promoter-GUS activity was observed in leaves, roots and flowers of plants from diverse developmental stages (Figure 3, Figure 4, Appendix A Figure A3). However, for cPK4 and cPK5 a quite strong GUS signal was observed during seedling stage, and the same lines showed no or very low expression in tissues of mature plants under standard growth conditions.

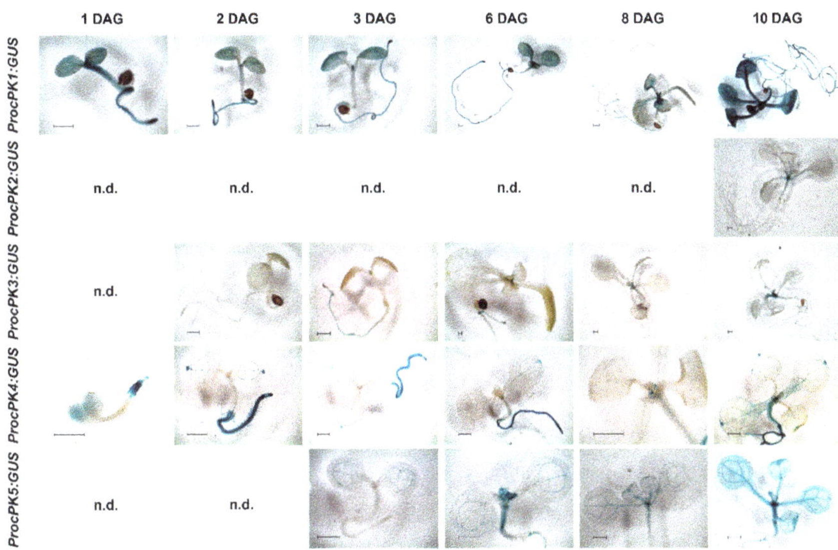

Figure 3. Histochemical β-glucuronidase (GUS) localization in seedlings of the *cPK1, cPK2, cPK3, cPK4* and *cPK5* promoter-GUS constructs, one day after germination (DAG) up to ten DAG. Representative images of one out of three individual transgenic lines are shown. n.d. indicates developmental stages without detectable GUS activity. Bars indicate 0.5 mm.

Analysis of the first ten days of seedling development revealed clear differences in the expression of selected *cPK* isogenes (Figure 3). While the *cPK1* promoter lead to broadly abundant GUS activity in roots and leaves of seedlings from the first day on, respective lines for the *cPK2* and *cPK3* promoters showed comparable GUS signals only at later stages. Notably, *cPK1* promoter GUS lines displayed strong staining in the shoot apical meristem (Figure 3) and in the basal part of young leaves (Figure 2 promcPK1::GUS, 8 DAG). *cPK2* promoter GUS expression did not initiate in the rosette axis until ten days after germination. The *cPK3* promoter-induced expression was restricted to the root tip and was already visible on the first day after germination. From the third day on, the signal became abundant in the entire root and was also observed in the leaf axis and spatial related trichomes. In *cPK4*-promoter plants expression was observed from the first day on in the entire root and the leaf vasculature, and later in hydathodes and at the sides of emerging leaf buds (Figure 3, promcPK4::GUS, 8 DAG). In contrast, expression of *cPK5* started later and was mainly restricted to the cotyledon, leaf and root vasculature. As indicated before, promoter-GUS constructs for *cPK1, 2* and *3* were expressed in the vasculature and mesophyll of rosette leaves with varying intensity depending on developmental stage of the leaf (Appendix A, Figure A3). In roots, *cPK1* appeared to be ubiquitously expressed, whereas GUS activity for *cPK2* and *cPK3* promoters was limited to confined areas. The *cPK2* promoter GUS lines showed blue stains restricted to primary roots (Figure 4N), while the *cPK3* promoter primarily lead to GUS activity in younger roots and root tips (Figure 4M). Promoters of both *cPK2* and *cPK3* drive GUS expression at sites of emerging lateral roots (Figure 4P,Q). In young flowers of *cPK1* promoter-GUS lines, no signal was observed, whereas fully developed flowers appeared to be ubiquitously stained in sepals, petals, filaments of stamen, style and the stigma tissue (Figure 4B,E,H). *cPK2* was also not expressed in young flowers, whereas in later stages, GUS-staining was limited to sepals and style (Figure 4C,F,I). In contrast, *cPK3* promoter-dependent GUS expression was already detected in the abscission zones of young flower buds (A), and later expression became apparent in the anthers as well (G).

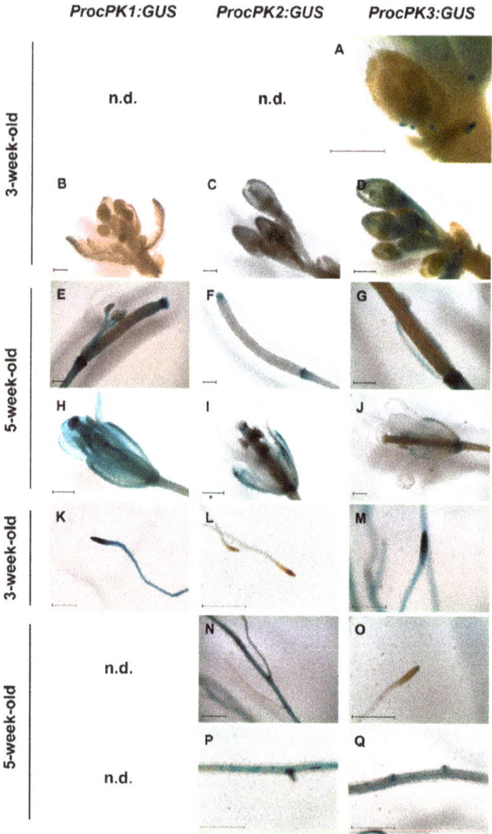

Figure 4. Histochemical GUS localization in flower organs (**A–G**) and in roots (**K–Q**), for the *cPK1*-promoter-GUS (**B,E,H,K**), the *cPK2*-promoter-GUS (**C,F,I,L,N,P**) and the *cPK3*-promoter-GUS (**A,D,G,J,M,O,Q**) in 3-week-old plants (**A–D**) and 5-week-old plants (**E–J**), is illustrated in developing flowers (**B–D**) and in fully developed carpels (**E–G**) with stamen as well as entire flowers (**H–J**) of 5-week-old plants. Pictures show root tips of 3-week-old plants (**K–M**) and GUS expression in 5-week-old plants, restricted to primary roots for the *cPK2*-promoter-GUS (**N**) and stem cells in root tips for the *cPK3*-promoter-GUS (**O**). Expression at sites of emerging secondary roots for the *cPK2*- and *cPK3*-promoter-GUS (**P,Q**). Representative images of one out of three different transgenic lines are shown. n.d. indicates plant organs without detectable GUS activity. Bars indicate 0.5 mm.

2.3. Cytosolic Pyruvate Kinases Respond to Cold Stress

In contrast to the plastid-localized pyruvate kinases, the analysis of public available transcript data [15] revealed that the cytosolic PKs are induced in response to cold treatment (Figure 4A). According to these data, *cPK1* and *cPK2* expression is induced, whereas *cPK3* transcript levels are not altered. The strongest induction upon cold treatment was found for the isogenes *cPK4* and *cPK5*. However, regarding the microarray data, *cPK4* and *cPK5* expression cannot be considered separately, since respective data do not discriminate between both genes. Thus, we validated these data by using *Arabidopsis* lines carrying the respective GUS-promoter constructs that were sampled after cold treatment (Figure 5B). In *cPK4*-lines, the GUS signal was very low in untreated control plants, and it was strongly induced in the leaf vasculature of cold treated plants. *cPK5* expression appeared to also be increased; however, the observed effect was not as strong as in the *cPK4*-GUS line. Therefore, our results indicate that cytosolic pyruvate kinases are induced during cold treatment.

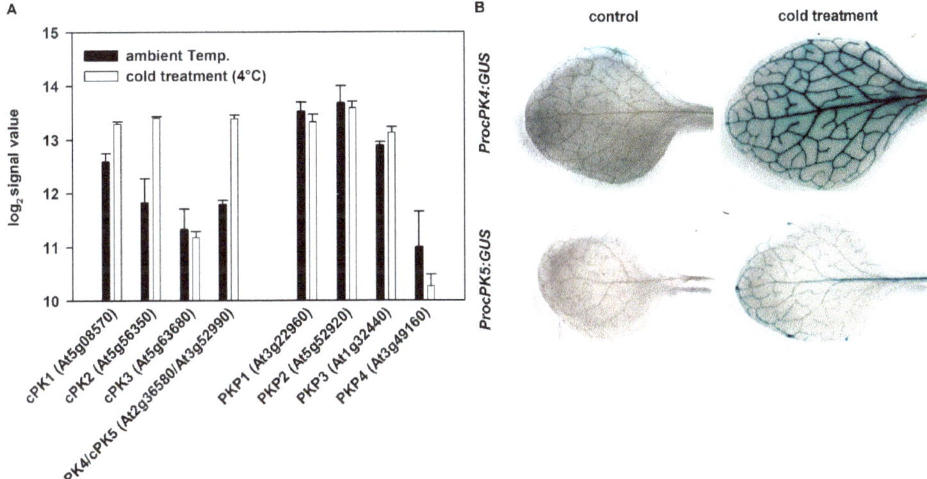

Figure 5. (**A**) Expression data of cytosolic and plastidic pyruvate kinases after cold treatment obtained from publicly available data source (https://genevestigator.com). Sixteen-day-old plants were transferred to the cold room (4 °C), and plant tissues were harvested at 24 h after onset of treatment. (**B**) To validate cold-induced expression of *cPK4* and *cPK5*, twenty-day-old *cPK4* and *cPK5* promoter GUS plants grown on soil were transferred to the cold room (4 °C) for three days. Control plants were grown under normal growth conditions. The oldest leaves of control and cold-treated plants were sampled, and the GUS activity was determined by histological staining. Representative images of one out of three different transgenic lines are shown.

2.4. Pyruvate Kinase Expression Follows a Diurnal Course

Carbon metabolism is exposed to diurnal changes, since photosynthesis cannot maintain energy provision during the dark period. Breakdown of starch and soluble sugars during the night leads to increased activity of glycolytic enzymes. In order to assess cPK activity, which is a key enzyme in glycolysis, in the course of the day/night cycle, *cPK* promoter GUS plants were sampled at the end of the night (after 8 h darkness) and at the end of the day (after 16 h light). Apart from differences in signal intensity between the lines, *cPK1*, *cPK2* and *cPK3* promoters lead to increased GUS activity when sampled after 8 h of darkness compared to plants sampled after 16 h in the light (Figure 6). In contrast, *cPK4* and *cPK5* promoter-GUS lines showed no detectable GUS activity in mature plants used for the day/night cycle experiment (data not shown).

Figure 6. Histochemical GUS expression in dependence on the daytime in 4-week-old plants expressing the *cPK1*-promoter-GUS, *cPK2*-promoter-GUS and *cPK3*-promoter-GUS constructs, grown on soil sampled after 16 h light and after 8 h darkness. Representative images of one out of three different transgenic lines are shown.

2.5. Kinetic Characterization of Cytosolic Pyruvate Kinases

To assess the kinetic parameters of cytosolic PK enzymes, coding sequences were cloned into the pET16B expression vector mediating N-terminal 6x-His tag. PK isoenzymes were expressed heterologously in *E. coli* and purified in order to obtain a sufficient yield (Appendix A Figure A4). Enriched isoenzymes were characterized in vitro applying a lactate dehydrogenase coupled enzyme assay according to [18]. Plots showing cPK activity versus linear dilution series of ADP and PEP describe hyperbolic saturation curves; thus, biochemical properties like Michaelis constant *Km*, enzymes maximum rate *Vmax*, and turnover number *kcat* were calculated applying the Michaelis–Menten equation (Table 1; Appendix A Figure A5). The lowest specific activity was observed for cPK1 with a *Vmax* of 1.4U/mg. On the contrary, PK5 exhibited the highest specific activity (*Vmax*: 6.4 U/mg). The *Km* values for ADP were in about the same range varying between 0.07 mM (cPK4) and 0.34 mM (cPK5). cPK2 had a quite high *Km* for PEP (0.76 mM), compared to the other isoenzymes.

Table 1. In vitro catalytic properties of cPK enzymes heterologously expressed in *Escherichia coli*: Michaelis constant *Km*, enzymes maximum rate *Vmax* and turnover number *kcat*.

Isoform	*Vmax* (U/mg)	*kcat* (1/s)	*Km* (mM) ADP	*Km* (mM) PEP
cPK1	1.4	1.53	0.15	0.06
cPK2	2.8	3.05	0.03	0.76
cPK3	2.2	2.4	0.14	0.17
cPK4	3.9	4.0	0.07	0.17
cPK5	6.4	6.68	0.34	0.05

2.6. Allosteric Effects on Pyruvate Kinase Enzyme Activity

PK enzymes are tightly regulated by allosteric effectors as known from other organisms [6,8,18]. To assess putative regulatory impact of diverse selected metabolites on cytosolic PK enzymes from *Arabidopsis thaliana*, single enzyme extracts were tested under substrate saturating conditions. ATP showed a strong effect on cPK activity. In the case of cPK1, the specific activity was reduced by more than 90%, and for cPK3 a reduction of 73% was observed (Table 2). In contrast, ATP did not affect the specific activity of cPK5. Our data indicate that cPK1 can be positively affected by FBP, whereby glutamate and aspartate lead to a decrease in specific enzyme activity. Activities of other cPK isoenzymes were not significantly altered by the selected metabolites. Citrate is a central metabolite of the citrate cycle and has been applied as negative control in several previous kinetic analyses of pyruvate kinases [10]. Also, in our study citrate had a strong negative allosteric effect on all five isoforms.

Table 2. Allosteric effects on cPK enzyme activity after application of F1.6BP (1 mM), serine (0.2 mM), AMP (0.1 mM), ATP (2 mM), glutamate (0.2 mM), aspartate (0.2 mM) and citrate (4 mM) compared to control.

Isoform	F1.6BP	Serine	AMP	ATP	Glutamate	Aspartate	Citrate
cPK1	122%	99%	102%	7%	85%	69%	17%
cPK2	104%	112%	90%	76%	99%	97%	26%
cPK3	101%	99%	109%	27%	102%	94%	7%
cPK4	104%	100%	97%	84%	100%	97%	3%
cPK5	97%	101%	97%	95%	100%	98%	2%

2.7. Effects of Subgroup Complex Formation

It was shown for plastidial PKs from *Arabidopsis thaliana* that their activity is dependent on the formation of α- and β-subunits. This is based on the observation that in vitro catalytic efficiency is significantly higher for reconstituted subunit complexes in comparison to the respective subunit alone [5]. To test whether subgroup complex association also has an effect on the cytosolic isoenzymes, mixtures of equal ratios of cPK extracts were used for further kinetic experiments. Under substrate saturating conditions, specific activities were determined for mixtures of selected isoform pairs. In a mixture of cPK1 and cPK2, specific activity was strongly increased by about 70% compared to cPK2 alone (Figure 7). A positive effect on activity in the same range was observed in a mixture of cPK1 and cPK4 (75%). Combinations of cPK2 and cPK3 or cPK4 and cPK3 also increased specific activity compared to the single enzymes but only by about 20%. However, the highest impact was observed in a batch containing equal ratios of cPK4 and cPK5, here the activity was 180% higher than the activity of cPK4 alone.

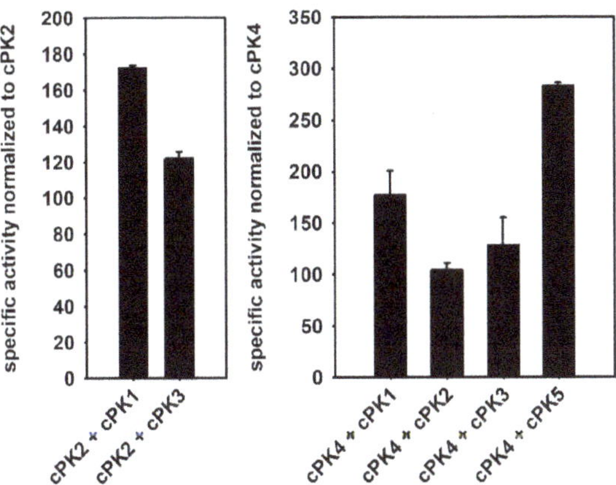

Figure 7. Effect of isoenzyme combinations on specific activities (U mg$_{protein}^{-1}$) of enzyme extracts containing equal amounts of two isoenzymes normalized to activity of extracts containing a single isoenzyme. Specific activity of single enzyme extracts of cPK2 and cPK4 were set to 100%.

3. Discussion

3.1. Arabidopsis thaliana Five Cytosolic Pyruvate Kinases Form Two Functional Subgroups

Various previous studies showed that plant PKs either localize to the cytosol or the plastid, whereby the subcellular distribution is of crucial relevance for enzyme activity. Plastidial PK isoforms are reported to be essential for biosynthesis of seed oil and the mobilization of storage compounds in *Arabidopsis thaliana* seedlings [5,19]. Cytosolic PKs on the contrary take over an important role in carbohydrate breakdown, cytosolic ATP biosynthesis and provision of pyruvate to the TCA cycle, as known from other plant and non-plant organisms [20]. Apart from their fundamental role in primary metabolism, reports on *Arabidopsis thaliana* PKs are limited to seed-expressed, plastid-localized isoforms [5]. In the present study, five potential cytosol-targeted PKs from *Arabidopsis thaliana* were identified based on phylogenetic studies and gene expression data [5] (Figure 1A). The prediction of cytosolic localization of cPK2 (At5g56350), cPK4 (At2g36580) and cPK5 (At3g52990) was confirmed by investigating YFP-fusion proteins in *Nicotiana benthamiana* leaf cells (Figure 2). Furthermore, PK activity was verified in vitro for all five isoforms since the isoenzymes were capable of hydrolyzing PEP into pyruvate and ATP (Table 1). Our observations suggest that within the subclade of cytosolic PKs, *cPK1*, *cPK2* and *cPK3* form a subgroup distinct from *cPK4* and *cPK5*, also displayed by structural characteristics of the genetic sequence. The specific roles for each subgroup were illustrated by GUS expression data, which showed entirely independent expression patterns for each isoform. Furthermore, the analysis of in vitro enzyme activity suggests regulatory properties of subunit complexes composed of cPK1/2, cPK2/3 or cPK4/5 isoforms, respectively (Figure 7).

3.2. Pyruvate Kinase Enzymes are Localized to the Cytosol

Three out of five PK candidates were localized to the cytosol via confocal microscopy of YFP-fusion proteins (Figure 2). This is in agreement with predictions based on a bona fide alignment of PK proteins from other plants [5]. Since consensus sequence predictions for cPK1 and cPK3 identified no target peptide for a specific organelle, we presume that both PKs also target to the cytosol [16]. Furthermore, formation of complexes with cPK2, cPK4 and cPK5 bears regulatory property of PK activity, assuming both isoforms of an enzyme pair to target to the same compartment.

3.3. Cytosolic Pyruvate Kinase Genes are Expressed in a Tissue-Specific and Developmental-Specific Manner

Our GUS studies show a distinct tissue-specific expression of the five *cPK* isogenes, also supported by publicly available microarray data [15]. Considering the fact that the cPK isoforms underlie distinct metabolic control, it can be concluded that plants are able to fine-tune cell and tissue-specific metabolism by coordinated *cPK* expression, in order to closely match the energy requirements under different conditions.

Cytosolic *PK1* appeared to be ubiquitously expressed in all tissues and in all developmental stages. This stands in contrast to *cPK2* and *cPK3* expression, which started at a later stage of seedling development and was initially restricted to meristematic tissues, such as the root tips (Figure 3) and to the proliferation zone of young leaves. However, all three isoforms were expressed in later developmental stages. In contrast, *cPK4* expression was pronounced in roots and cotyledons during the first days of development (Figure 3). At a later stage, *cPK4* together with *cPK5* was expressed only under stress conditions (Figure 5). Seedling establishment might be considered as a stress situation, since the plants face limited carbon supply because of an inactive or not yet fully activated photosynthetic machinery. Accordingly, during this stage no carbon assimilation takes place, and the plants depend on the degradation of compounds that were supplied by the mother plant [21]. During germination, triacylglycerol (TAG) is degraded via β-oxidation, and the released fatty acids are converted to sucrose through the glyoxylate cycle and gluconeogenesis. Thereby PEP is formed from oxaloacetate by the cytosolic PEP carboxykinase (PEPCK) [22]. Possibly, conditions favoring gluconeogenic respiration, for example high ATP levels, restrict some cPK isoforms more than others. Interestingly, cPK4 and cPK5 remained unaffected by ATP levels, while cPK1 and cPK3 were subjects of strong ATP inhibition (Figure 7). In general, gene expression is only one level of regulation and does not necessarily reflect the protein content in the respective tissue, since cPK has been reported also to be strongly regulated by protein degradation [14,23]. Furthermore, we demonstrated that full cPK activity depends on the formation of subgroup complexes (Figure 7). This leads to the possibility that a comparatively weakly expressed isoform that is part of a complex might still have an essential impact on total cPK activity. We induced the expression of cPK4 and cPK5-GUS via cold treatment, as we could not detect any expression under ordinary conditions. This result is in line with published microarray data [15]. Exposure to low temperatures enhances freezing tolerance, a versatile process involving various alterations in biochemical processes and global changes in gene expression that are accompanied by a shift in the composition of primary and secondary metabolites [24–27]. Under these conditions cell division and expansion is actively inhibited, leading to sink limitation and subsequent accumulation of carbohydrates [28]. The chilling response involves an acceleration in levels of γ-aminobutyric acid (GABA), proline and sucrose [29], metabolites that were shown to be involved in protecting membranes and proteins from freezing damage [30–32]. Under these conditions, cPKs have to maintain TCA cycle flux independent of the cytosolic ATP status, as GABA and proline depend on TCA cycle intermediates.

3.4. Cytosolic Pyruvate Kinases are Vital for Energy Allocation

Cytosolic PK is a key player in energy allocation, as it generates ATP in the cytosol, supplies pyruvate to the TCA cycle and thereby drives mitochondrial ATP synthesis. Cytosolic PK isoforms were strongly expressed in the leaf vasculature (Figure 6). This is expected since most cells of the vasculature are heterotrophic and depend on ATP generated via glycolysis and mitochondrial respiration. Furthermore, this enables fast access to carbohydrate substrates, for example sucrose, that is delivered from photosynthetic source tissues via the phloem sap. Our GUS experiments also revealed expression of *cPK1*, *cPK2* and *cPK3* in mesophyll cells. This expression became more pronounced at the end of the dark period, underlining the increased respiratory activity, which is normally present during the night. Furthermore, we observed distinct GUS expression in root tips, shoot apical meristems and leaf primordia (Figures 3 and 4A,K–Q), indicating a significant role of cPK isoforms in ATP and pyruvate provision to these young, developing and highly energy-consuming tissues.

3.5. Pyruvate Kinase Enzymes are Differently Regulated by Metabolites

Several glycolytic enzymes underlie negative allosteric regulation by TCA cycle intermediates [33]. A major issue is to control the cellular $NADH + H^+/NAD^+$ and ATP/ADP ratios in order to sustain tolerable physiological conditions. This is achieved by the pH-dependent balance between citrate and isocitrate formation, which determines the overall TCA cycle flux [34]. Our kinetic characterization demonstrated that all five cytosolic PK enzymes from *Arabidopsis thaliana* are strongly inhibited in the presence of citrate, confirming previous studies on PKs [33]. cPK1 underlies a further mechanism of control because its activity was affected by aspartate and glutamate (Table 2). This might provide a feedback control balancing the generation of carbon skeletons required for NH_4^+ assimilation in tissues highly active in amino acid biosynthesis. Furthermore, the activity of cPK1, cPK2 and cPK3 was negatively controlled by ATP, which suggests that these isoforms are inhibited in vivo under sufficient energy supply. Activities of cPK4 and cPK5 were not affected by ATP, indicating that both isoforms act independently of the energy status of the cell, for instance under stress-related conditions. In summary, our data show that *Arabidopsis thaliana* encodes five isoenzymes with individual regulatory properties, allowing a versatile system of cPK control.

3.6. Cytosolic Pyruvate Kinases are Regulated by Subgroup Association and Dissociation

Most glycolytic regulatory enzymes, including PFK, PEPC and PK, were shown to exist as oligomers, whereby many of them can reversibly dissociate after binding effector molecules. This is often accompanied by an altered enzyme activity and therefore provides a mechanism for regulation [33].

Our catalytic characterization of cPK enzymes supports this hypothesis, since single isoform activities are significantly lower than activities of equimolar mixtures of specific isoforms (Figure 7). For plastidial PK enzymes, it was even reported that single enzymes or their respective homo oligomers, have no activity at all [5]. Thus, enzyme oligomerization seems to be a common mechanism to regulated PK enzyme activity.

The determined enzyme pairs are based not simply on sequence homology, but they are more in agreement with the tissue-specific expression pattern of the respective isoforms observed in our GUS studies. Accordingly, the most pronounced effect on enzyme activity was observed when the isoforms cPK4 and cPK5 were combined (Figure 7), which both appear to be strongly induced by cold treatment. The observation that cPK2 and cPK3 have positive impact on the activity of each other is underlined by co-expression data, which show that expression of *cPK2* and *cPK3* is strongly correlated (ATTEDII, version 7.1), indicating both isoforms to be active in the same metabolic pathway simultaneously. Interestingly, cPK1 appears to take over a special regulatory role because it exhibited the lowest specific activity as single enzyme (Table 1), but it was able to increase the activity of other isoforms in different combinations (Figure 7). Furthermore, cPK1 seems to underlie the most multisided metabolic control (Table 2) and appears to be most generally expressed isoform, independent of tissue and developmental stages. Finally, this supports the hypothesis of a further regulatory aspect, as *cPK1* expression might be sufficient to meet energy demand alone under certain conditions, whereas a second isoform is expressed additionally at raised energy need.

4. Materials and Methods

4.1. Assaying Tissue-Specific Pyruvate Kinase Expression by Promoter-GUS Fusion

To analyze the localization of pyruvate kinase expression during plant growth and development β-glucuronidase (GUS) fusion constructs were stably transformed into *Arabidopsis thaliana*. The first 16 days after germination were investigated on seedlings grown on half strength MS (0.5% agarose), whereas samples of later developmental stages were taken from plants grown on soil in a growth chamber. For diurnal expression analysis plants were grown under short day conditions (16 h dark, 8 h light, 160 $\mu E \cdot m^{-2} \cdot s^{-1}$). Histochemical staining of transgenic plants was performed by vacuum infiltration with staining solution (0.1M $NaPO_4$ pH 7.2, 10 mM EDTA, 0.5 mM $K_3Fe[CN]_6$, 0.5 mM $K_4Fe[CN]_6$, 10% Triton x-100) supplied with 1 mM of the substrate of β-glucuronidase x-Gluc (5-bromo-4-chloro-3-indolyl s-D-glucuronic acid, solved in dimethylformamide) in an evacuated exsiccator according to the protocol established by Jefferson et al. (1987) [35]. For staining, the samples were incubated for 37 °C overnight and subsequently relieved from chlorophyll with 80% EtOH at 60 °C. Stained younger plants were documented with a Leica binocular (S8APO, equipped with an EC3 camera, Leica), and pictures were processed with the compatible LAS EZ imaging software (Leica). Older plants were photographed with a digital single-lens reflex camera (Sony α330). GUS-stained tissues and plants shown in this work represent the representative results of at least four independent lines for each construct.

4.2. Determination of Protein Localization in Nicotiana benthamiana

For transient transformation of *Nicotiana benthamiana* plants, *Agrobacterium tumefaciens* cells of the strain GV3101 pMP90 harboring the respective c-terminal YFP-tagged PK-CDS were grown overnight in a 25 mL culture (YEB, 100 ng/μL carbenicillin, 25 ng/μL gentamycin, 12 ng/μL kanamycin, 100 ng/μL rifampicin). Cells were harvested by centrifugation at 4000× *g* for 10 min and 4 °C, re-suspended in 1 mL infiltration medium (5% *w/v* sucrose, 0.01% *v/v* Silwet l-77, 2 mM $MgSO_4$, 0.5% *w/v* glucose, 450 μM acetosyringone), and OD_{600} was adjusted to 0.4. After incubation for at least 1 h on ice, and subsequent warming to room temperature, the suspension was infiltrated at the lower side of the tobacco leaves using a 1 mL blunt end tip syringe. For determination of enzyme localization, leaf samples were taken three to four days after infiltration and analyzed by confocal laser scanning microscopy using a Zeiss

LSM 700 microscope. LAS AF imaging software (Leica Application Suite Advanced Fluorescence, Leica) was used for image processing and documentation.

4.3. Generation of Promoter-ß-Glucuronidase Fusion Constructs

Promoter-β-glucuronidase (GUS) fusion constructs were generated in order to localize the expression of selected genes involved in growth and development of *Arabidopsis thaliana*. The 5′-flanking regions of cytosolic PK genes were cloned into the gateway binary vector pGWB3, which mediates GUS fusion [36]. Initially, ~1800 bp of the promoter regions were amplified from *Arabidopsis thaliana* genomic DNA by preparative PCR and introduced into gateway pENTR vectors. For the cPK1 promoter, the *Nco*I site was inserted into the 5′ end forward primer prom_cPK1_*Nco*I_for and the *Xma*I site into the 3′ end reverse primer prom_cPK1_*Xma*I_rev to facilitate T4-ligase (New England Biolabs)-mediated ligation into pENTR4 (Invitrogen, Carslbad, USA). Oligonucleotide sequences for cloning are provided in the Appendix A (Appendix A Table A1).

All other constructs were generated by TOPO cloning into the pENTR/D-TOPO vector (Invitrogen) according to the manufacturer's instructions using the following primer pairs for DNA amplification: for cPK2, prom_cPK2_TOPO_for and prom_cPK2_TOPO_rev; for cPK3, prom_cPK3_TOPO_for and prom_cPK3_TOPO_rev; for cPK4, prom_cPK4_TOPO_for and prom_cPK4_TOPO_rev; for cPK5, prom_cPK5_TOPO_for and prom_cPK5_TOPO_rev. The resulting pENTR constructs were transformed into the *E. coli* strain DH5α. After transformation, the *E. coli* were plated onto LB agar containing the respective antibiotics in order to select for successfully transformed colonies. Transformed colonies were confirmed via PCR and restriction digestion of isolated plasmid DNA as well as double-strand sequencing. The promoter fragments were cloned into the vector pGWB3 by LR reaction (LR-clonase, Invitrogen, Carslbad, CA, USA). The final plasmids were stably introduced into *Arabidopsis thaliana* plants by *Agrobacterium tumefaciens* (*A. tumefaciens*)-mediated transformation, and transformants were isolated following selection with the respective antibiotics on half-strength MS agar (Duchefa, 50 µg/mL kanamycin, 50 µg/mL hygromycin, 0.5% agarose).

4.4. Generation of 5x His-Tagged Fusion Constructs

For heterologous expression of cytosolic PK proteins in *E. coli*, full-length cDNA was cloned into the pET16b vector (Novagen, Darmstadt, Germany) mediating N-terminal His-tag fusion. The cDNA of cPK2 and cPK3 was amplified with the following oligonucleotides carrying the restriction enzyme recognition sites for subsequent classical ligation into the destination vector for cPK2 cPK2_*Nde*I_for and cPK2_*Nde*I_rev, for cPK3 cPK3_*Nde*I_for and cPK3_*Bam*HI_rev. In order to increase the DNA yield of cPK1, cPK4 and cPK5 amplicons, the cDNA was subcloned into the pENTR/D-TOPO vector (Invitrogen, Carslbad, USA) by TOPO cloning using primers cPK1_TOPO_for and cPK1_TOPO_rev for cPK1, cPK4_TOPO_for and cPK4_TOPO_rev for cPK4, and cPK5_TOPO_for and cPK5_TOPO_rev for cPK5. Oligonucleotide sequences for cloning are provided in the Appendix A (Appendix A Table A1). Amplified cDNA was digested with the indicated enzymes, purified by gel extraction and cloned into pET-16b by T4 Ligase (New England Biolabs, Ipswich, USA). Obtained plasmids were transformed into the *E. coli* strain BLR21, and the constructs were verified by colony PCR and restriction digestion.

4.5. Protein Purification by Metal Chelate Affinity Chromatography

After harvest and cell lysis, heterologously expressed 5xHis-PK proteins were purified by Ni-NTA affinity chromatography in a batch procedure. The cleared lysate was mixed for 1 h at 4 °C together with 2 mL Ni-NTA agarose beads (Macherey Nagel, Düren, Germany), which had been washed earlier with lysis buffer (50 mM TRIS-HCl pH 8, 300 mM NaCl, 1 mM imidazole). The supernatant obtained by centrifugation at 300× *g* for 5 minutes at 4 °C was removed. Afterwards, the Ni-NTA beads were resuspended in 4 mL wash buffer (50 mM TRIS-HCl pH 8, 300 mM NaCl, 2 mM imidazole) and added to a self-made column plugged with cotton wool. The flow through was collected and a further washing step followed. Matrix-bound proteins were eluted stepwise by application of four times

0.5 mL elution buffer (50 mM TRIS-HCl pH 8, 300 mM NaCl, 250 mM imidazole). The obtained elution fractions were desalted afterwards.

4.6. Desalting of Purified Protein by Size Exclusion Chromatography

After Ni-NTA chromatography, obtained elution fractions were desalted on Sephadex G-25 columns (NAP-5, GE-Healthcare). Columns were equilibrated 3 times with 2 mL PK storage buffer (100 mM TRIS-HCl pH 8, 1 mM dithiothreitol, 1 mM EDTA, 5 mM $MgCl_2$) before the elution fractions were added and PK protein was eluted in 1 mL PK storage buffer. Desalted protein was stored on ice until protein quantification, SDS-page (Appendix A Figure A4) and kinetic characterization studies (Appendix A Figure A5) followed.

4.7. Kinetic Characterization of Pyruvate Kinases

The activity of purified cPK isoenzymes was assayed in a coupled reaction system involving lactate dehydrogenase (LDH) according to a previously described method by Plaxton [19], based on photometric detection of NAD+ formation following LDH-mediated NADH oxidation. PK activity measurement was performed in a total volume of 200 µL reaction solution (50 mM TRIS-HCl, pH 7.5, 50 mM KCl, 10 mM $MgCl_2$, 5% *w/v* PEG 200, 1 mM DTT, 0.15 mM NADH) containing LDH (20 U/mL LDH, Roche) applying 5-10 µL of freshly purified PK protein in diverse concentrations. The shift in absorption was detected at $\lambda = 340$ nm in a 96-well format in a microplate reader (Infinite M200, Tecan, Männedorf, Schweiz) at 25 °C. To start reactions, substrates and potential effectors of PK enzymes were added to all reaction batches simultaneously applying a house made applicator with 96 spatulas.

Author Contributions: For research articles with sever Conceptualization, S.K.; methodology and investigation, S.W., S.S. and S.K.; validation and interpretation, S.W., and S.K.; formal analysis, S.S. and S.K.; resources, S.K.; writing (original draft), S.W. and S.K.; writing (review and editing), all co-authors; supervision, S.K., All authors have read and agreed to the published version of the manuscript.

Funding: This research was funded by German Science Foundation, grant number Kr4245/1-1 and Kr4245/2-1.

Acknowledgments: We acknowledge Diana Vogelmann for technical assistance and Rainer Schwacke (Research Centre, Jülich) for critical comments and discussion.

Conflicts of Interest: The authors declare no conflict of interest.

Appendix A

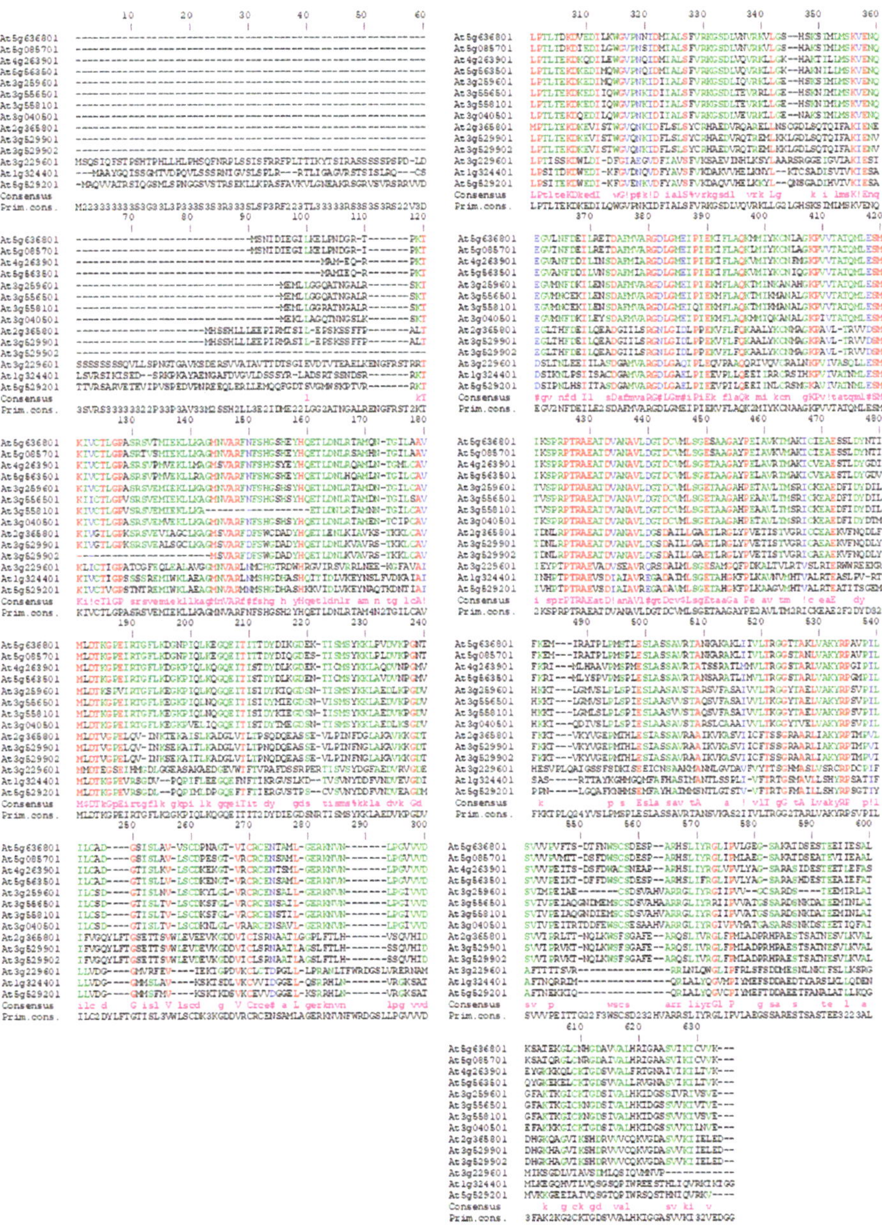

Figure A1. Multiple sequence alignment of *Arabidopsis thaliana* pyruvate kinases.

		cPK4	cPK5	cPK5					cPK3	cPK1		cPK2		
	At3g49160.1	At2g36580.1	At3g52990.1	At3g52990.2	At3g04050.1	At3g25960.1	At3g55650.1	At3g55810.1	At5g63680.1	At5g08570.1	At4g26390.1	At5g56350.1	At3g22960.1	At1g32440.1	At5g52920.1
	At3g49160.1	100	24	24	24	31	30	30	30	31	30	30	29	24	27	25
cPK4	At2g36580.1	24	100	94	95	40	41	41	40	43	43	42	43	25	27	25
cPK5	At3g52990.1	24	94	100	100	40	41	41	40	42	42	42	44	26	27	25
cPK5	At3g52990.2	24	95	100	100	40	41	40	39	42	42	41	43	26	27	25
	At3g04050.1	31	40	40	40	100	86	85	85	70	70	72	73	32	34	35
	At3g25960.1	30	41	41	41	86	100	91	91	71	70	71	74	32	34	36
	At3g55650.1	30	41	41	40	85	91	100	97	69	69	69	71	32	34	35
	At3g55810.1	30	40	40	39	85	91	97	100	68	68	69	71	31	33	35
cPK3	At5g63680.1	31	43	42	42	70	71	59	68	100	93	77	79	31	36	33
cPK1	At5g08570.1	30	43	42	42	70	70	59	68	93	100	77	79	31	36	34
	At4g26390.1	30	42	42	41	72	71	69	69	77	77	100	90	30	34	35
cPK2	At5g56350.1	29	43	44	43	73	74	71	71	70	79	90	100	31	34	34
	At3g22960.1	24	25	26	26	32	32	32	31	31	31	30	31	100	43	44
	At1g32440.1	27	27	27	27	34	34	34	33	36	36	34	34	43	100	66
	At5g52920.1	25	25	25	25	35	36	35	35	33	34	35	34	44	66	100

percent identity | 100 | 50 | 26

Figure A2. Percentages of identity between *Arabidopsis thaliana* PK isoenzymes based on multiple sequence alignments of the respective amino acid sequences. The percentage of identity is shown in numbers and visualized by a green color gradient.

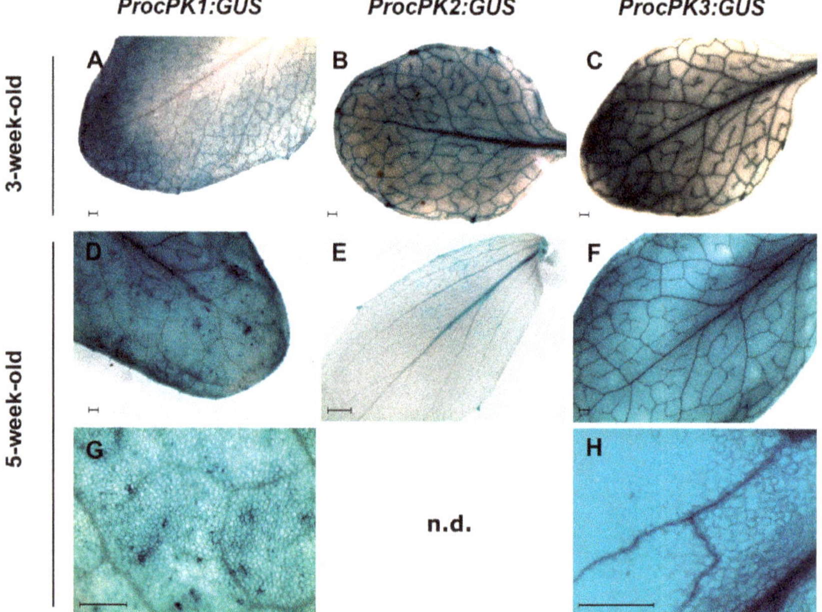

Figure A3. Histochemical GUS localization in oldest rosette leaves for the *cPK1* promoter (**A,D,G**), the *cPK2* promoter (**B,E**) and the *cPK3* promoter (**C,F,H**) in 3-week-old (**A–C**) and 5-week- old plants (**D–H**). Bars indicate 0.5 mm. G and H are magnifications of D and F, respectively. Representative images of one out of three different transgenic lines are shown.

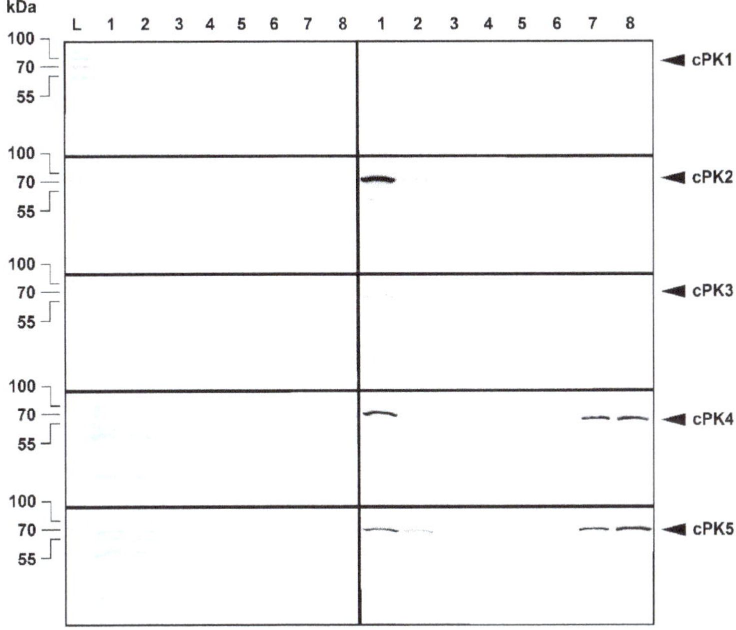

Figure A4. Affinity purification of heterologous expressed cytosolic pyruvate kinases. Protein extracts (2 µL) of different fractions (1, cell extract; 2, flow through; 3, wash fraction I; 4, wash fraction II; 5-8, elution fraction I-IV) were separated on SDS-polyacrylamide gels and stained either by Coomassie protein stain or blotted and detected by Western blot analysis using anti-His antibodies.

Table A1. Oligonucleotide sequences for cloning.

Name	Sequence
prom_cPK1_*Nco*I_for	ggccatggaatgaatgtttttgcagtat
prom_cPK1_*Xma*I_rev	ccccgggttttttctccttctcaagtt
prom_cPK2_TOPO_for	cacctactatcagctattagaattatatatc
prom_cPK2_TOPO_rev	cgctaagtgagaaaaaaaacag
prom_cPK3_TOPO_for	caccctctgtgatggatccaagtag
prom_cPK3_TOPO_rev	gtaaatctccaaaaacctaatc
prom_cPK4_TOPO_for	caccatttaccgaagtgggttagatcgg
prom_cPK4_TOPO_rev	agttgctgatcagaattcggaga
prom_cPK5_TOPO_for	cacctccttctagttttacgaaca
prom_cPK5_TOPO_rev	cttcggtgacggaagggagagagatc
cPK2_*Nde*I_for	gggcatatgatggcgatgatagagcaaagg
cPK2_*Nde*I_rev	ccccatatgtcacttgacggtcaagatcttgatca
cPK3_*Nde*I_for	gggcatatgatgtcgaacatagacatagaag
cPK3_*Bam*HI_rev	cccggatcctacttcaccacacagatcttg
cPK1_TOPO_for	caccacccatatgatgtcgaacatagacatagaaggg
cPK1_TOPO_rev	ggtcatatgtcacttaaccacacagatcttaa
cPK4_TOPO_for	caccaaccatatgatgcattcaagtcatctcc
cPK4_TOPO_rev	aacggatccctaatcctctagctcgatgattt
cPK5_TOPO_for	caccaaccatatgatgcattccagtcatcttct
cPK5_TOPO_rcv	gttggatccttaatcctcaagctcaatga

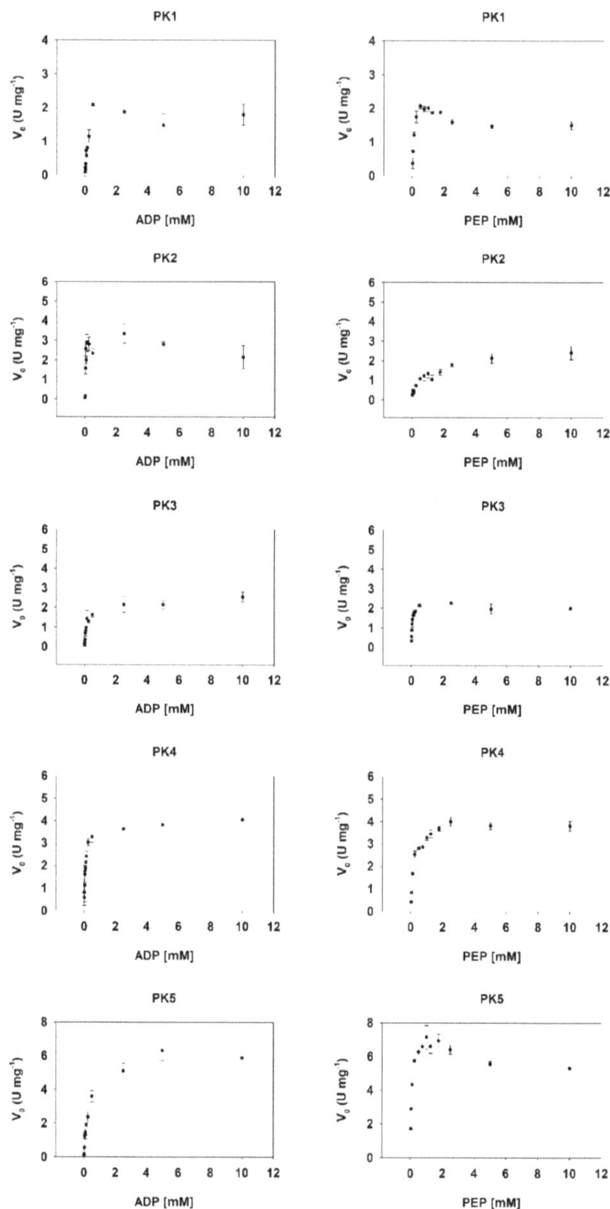

Figure A5. Biochemical characterization of cytosolic pyruvate kinase isoenzymes. Km values of cPK1, cPK2, cPK3, cPK4 and cPK5 for ADP (left panel) and PEP (right panel) were determined by nonlinear regression based on the Michaelis–Menten equation. Linear dilutions of ADP (0.005, 0.01, 0.025, 0.05, 0.075, 0.1, 0.15, 0.25, 0.5, 2.5, 5.0 and 10.0 mM) and PEP (0.025, 0.05, 0.1, 0.25, 0.5 0, 0.75, 1.0, 1.25, 1.75, 2.5, 5.0 and 10.0 mM) were used. Data are mean ± SD of four replicates.

References

1. Schwender, J.; Ohlrogge, J.B.; Shachar-Hill, Y. A flux model of glycolysis and the oxidative pentosephosphate pathway in developing Brassica napus embryos. *J. Biol. Chem.* **2003**, *278*, 29442–29453. [CrossRef]
2. Flugge, U.-I. Phosphate Translocators in Plastids. *Annu. Rev. Plant Physiol. Plant Mol. Biol.* **1999**, *50*, 27–45. [CrossRef]
3. Weber, A.P.M. Solute transporters as connecting elements between cytosol and plastid stroma. *Curr. Opin. Plant Biol.* **2004**, *7*, 247–253. [CrossRef]
4. Flügge, U.-I.; Häusler, R.E.; Ludewig, F.; Gierth, M. The role of transporters in supplying energy to plant plastids. *J. Exp. Bot.* **2011**, *62*, 2381–2392. [CrossRef]
5. Andre, C.; Froehlich, J.E.; Moll, M.R.; Benning, C. A heteromeric plastidic pyruvate kinase complex involved in seed oil biosynthesis in Arabidopsis. *Plant Cell* **2007**, *19*, 2006–2022. [CrossRef] [PubMed]
6. Jurica, M.S.; Mesecar, A.; Heath, P.J.; Shi, W.; Nowak, T.; Stoddard, B.L. The allosteric regulation of pyruvate kinase by fructose-1,6-bisphosphate. *Structure* **1998**, *6*, 195–210. [CrossRef]
7. Turner, W.L.; Knowles, V.L.; Plaxton, W.C. Cytosolic pyruvate kinase: Subunit composition, activity, and amount in developing castor and soybean seeds, and biochemical characterization of the purified castor seed enzyme. *Planta* **2005**, *222*, 1051–1062. [CrossRef] [PubMed]
8. Hu, Z.; Plaxton, W.C. Purification and characterization of cytosolic pyruvate kinase from leaves of the castor oil plant. *Arch. Biochem. Biophys.* **1996**, *333*, 298–307. [CrossRef] [PubMed]
9. Singh, D.K.; Malhotra, S.P.; Singh, R. Purification and characterization of cytosolic pyruvate kinase from developing seeds of Brassica campestris L. *Indian J. Biochem. Biophys.* **2000**, *37*, 51–58. [PubMed]
10. Smith, C.R.; Knowles, V.L.; Plaxton, W.C. Purification and characterization of cytosolic pyruvate kinase from Brassica napus (rapeseed) suspension cell cultures: Implications for the integration of glycolysis with nitrogen assimilation. *Eur. J. Biochem.* **2000**, *267*, 4477–4485. [CrossRef] [PubMed]
11. Cai, Y.; Li, S.; Jiao, G.; Sheng, Z.; Wu, Y.; Shao, G.; Xie, L.; Peng, C.; Xu, J.; Tang, S.; et al. OsPK2 encodes a plastidic pyruvate kinase involved in rice endosperm starch synthesis, compound granule formation and grain filling. *Plant Biotechnol. J.* **2018**, *16*, 1878–1891. [CrossRef] [PubMed]
12. Ye, J.; Mancuso, A.; Tong, X.; Ward, P.S.; Fan, J.; Rabinowitz, J.D.; Thompson, C.B. Pyruvate kinase M2 promotes de novo serine synthesis to sustain mTORC1 activity and cell proliferation. *Proc. Natl. Acad. Sci. USA* **2012**, *109*, 6904–6909. [CrossRef] [PubMed]
13. Giegé, P.; Heazlewood, J.L.; Roessner-Tunali, U.; Millar, A.H.; Fernie, A.R.; Leaver, C.J.; Sweetlove, L.J. Enzymes of glycolysis are functionally associated with the mitochondrion in Arabidopsis cells. *Plant Cell* **2003**, *15*, 2140–2151. [CrossRef] [PubMed]
14. Tang, G.-Q.; Hardin, S.C.; Dewey, R.; Huber, S.C. A novel C-terminal proteolytic processing of cytosolic pyruvate kinase, its phosphorylation and degradation by the proteasome in developing soybean seeds. *Plant J.* **2003**, *34*, 77–93. [CrossRef]
15. Winter, D.; Vinegar, B.; Nahal, H.; Ammar, R.; Wilson, G.V.; Provart, N.J. An "Electronic Fluorescent Pictograph" browser for exploring and analyzing large-scale biological data sets. *PLoS ONE* **2007**, *2*, e718. [CrossRef]
16. Schwacke, R.; Fischer, K.; Ketelsen, B.; Krupinska, K.; Krause, K. Comparative survey of plastid and mitochondrial targeting properties of transcription factors in Arabidopsis and rice. *Mol. Genet. Genomics* **2007**, *277*, 631–646. [CrossRef]
17. Huson, D.H.; Scornavacca, C. Dendroscope 3: An interactive tool for rooted phylogenetic trees and networks. *Syst. Biol.* **2012**, *61*, 1061–1067. [CrossRef]
18. Turner, W.L.; Plaxton, W.C. Purification and characterization of cytosolic pyruvate kinase from banana fruit. *Biochem. J.* **2000**, *352 Pt 3*, 875–882. [CrossRef]
19. Plaxton, W.C. Purification of Pyruvate Kinase from Germinating Castor Bean Endosperm 1. *Plant Physiol.* **1988**, *86*, 1064–1069. [CrossRef]
20. Fernie, A.R.; Carrari, F.; Sweetlove, L.J. Respiratory metabolism: Glycolysis, the TCA cycle and mitochondrial electron transport. *Curr. Opin. Plant Biol.* **2004**, *7*, 254–261. [CrossRef]
21. Penfield, S.; Graham, S.; Graham, I.A. Storage reserve mobilization in germinating oilseeds: Arabidopsis as a model system. *Biochem. Soc. Trans.* **2005**, *33*, 380–383. [CrossRef] [PubMed]

22. Leegood, R.C.; Walker, R.P. Regulation and roles of phosphoenolpyruvate carboxykinase in plants. *Arch. Biochem. Biophys.* **2003**, *414*, 204–210. [CrossRef]

23. Auslender, E.L.; Dorion, S.; Dumont, S.; Rivoal, J. Expression, purification and characterization of Solanum tuberosum recombinant cytosolic pyruvate kinase. *Protein Expr. Purif.* **2015**, *110*, 7–13. [CrossRef] [PubMed]

24. Guy, C. Cold Acclimation and Freezing Stress Tolerance: Role of Protein Metabolism. *Annu. Rev. Plant Physiol.* **1990**, *41*, 187–223. [CrossRef]

25. Guy, R.D.; Vanlerberghe, G.C.; Turpin, D.H. Significance of Phosphoenolpyruvate Carboxylase during Ammonium Assimilation: Carbon Isotope Discrimination in Photosynthesis and Respiration by the N-Limited Green Alga Selenastrum minutum. *Plant Physiol.* **1989**, *89*, 1150–1157. [CrossRef]

26. Renaut, J.; Hausman, J.F.; Wisniewski, M. Proteomics and low-temperature studies: Bridging the gap between gene expression and metabolism. *Physiol. Plant.* **2006**, *126*, 97–109. [CrossRef]

27. Wang, X.; Li, W.; Li, M.; Welti, R. Profiling lipid changes in plant response to low temperatures. *Physiol. Plant.* **2006**, *126*, 90–96. [CrossRef]

28. Muller, B.; Pantin, F.; Génard, M.; Turc, O.; Freixes, S.; Piques, M.; Gibon, Y. Water deficits uncouple growth from photosynthesis, increase C content, and modify the relationships between C and growth in sink organs. *J. Exp. Bot.* **2011**, *62*, 1715–1729. [CrossRef]

29. Kaplan, F.; Kopka, J.; Sung, D.Y.; Zhao, W.; Popp, M.; Porat, R.; Guy, C.L. Transcript and metabolite profiling during cold acclimation of Arabidopsis reveals an intricate relationship of cold-regulated gene expression with modifications in metabolite content. *Plant J.* **2007**, *50*, 967–981. [CrossRef]

30. Carpenter, J.F.; Crowe, J.H. The mechanism of cryoprotection of proteins by solutes. *Cryobiology* **1988**, *25*, 244–255. [CrossRef]

31. Heber, U.; Tyankova, L.; Santarius, K.A. Stabilization and inactivation of biological membranes during freezing in the presence of amino acids. *Biochim. Biophys. Acta* **1971**, *241*, 578–592. [CrossRef]

32. Yancey, P.H.; Clark, M.E.; Hand, S.C.; Bowlus, R.D.; Somero, G.N. Living with water stress: Evolution of osmolyte systems. *Science* **1982**, *217*, 1214–1222. [CrossRef]

33. Plaxton, W.C. The Organization and Regulation of Plant Glycolysis. *Annu. Rev. Plant Physiol. Plant Mol. Biol.* **1996**, *47*, 185–214. [CrossRef] [PubMed]

34. Popova, T.N.; Pinheiro de Carvalho, M.A. Citrate and isocitrate in plant metabolism. *Biochim. Biophys. Acta* **1998**, *1364*, 307–325. [CrossRef]

35. Jefferson, R.A.; Kavanagh, T.A.; Bevan, M.W. GUS fusions: Beta-glucuronidase as a sensitive and versatile gene fusion marker in higher plants. *EMBO J.* **1987**, *6*, 3901–3907. [CrossRef] [PubMed]

36. Nakagawa, T.; Kurose, T.; Hino, T.; Tanaka, K.; Kawamukai, M.; Niwa, Y.; Toyooka, K.; Matsuoka, K.; Jinbo, T.; Kimura, T. Development of series of gateway binary vectors, pGWBs, for realizing efficient construction of fusion genes for plant transformation. *J. Biosci. Bioeng.* **2007**, *104*, 34–41. [CrossRef] [PubMed]

Article

Growth under Fluctuating Light Reveals Large Trait Variation in a Panel of Arabidopsis Accessions

Elias Kaiser [1,2,*], Dirk Walther [1] and Ute Armbruster [1,*]

[1]   Max Planck Institute of Molecular Plant Physiology, Wissenschaftspark Golm, Am Mühlenberg 1, 14476 Potsdam, Germany; walther@mpimp-golm.mpg.de

[2]   Horticulture and Product Physiology, Wageningen University, Droevendaalsesteeg 1, 6708 PB Wageningen, The Netherlands

*   Correspondence: elias.kaiser@wur.nl (E.K.); armbruster@mpimp-golm.mpg.de (U.A.)

Received: 16 December 2019; Accepted: 18 February 2020; Published: 3 March 2020

Abstract: The capacity of photoautotrophs to fix carbon depends on the efficiency of the conversion of light energy into chemical potential by photosynthesis. In nature, light input into photosynthesis can change very rapidly and dramatically. To analyze how genetic variation in *Arabidopsis thaliana* affects photosynthesis and growth under dynamic light conditions, 36 randomly chosen natural accessions were grown under uniform and fluctuating light intensities. After 14 days of growth under uniform or fluctuating light regimes, maximum photosystem II quantum efficiency (F_v/F_m) was determined, photosystem II operating efficiency (Φ_{PSII}) and non-photochemical quenching (NPQ) were measured in low light, and projected leaf area (PLA) as well as the number of visible leaves were estimated. Our data show that Φ_{PSII} and PLA were decreased and NPQ was increased, while F_v/F_m and number of visible leaves were unaffected, in most accessions grown under fluctuating compared to uniform light. There were large changes between accessions for most of these parameters, which, however, were not correlated with genomic variation. Fast growing accessions under uniform light showed the largest growth reductions under fluctuating light, which correlated strongly with a reduction in Φ_{PSII}, suggesting that, under fluctuating light, photosynthesis controls growth and not vice versa.

Keywords: acclimation; chlorophyll *a* fluorescence; fluctuating light; natural variation; photosynthesis

---

1. Introduction

In nature, light energy supply for plant photosynthesis varies strongly in both amplitude and frequency. How plants respond to dynamic light environments is still poorly understood. This has many reasons, among them being that for most experiments, plants are grown under standard, highly controlled, and uniform light regimes (U). However, plant responses strongly depend on the environment that plants have acclimated to, as they adjust their metabolism to cope most efficiently with the prevailing condition. Several recent studies focused on the model plant *Arabidopsis thaliana* (hereafter: Arabidopsis) under dynamic light [1–3], but these were restricted to few accessions only and thus did not assess the effects of genetic factors. The study of a larger set of genotypes is warranted, also because photosynthesis shows great intraspecific variation in Arabidopsis [4,5], and variation on the genetic level has been shown to translate more strongly into phenotypic variation under fluctuating light regimes (FL; [6]). Using genome-wide association mapping, natural variation in the photosynthetic response to high light could be linked to several quantitative trait loci [5], supporting the notion that in Arabidopsis, traits linked to photosynthesis are heritable.

Compared to U, FL with the same average intensity often reduces plant growth [7–10], although exceptions exist [11]. There are at least two reasons for this reduction. Firstly, photosynthesis responds nonlinearly to light, i.e., at higher light intensities the rate of photosynthesis is limited by its capacity for CO_2 fixation, in turn leading to the activation of photoprotective mechanisms that dissipate absorbed light energy as heat, and resulting in a decrease in photosynthetic efficiency (reviewed by [12]). Leaves under FL, in contrast to U, are generally exposed to light periods during which photosynthesis is saturated. Secondly, photosynthesis generally lags behind rapid changes in light intensity, but the loss in CO_2 fixation after an increase in light intensity typically exceeds any gains in CO_2 fixation after a decrease (though see [13] for a further discussion on the role of post-illumination CO_2 fixation). The latter is partially connected to the slow relaxation of photoprotective mechanisms in low light [14].

Theoretically, plant growth is linked to how quickly photosynthesis can switch between protective mechanisms in high light and highly efficient light capture and conversion in shade periods [14]. Insufficient protection in high light causes photooxidative damage [15], while overprotection can result in low rates of photosynthesis through a reduction in the operating efficiency of PSII (Φ_{PSII}), particularly in shade periods when light availability limits the rate of photosynthesis [16]. Non-photochemical quenching (NPQ) is a central photoprotective mechanism in plants [17]. Most NPQ is rapidly reversible, and is controlled by the proton concentration of the lumen (reviewed by [12]). Light-driven electron transport from water to NADPH along the thylakoid localized electron transport chain is coupled to the transfer of protons from the stroma into the lumen. Protons then exit the lumen via the ATP synthase, thereby providing the energy required for ATP synthesis. In high light conditions, when downstream metabolic reactions are limiting, the proton concentration in the lumen rises, as efflux via the ATP synthase is restricted [18,19]. Above a threshold, the proton concentration in the lumen induces a reorganization of the PSII supercomplex via protonating key amino acid residues of the PsbS protein and activates the violaxanthin-deepoxidase (VDE), both of which are important for maximum pH-dependent quenching, which is also referred to as energy-dependent quenching (qE, reviewed by [12]). Under prolonged stress conditions, photoinhibitory quenching (qI) is induced, which coincides with oxidative damage to the D1 protein of photosystem II [17]. Such damage also causes an increased Chl *a* fluorescence of dark acclimated plants and is reflected as a decrease in F_v/F_m, which is a measure of the maximum quantum efficiency of PSII. However, reductions of F_v/F_m are often only observed under relatively harsh conditions, whereas NPQ and Φ_{PSII} already respond to milder stresses.

While the rapid response of photosynthesis to high light is comparably well studied on the molecular level (reviewed by [12]), much less is known about molecular mechanisms that allow photosynthesis to rapidly adjust to low light periods. Only recently, it was shown that plants contain at least one molecular player that accelerates the response of photosynthesis to shade periods [16,19]. Furthermore, the capacity for NPQ can be upregulated in Arabidopsis acclimated to FL [1,20], but it is not known whether (i) such an upregulation might reduce Φ_{PSII}, particularly in shade periods, during which this reduction could reduce photosynthetic efficiency and growth and (ii) whether natural genetic variation exists for these phenomena. We hypothesized that (i) NPQ would be upregulated and Φ_{PSII} would be decreased in low light in plants grown under FL compared to U, (ii) dark-adapted F_v/F_m would be largely unaffected by FL relative to U, (iii) growth would be reduced under FL compared to U, and that (iv) large differences for the extent of these changes between FL and U would become apparent between genotypes. To test these hypotheses, we grew 36 natural accessions of Arabidopsis (Table 1) under U and FL and next to their growth and development assessed Φ_{PSII} and NPQ under low light, as well as dark-adapted F_v/F_m.

Table 1. Full names and abbreviations (as used in the figures) of Arabidopsis accessions. Accessions are sorted by the fluctuating light (FL) experiment they were used in. Information on country of origin, latitude, and number of leaves at flowering was accessed on the 1001 genomes website (https://1001genomes.org/). nd = not determined.

Experiment	Name	Abbreviation	Country of Origin	Latitude	#Leaves Flowering
1	Be-1	Be	Germany	49.68	nd
1	Cen-0	Cen	France	49.00	nd
1 & 2	Col-0	Col	USA	nd	16.00
1	Da(1)-12	Da	Czech Republic	49.85	17.50
1	DraIV6-16	Dra	Czech Republic	49.41	nd
1	Fei-0	Fei	Portugal	40.50	22.25
1	Ge-0	Ge0	Switzerland	46.50	60.00
1	Hn-0	Hn	Germany	51.35	21.25
1	Hs-0	Hs	Germany	52.24	17.50
1	Kelsterbach-2	Kel	Germany	50.07	nd
1	Ler-1	Ler	Germany	nd	11.25
1	Mnz-0	Mnz	Germany	50.00	23.50
1	PHW-34	P34	France	48.61	43.50
1	PHW-37	P37	France	48.61	nd
1	Pla-0	Pla	Spain	41.50	40.50
1	Sha	Sha	Tajikistan	38.35	nd
1	TOU-H-13	TOU	France	46.67	nd
1	Tsu-0	Tsu	Japan	34.43	36.75
1	Ws-0	Ws	Russia	nd	42.25
1	Yo-0	Yo	USA	37.45	56.33
1	ZdrI2-24	Zdr	Czech Republic	49.39	nd
2	Amel-1	Ame	Netherlands	53.45	41.50
2	Db-0	Db	Germany	50.31	nd
2	Ge-1	Ge1	Switzerland	46.50	nd
2	Gel-1	Gel	Netherlands	51.02	20.0
2	HSm	HSm	Czech Republic	49.33	41.66
2	Kin-0	Kin	USA	44.46	nd
2	LDV-58	LDV	France	48.52	nd
2	MNF-Che-2	MNF	USA	43.53	46.00
2	PAR-3	PAR	France	46.65	nd
2	Sapporo-0	Sap	Japan	43.06	nd
2	Ta-0	Ta	Czech Republic	49.50	35.75
2	Tad01	Tad	Sweden	62.87	nd
2	TDr-3	TDr	Sweden	55.77	nd
2	UKNW06-060	UKN	UK	54.40	nd
2	VOU-2	VOU	France	46.65	nd

2. Results

2.1. Chlorophyll a Fluorescence

Twenty-six accessions (i.e., 72%) showed significantly reduced values of Φ_{PSII} when grown under FL as compared to growth under U (Figure 1A). No accession showed increased Φ_{PSII} under FL. NPQ was significantly increased in 13 (36%) of the FL-grown accessions (Figure 1B), and none of the FL-grown accessions showed significantly reduced NPQ. Maximum quantum efficiency of photosystem II (F_v/F_m), measured in dark-adapted leaves, showed a very different pattern compared to Φ_{PSII} and NPQ: of the ten accessions showing a significant effect of FL on F_v/F_m, six FL-grown accessions showed significantly enhanced F_v/F_m, while four others showed significantly reduced F_v/F_m (Figure 1C).

2.2. Growth and Development

Projected leaf area (PLA) and the number of visible leaves were assessed as proxies for growth and development, respectively. PLA of plants was significantly reduced when grown under FL as compared to U in 22 (61%) accessions (Figure 2A). On average, PLA was reduced by 36% across all accessions, meaning that the reduction in PLA was not only significant for many accessions, but also substantial (although PLA showed a large coefficient of variation: 42.3%). However, one accession, Hs-0, had a strongly increased PLA (74%) under FL compared to U (Figure 2A). Of the 12 accessions that showed significant treatment effects on the number of visible leaves, FL decreased leaf number in nine cases, but increased it in another three cases (among which was Hs-0; Figure 2B). Finally, to account for the possibility that differences in projected leaf area may be caused by differences in the number of leaves, average leaf size was determined by dividing projected leaf area by leaf number (Figure 2C). When expressed this way, 24 (67%) FL-grown accessions showed significant reductions in average leaf size, most of which overlapped with those showing reduced PLA. Hs-0 again displayed a significant increase in average leaf size when grown under FL. Two other accessions, Tsu-0 and Cen-0, showed larger growth- and development-related parameters under FL, albeit to a lesser extent than Hs-0. While for Tsu-0 this did not correlate with treatment effects on Chl *a* fluorescence data, Cen-0, like Hs-0, showed positive growth and development despite reduced Φ_{PSII} and increased NPQ in FL as compared to U (Figures 1 and 2). Accession Col-0 had been grown in both experiments: Φ_{PSII}, leaf area, leaf number and average leaf area were all significantly reduced in FL-grown Col-0, compared to Col-0 under uniform irradiance, in both experiments (Figures 1 and 2). NPQ was significantly increased under FL in Col-0 in Exp. 1 but not in Exp. 2, although data showed the same tendency (Figure 1B). Similarly, F_v/F_m was significantly reduced in FL-treated Col-0 in Exp. 1 but not in Exp. 2, however it tended to be reduced (Figure 1C). Altogether, these data suggest that similar conclusions could be drawn from both FL experiments, reassuring us of the repeatability of the experimental setup used.

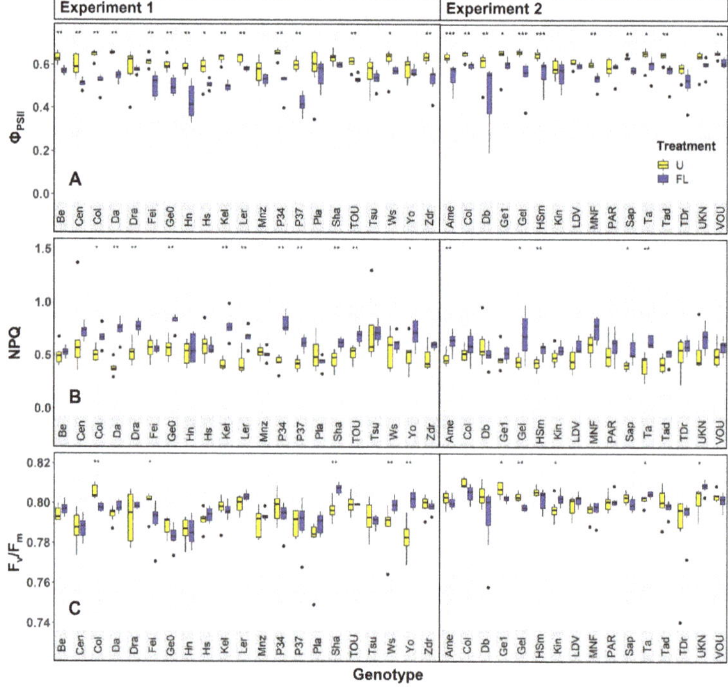

Figure 1. Chlorophyll *a* fluorescence analysis of 36 Arabidopsis accessions acclimated to uniform (U; yellow boxes) and fluctuating light intensities (FL; blue boxes), grown in two different experiments. (**A**) photosystem II operating efficiency (Φ_{PSII}), (**B**) non-photochemical quenching (NPQ), and (**C**) photosystem II maximum quantum efficiency (F_v/F_m). Φ_{PSII} and NPQ were measured at 90 µmol m^{-2} s^{-1}, whereas F_v/F_m was measured on dark-adapted leaves. Bars depict interquartile range (IQR; 25th–75th percentile) and median (thick line inside bar), whiskers depict data up to 1.5 × IQR, dots outside whiskers depict outliers (>1.5 × IQR). In the case of significant differences between average values under U and FL, these are shown for a given accession as: *** = $p < 0.001$, ** = $p < 0.01$ and * = $p < 0.05$ (n = 5–7).

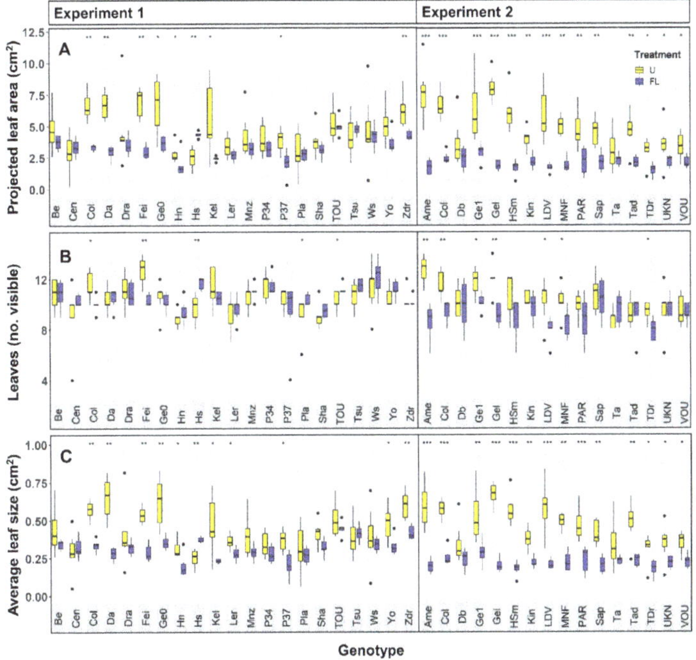

Figure 2. Growth and development of 36 Arabidopsis accessions acclimated to uniform (U; yellow boxes) and fluctuating light intensities (FL; blue boxes), grown in two different experiments. (**A**) Projected leaf area, (**B**) number of visible leaves and (**C**) average leaf size. Values were obtained from chlorophyll *a* fluorescence pictures. Bars depict interquartile range (IQR; 25th–75th percentile) and median (thick line inside bar), whiskers depict data up to 1.5 × IQR, dots outside whiskers depict outliers (>1.5 × IQR). In the case of significant differences between average values under U and FL, these are shown for a given accession as: *** = $p < 0.001$, ** = $p < 0.01$ and * = $p < 0.05$ (n = 5–7).

We observed a large trait variation across accessions and treatments (Figures 1 and 2). For example, projected leaf area for a given accession ranged from 1.4 to 8.0 cm^2 (Figure 2A), resulting in a coefficient of variation (CV) of 42.3%. Average leaf size varied to a slightly lesser degree (35.9% CV; Figure 2C), while leaf number per plant displayed a comparably low CV of 10.3% (Figure 2B). Of the chlorophyll *a* fluorescence traits, NPQ showed the largest variation (0.38–0.82; 19.2% CV; Figure 1B) while Φ_{PSII} (0.42–0.66; 9.0% CV; Figure 1A) varied comparably less. In contrast to these values, F_v/F_m showed a very small variation of only 0.8% CV (Figure 1C).

2.3. Principle Component Analysis

For each trait and accession, we calculated a response ratio by dividing the value of the trait due to growth under FL by its value under U. Based on the log−2 transformed response ratios of all phenotypes to FL relative to U, a principal component analysis was constructed (Figure 3). Principal component 1 accounted for 87.9% of the total variance, and was characterized by large loadings for traits in growth and development (PLA, leaf size, leaf number). Principal component 2 accounted for another 9.3%, and was dominated by a large loading for NPQ. Given that principal component 1 accounted for most of the variation and was dominated by traits related to growth, the PCA reveals that variation in growth was the biggest determinant for the overall variation in our data. Generally, there was little clustering of accessions based on any of the two principal components (Figure 3), suggesting large genetic variation for both groups of traits among the 36 accessions studied here.

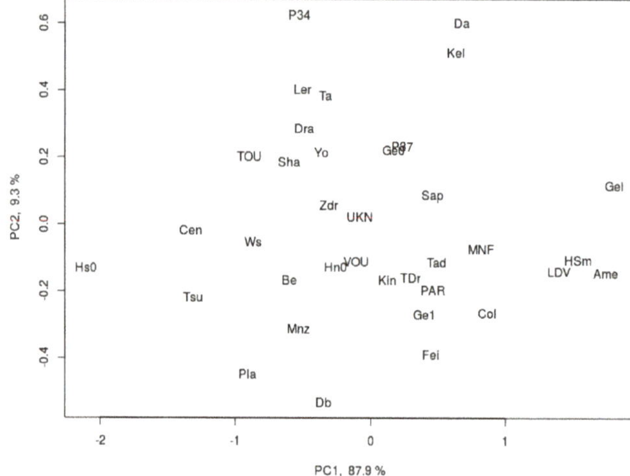

Figure 3. Principal component analysis of 36 Arabidopsis accessions, based on differences in phenotypical response ratio under fluctuating vs. uniform light (log−2 transformed FL/U ratio). PC1 accounts for 87.9% and PC2 accounts for 9.3% of the total variation.

2.4. Phenotypic and Genomic Cluster Analysis

Accessions were clustered based on values of the log−2 transformed response ratios (Figure 4A). Accessions were most strongly divided into different groups based on growth (PLA, leaf size), with Hs-0 being distinct from all other accessions. Another group of accessions, LDV-58, HSm, Gel-1, and Amel-1 could be defined as FL-sensitive in that these accessions showed the strongest reductions in growth and number of leaves, while also showing some reductions in Φ_{PSII} and relatively large increases in NPQ (Figure 4A). When clustered based on their genomic differences, on the other hand, accessions showed an entirely different pattern (Figure 4B). For example, Col-0 was suggested to have the largest genomic distance from all other accessions, while phenotypic clustering suggested its response to FL to be close to that of, e.g., Fei-0 and MNF-Che-2 (Figure 4A). Another example for the incongruence between genomic and phenotypic clustering is a group of closely related accessions that all originated in the Czech Republic: HSm, Ta-0, DraIV6-16, ZdrI2-24 and Da(1)-12 (Figure 4B, Table 1). The phenotypic analysis, on the other hand, did not indicate a close link between these accessions (Figure 4A). Indeed, no significant correlation was found between log−2 transformed response ratios and corresponding genomic distances between all pairwise comparisons among the 36 accessions (36 × 35/2 pairs; Pearson correlation coefficient = −0.02; p = 0.56). Together, these results suggest that the genomic distance alone cannot be used to predict behavior under FL compared to U, indicating that genetic variation in specific genes may account for the difference.

2.5. Correlation Analysis

For a more detailed view of the interrelations between chlorophyll *a* fluorescence, growth, and development data, we constructed a correlation matrix between the average values for each accession and the measured variable, and we included latitude of origin and the number of leaves formed until flowering (Table 1). Strong correlations between most chlorophyll *a* fluorescence parameters (except for the relationship between NPQ and F_v/F_m) and all growth and developmental parameters were found (Table 2). Average leaf size correlated strongly and positively with Φ_{PSII}, and negatively with NPQ (Figure 5), suggesting that light use efficiency had positive effects on leaf growth. These correlations are especially apparent for plants grown under U (yellow symbols in Figure 5). Both projected leaf area and the number of visible leaves correlated positively with F_v/F_m (Figure 6A,B). Also, F_v/F_m correlated negatively with the number of leaves formed until flowering (Figure 6C).

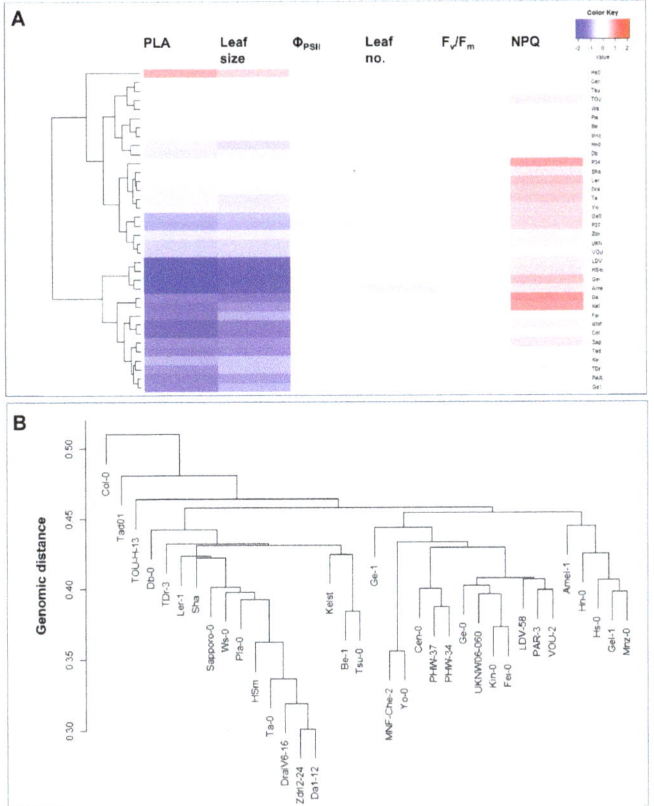

Figure 4. Clustering of 36 Arabidopsis accessions based on (**A**) differences in phenotypical logarithmic response ratios under fluctuating vs. uniform light (FL/U ratio; plot produced using heatmap.2 function with default settings of R package gplot) and (**B**) genomic distances, based on published SNP data (single linkage).

Table 2. Correlation matrix for traits observed in plants grown under uniform and fluctuating light. Blue colored backgrounds indicate a positive correlation, red indicates negative; the more strongly colored the background, the steeper the slope of the correlation. Statistically significant correlations ($p < 0.05$) are marked in bold. Numbers indicate Spearman's ρ, stars indicate the significance of the correlation, as: *** = $p < 0.001$, ** = $p < 0.01$ and * = $p < 0.05$ (n = 15–72). Lat., latitude of origin (°), #leaves flowering, number of leaves at flowering.

Trait	#Leaves at Flowering	Φ_{PSII}	NPQ	F_v/F_m	PLA	#Leaves	Leaf Size
Lat.	−0.51	−0.03	−0.09	−0.05	−0.18	−0.16	−0.17
#leaves at flowering		−0.13	0.16	**−0.41 ***	0.16	0.10	0.10
Φ_{PSII}			**−0.64 ***	**0.38 ***	**0.52 ***	0.12	**0.58 ***
NPQ				0.00	**−0.41 ***	−0.08	**−0.47 ***
F_v/F_m					**0.31 ***	**0.38 ***	**0.27 ***
PLA						**0.72 ***	**0.98 ***
#leaves							**0.59 ***

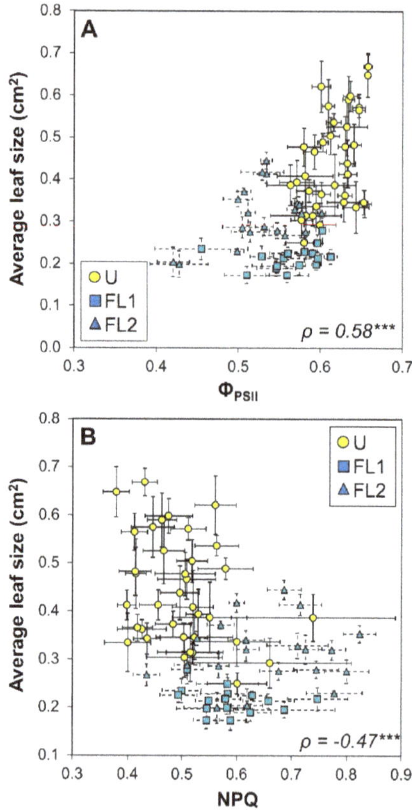

Figure 5. Relationships between average leaf size and (**A**) photosystem II operating efficiency (Φ_{PSII}) and (**B**) non-photochemical quenching (NPQ) in 36 Arabidopsis accessions acclimated to uniform (U) and fluctuating light intensities (FL). Data from plants grown under FL are sorted by experiment 1 (squares) and experiment 2 (triangles). Averages ± SE (n = 5–7). Spearman's ρ and the significance of a linear correlation through all points is shown (*** = $p < 0.001$).

Next, we tested whether response ratios of parameters derived from FL over U grown plants (FL/U) correlated at the trait level, as well as with latitude of origin and number of leaves until flowering (Table 3). This analysis showed a strong, positive correlation between $\Delta\Phi_{PSII}$ and $\Delta F_v/F_m$ (Table 3), suggesting that accessions with a strong reduction in Φ_{PSII} also showed a stronger reduction in F_v/F_m under FL. As might be expected, the response ratio of projected leaf area correlated strongly and positively with the response ratios of leaf number and leaf size (Table 3). We tested whether response ratios derived from either FL experiment 1 or FL experiment 2 yielded similar results, by repeating the same correlation analysis as shown in Table 3 for these two subsets of data (Tables S1 and S2). Both subsets yielded highly similar correlation coefficients, which themselves showed a strong linear correlation (Figure S1, $p < 0.001$). Correlation coefficients from each FL experiment subset also correlated strongly with those derived from the total dataset as shown in Table 3 ($p < 0.001$ in both cases, plots not shown). These results strongly suggest that the effects of fluctuating growth light on plants were repeatable within our experimental setup, and that hence similar conclusions can be drawn from both FL experiment 1 and FL experiment 2, further validating our findings.

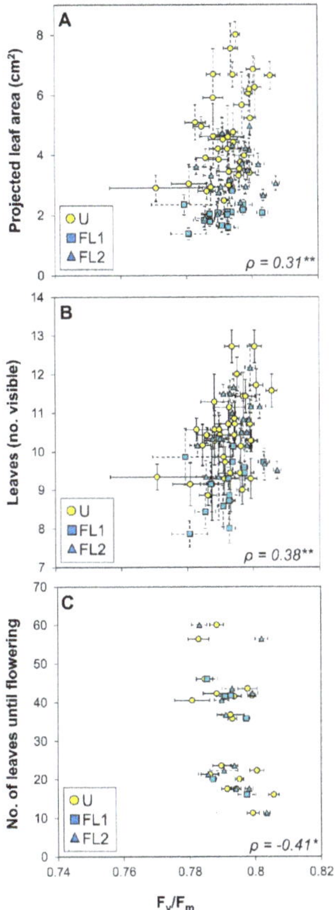

Figure 6. Relationships between photosystem II maximum quantum efficiency (F_v/F_m) and (**A**) projected leaf area, (**B**) the number of visible leaves and (**C**) the number of leaves required until flowering, in 36 Arabidopsis accessions acclimated to uniform (U) and fluctuating light intensities (FL). Data from plants grown under FL are sorted by experiment 1 (squares) and experiment 2 (triangles). Averages ± SE (n = 5–7). Spearman's ρ and the significance of a linear correlation through all points is shown (** = $p < 0.01$, * = $p < 0.05$).

Table 3. Correlation matrix for response ratio in traits under fluctuating light divided by those under uniform light (Δ = FL/U). Blue colored backgrounds indicate a positive correlation, red indicates negative; the more strongly colored the background, the steeper the slope of the correlation. Statistically significant correlations ($p < 0.05$) are marked in bold. Numbers indicate Spearman's ρ, stars indicate the significance of the correlation, as: *** = $p < 0.001$, ** = $p < 0.01$ and * = $p < 0.05$ (n = 15–36). Lat., latitude of origin (°), #leaves flowering, number of leaves at flowering.

Trait	#Leaves Flowering	$\Delta\Phi_{PSII}$	ΔNPQ	$\Delta F_v/F_m$	ΔPLA	Δ#Leaves	ΔLeaf Size
Lat.	−0.51	**−0.36** *	0.04	−0.03	−0.17	−0.07	−0.16
#leaves at flowering		0.07	0.05	0.13	−0.01	−0.01	0.06
$\Delta\Phi_{PSII}$			−0.16	**0.57** ***	0.17	0.16	0.14
ΔNPQ				0.05	−0.29	−0.19	**−0.35** *
$\Delta F_v/F_m$					**0.42** *	**0.48** **	**0.38** *
ΔPLA						**0.90** ***	**0.98** ***
Δ#leaves							**0.84** ***

The response ratio (FL/U) of PLA correlated strongly and negatively with PLA of plants grown under U (Figure 7A), suggesting that the reduction in PLA under FL was strongest in plants that showed high growth under U. The response ratio of average leaf size correlated positively with $\Delta F_v/F_m$ and negatively with ΔNPQ (Figure 7B,C), again suggesting that leaf growth was directly related to rates of photoprotection and photoinhibition. Interestingly, the latitude of origin correlated negatively with $\Delta\Phi_{PSII}$ (Figure 8), revealing a trend for photosynthesis of accessions collected further north on the globe to be more negatively affected by FL. Lastly, both the response ratio of projected leaf area and of number of visible leaves correlated positively with $\Delta F_v/F_m$, but this correlation was less meaningful given the large uncertainty around the mean for values of single accessions (Figure S2).

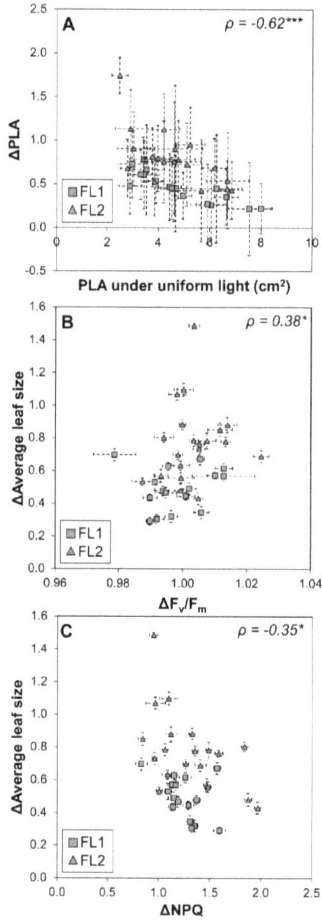

Figure 7. Changes in growth and chlorophyll *a* fluorescence in 36 Arabidopsis accessions grown under fluctuating compared to uniform light. (**A**) Relationship between projected leaf area (PLA) under uniform light and the response ratio of PLA under fluctuating light divided by PLA under uniform light (Δ = FL/U), (**B**) relationship between the response ratio of average leaf size and the response ratio of photosystem II maximum quantum efficiency ($\Delta F_v/F_m$), and (**C**) relationship between the response ratio of average leaf size and the response ratio of non-photochemical quenching (ΔNPQ). Data are sorted by FL experiment 1 (squares) and FL experiment 2 (triangles). Averages ± SE (n = 5–7). Spearman's ρ and the significance of a linear correlation through all points is shown (*** = p <0.001, * = p< 0.05).

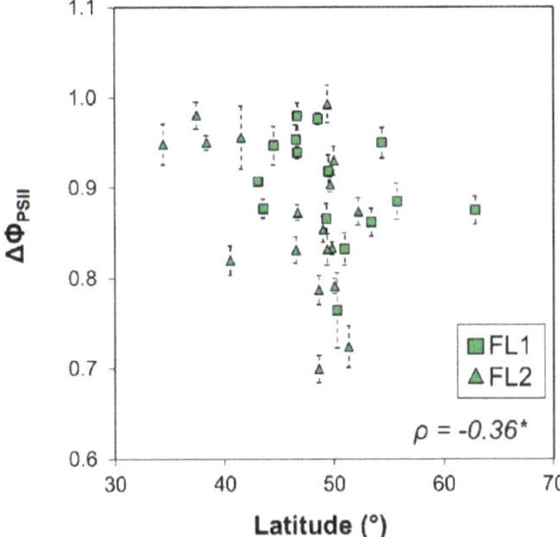

Figure 8. Relationship between latitude of origin of 36 Arabidopsis accessions and the response ratio in photosystem II operating efficiency ($\Delta\Phi_{PSII}$) between plants grown under fluctuating light intensities (FL) divided by values from plants grown under uniform light intensities (Δ = FL/U). Data are sorted by FL experiment 1 (squares) and FL experiment 2 (triangles). Averages ± SE (n = 5–7). Spearman's ρ and the significance of a linear correlation through all points is shown ($p < 0.05$).

3. Discussion

As photoautotrophs, plants interact with light in a direct manner. While our knowledge on rapid responses to changes in light intensity in the range of seconds to minutes is well advanced of plants grown under uniform light regimes (U; reviewed in [12,21–23]), less is known about the long-term response, i.e., acclimation occurring within days to fluctuating light (FL; [1,2]), and how this affects short-term responses to FL [2]. Here, we found that PLA and Φ_{PSII} generally decrease, NPQ generally increases, and F_v/F_m as well as number of leaves per plant generally remain unchanged, when plants are grown under FL compared to U. This, together with the large phenotypic variation observed, broadly confirms our hypotheses. However, we acknowledge that because (i) plants grown under U and FL received different light sums and day lengths during the first 14 days after sowing and (ii) the FL experiment was run twice, with different groups of accessions each, further experimental work should be conducted to confirm the robustness of these results. Additionally, it will be desirable to increase the number of accessions for future analyses.

3.1. Fluctuating Light Reduces Visible Leaf Area Most Strongly in Plants with High PLA Under Uniform Light

Most (61%) accessions showed a reduction in projected leaf area under FL, and this reduction was substantial (36% reduction in PLA across accessions). This compares well to previous studies [7–9] in which a reduction of 30–58% in biomass across several FL regimes and species was shown. In our study, accessions with the largest PLA under U showed a much smaller PLA and a reduced number of leaves under FL, e.g., Fei-0, Amel-1, Col-0, and Gel-0 (Figure 2). That these accessions showed high PLA and leaf number in U suggests that they were not generally restricted in their capacity for growth. That these accessions had a decreased PLA and number of leaves under FL suggests that reduced photosynthesis under FL is the primary cause for the reduction in these two growth proxies. Here, we acknowledge that possible effects of FL on leaf angles and leaf thickness could be confounding factors when trying to correlate PLA with biomass, the direct read out for growth capacity. However, the decrease in PLA correlates with lower number of leaves, together suggesting that growth is negatively affected in FL. In nature, the reverse is often the case, i.e., plant growth rate constrains photosynthesis, mostly due to suboptimal temperatures and/or nutrient or water availability [24]. In our experiment, both scenarios of temperature- or nutrient-constrained growth seem unlikely for most accessions, as there was a strong negative correlation between PLA of plants grown under U and the response ratio of the same trait (Figure 7A). This correlation suggests that "fast growers" under U showed strong reductions under FL, and that consequently under U they were not restricted in their rate of leaf area expansion by factors other than those directly related to photosynthesis. Conversely, for most accessions that showed low PLA under U (Hs-0, Cen-0, Pla-0, Tsu-0; Figure 2), PLA, leaf number as well as Φ_{PSII} and NPQ were less strongly affected by FL (Figure 4A).

3.2. Large Natural Genetic Variation for PLA and Chlorophyll a Fluorescence Traits

The coefficient of variation in accession-specific phenotypes ranged from 9 to 42%, suggesting that there was significant natural genetic variation (Figure 1, Figure 2, Figure 5, and Figure 6). Also, a large spread of accessions along principal components in the PCA was shown (Figure 3), suggesting that instead of distinct groups there was a continuum of responses. A similar range of genetic variation for traits related to photosynthesis and growth was previously found in populations of Arabidopsis [4,5], wheat [25,26] and rice [27]. The only exception to this observation was dark-adapted F_v/F_m with a CV of only 0.8%, suggesting that it was not strongly affected by either treatment or accession. This is in agreement with previous data from Arabidopsis [4] and rice [27], in which the photosynthetic trait with the smallest genetic variation was F_v/F_m, but other traits varied considerably. The range in Φ_{PSII} values observed here (0.42–0.66; Figure 5A) compares very well to data reported by van Rooijen et al. [5], a study which in a larger panel of Arabidopsis accessions found a spread of 0.47–0.66 for Φ_{PSII} when this was determined in plants acclimated to and measured at 100 μmol m^{-2} s^{-1}.

3.3. Correlations between Growth and Fluorescence: A Case for Rapid Phenotyping

Our data showed many interrelations between chlorophyll *a* fluorescence and growth-related traits (PLA, no. of visible leaves, average leaf size), suggesting that light use efficiency of photosystem II electron transport generally correlated with growth (Tables 2 and 3; Figures 5–7). Also, a strong negative and highly significant correlation ($p < 0.001$) between Φ_{PSII} and NPQ across treatments and accessions supports a link between photosynthetic efficiency and photoprotection (Table 2). This is remarkable, as this link is not often apparent at the relatively low light intensities that these measurements were conducted at (90 μmol m^{-2} s^{-1}).

Our data again emphasize chlorophyll *a* fluorescence imaging as a powerful tool for rapid plant phenotyping, enabling the analysis of photosynthesis-related traits in many accessions under multiple (and sometimes rapidly changing) environmental conditions [6,28,29]. Rapid plant phenotyping aims to simultaneously and repeatedly determine a large number of traits on a large number of plants. To that end, several weighing and imaging tools exist to determine, e.g., whole shoot and root growth [30], plant architecture, relative or absolute transpiration rates, as well as leaf temperature, photosynthetic capacity, spectral absorptivity [31], thickness, pigmentation and sugar concentration [32]. Since chlorophyll *a* fluorescence is often closely related to actual photosynthesis rates, and since photosynthesis reacts in a highly sensitive manner to intrinsic and extrinsic (e.g., environmental) factors, chlorophyll *a* fluorescence is a great tool to determine differences in photosynthesis within plants, across plants, and over time.

3.4. Can Evolutionary Adaptation Explain the Phenotype of an Accession?

Adaptation to a specific ecological niche shapes an organism's genome and its response to changes in its surroundings [33]. Correlation analysis revealed a greater reduction in Φ_{PSII} under FL the further north on the globe the accession had originated from (Figure 8). This finding could potentially be explained by the following factors, considering that the main phase of vegetative growth of most Arabidopsis accessions is in early spring: (i) lower temperatures with increasing latitude, thus, accessions from further north may have optimized their photosynthesis to lower temperatures than those in our experiment (16 °C night, 21 °C day); (ii) differences in day length, because changes in gene expression that drive acclimation in response to FL have been shown to intercept with the circadian clock [3]; and/ or (iii) light intensity and availability.

A negative correlation was also found between F_v/F_m and number of leaves until flowering (Figure 7C). The number of leaves until flowering is an indicator for whether Arabidopsis accessions germinate in spring or whether they germinate in autumn and require a long cold period in winter (vernalization) before flowering. Thus, it is tempting to speculate that species which form many leaves before flowering germinate in autumn and are thus better adapted to photosynthesis under cold temperatures, as they overwinter. A decrease in F_v/F_m is associated with sustained photoinhibitory quenching (qI) that accompanies PSII damage. It has been speculated that slowly reversible qI may have a photoprotective function under some conditions [34]. Thus, qI may be prematurely switched on in winter-grown accessions in order to protect against cold-stress.

4. Materials and Methods

4.1. Plant Material, Growth Conditions and Treatments

Thirty-five accessions of *Arabidopsis thaliana* were randomly selected from a collection of 330 accessions, and Col-0 was additionally selected as a reference genotype (36 accessions were used in total) due to its use as a wildtype in many reverse genetics studies. Information on country of origin, longitude, and the number of leaves until flowering (Table 1) was accessed on the 1001 genomes website (https://1001genomes.org/; [35]). In some cases, the Google Maps Arabidopsis viewer was additionally used to locate an accession. Latitudes for Col-0, Ler-1, and Ws-0 were omitted from Table 1 and correlation analyses (see below) as these genotypes have been cultivated in laboratories for decades and may not anymore be representative of the original accessions. All accessions are from the Northern Hemisphere, 28 (78%) were initially collected in Europe and another eight in Asia and North America (Table 1).

Seeds were sown on substrate prepared for Arabidopsis ('Arabidopsis substrate'; 70% white peat, 20% vermiculite, 10% sand; Stender, Schermbeck, Germany) which was enriched with 1 g L^{-1} each of two standard fertilizers: Osmocote Start® (ICL Specialty Fertilizers, Tel Aviv, Israel; composition: 11% N, 11% P_2O_5, 17% K_2O, 2% MgO, 0.38% Fe, 0.05% Mn, 0.01% B, 0.09% Cu, 0.009% Mo, 0.014% Zn) and Triabon® (Combo Expert, Münster, Germany; composition: 16% N, 8% P_2O_5, 12% K_2O, 4% MgO, 9% S, 0.02% B, 0.04% Cu, 0.1% Fe, 0.1% Mn, 0.02% Mo, 0.01% Zn; [36]). For the first 14 days, plants that were later used in the uniform light treatment were grown under a 16 h photoperiod, at ~150 µmol m^{-2} s^{-1} photosynthetically active radiation (PAR; 400–700 nm), while plants later used for the fluctuating light treatment were grown in a 12 h photoperiod, at 250 µmol m^{-2} s^{-1} PAR, for the first 14 days. This means that FL grown plants were initially exposed to a higher daily light integral than those in U (10.8 mol photons d^{-1} and 8.6 mol photons d^{-1}, respectively). Then, single plants were placed in a 0.11 L pot containing Arabidopsis substrate, and exposed to the light treatments until they were 28 days old. Day/night temperatures and relative humidity were 20/16 °C and 60/75% in all cases (for light and temperature recordings in the growth chambers, see Figure S3).

The treatments were as follows: uniform light (U) of 250 µmol m^{-2} s^{-1} PAR, and fluctuating light (FL) of alternating cycles of 900 µmol m^{-2} s^{-1} and 90 µmol m^{-2} s^{-1} PAR for one and four minutes, respectively (average light intensity: 252 µmol m^{-2} s^{-1} PAR). In both treatments, the photoperiod was 12 h, totaling 144 high/low light cycles in the FL treatment. Changes between the two light intensities in the FL experiment were very rapid and accurate (Figure S3). Three types of LED lamps (Roschwege, Greifenstein, Germany) were used in both treatments: white (LED-Star 2700 K 10 W), red (LED-Star DR 660 nm 5 W) and blue (LED-Star DB 460 nm 5 W). The output setpoints of all LED lamps were kept identical for any given light intensity, ensuring that there were no changes in light spectrum as light intensity changed. Accessions under the FL treatment were grown in two separate experiments: 20 accessions were grown in experiment 1, 15 different accessions were grown in experiment 2, and Col-0 was grown in both experiments (Table 1). Utmost care was taken that conditions in both FL experiments were identical. In the U treatment, all accessions were grown in one experiment. Plants were watered 2–3 times per week depending on substrate wetness to the touch (as per usual practice in the institute). To account for position effects in the climate chamber, plants were randomized during the treatment period.

4.2. Chlorophyll a Fluorescence

Images of chlorophyll *a* fluorescence values were obtained using the IMAG-MAX/L imaging PAM system (Heinz Walz GmbH, Effeltrich, Germany), under which six plants were measured simultaneously. Plants that had been light-adapted under growth conditions were first dark-adapted for $\geq$20 min, after which minimal (F_o) and maximal (F_m) chlorophyll *a* fluorescence emissions were measured. Plants were then adapted to growth light conditions ($\geq$30 min) for complete photosynthetic induction, after which they were shade-adapted at 90 μmol m^{-2} s^{-1} PAR in the imaging PAM for four minutes before chlorophyll *a* fluorescence emission under actinic light (F_s) and maximal fluorescence from the light-adapted leaf (F_m') were determined. Saturating beam duration was 0.7 s and saturating beam intensity was ~1300 μmol m^{-2} s^{-1} when determining F_m and ~2700 μmol m^{-2} s^{-1} in the case of F_m'. All measurements were conducted between 8:30 h and 14:00 h (growth lights switched on at 7:00).

From chlorophyll *a* fluorescence images, in ImagingWin (v2.47, Heinz Walz GmbH) four circular areas of interest (AOI) were selected per plant. In ImagingWin, AOI with several pre-defined diameters can be selected. AOI were chosen to cover as much total plant leaf area as possible while avoiding parts of the picture not covered by plant material; consequently, AOI were typically chosen to cover parts of the largest leaves of a plant (Figure S4A). From each area of interest, an average value of F_o, F_s, F_m and F_m' was obtained. The four values were later averaged to represent one biological replicate. Photosystem II maximum quantum efficiency was calculated as $F_v/F_m = (F_m - F_o)/F_m$, photosystem II operating efficiency was calculated as $\Phi_{PSII} = (F_m' - F_s)/ F_m'$, and non-photochemical quenching was calculated as NPQ = $(F_m - F_m')/F_m'$.

4.3. Growth and Development

Projected leaf area (PLA; i.e., total visible leaf area) was used as a proxy for growth, while the number of visible leaves was used as a proxy for development. For both variables, images of F_m, obtained with the imaging PAM, were used (Figure S4B). ImageJ (https://imagej.nih.gov/ij/) was used to calculate PLA from the number of pixels per plant; pot diameter was used to scale pixel number to actual plant size. Average leaf size was obtained by dividing PLA by the number of visible leaves.

4.4. Clustering of Accessions by Phenotypic Differences

Accessions were clustered into different categories depending on log$-$2-transformed values of the response ratio (ΔP), which was calculated for each parameter (P) as ΔP = PFL/PU, where PU and PFL are the average values of the parameter for each accession. Cluster 3.0 (http://www.falw.vu/~{}huik/cluster.htm) with Euclidean distance as similarity metric and average linkage was used. Based on this result, categories were manually adjusted to fit the specified criteria.

4.5. Principal Component Analysis

A principal component analysis (PCA) was performed on the log$-$2-transformed response ratios (ΔP) using the prcomp function as implemented in R (https://www.r-project.org/).

4.6. Genomic Distances

Genomic distances between accessions were computed as Kimura distances using the program dnadist (v3.698, [37]). Genomic sequences of all accessions were taken as those composed of the 178,083 sites that were detected polymorphic in at least one of the 36 accessions. Original sequences were taken from [38] with exception of the sequence information of accession PHW-34, which was not included in the set, and was taken from [35].

4.7. Statistical Analysis

Means of each accession grown under the two treatments were compared using the nonparametric Wilcoxon's rank-sum test, and *p*-values were adjusted for multiple comparisons using the Benjamini-Hochberg procedure [39]. Correlation analysis was performed on average responses per accession using Spearman's rank correlation coefficient analysis. When correlation analysis was performed on response ratios, log−2 transformed values were used. Data were analyzed in R, using the packages 'ggplot2' by Hadley Wickham, 'Hmisc' by Frank Harrell and 'ggpubr' by Alboudakel Kassambra. The number of biological replicates per genotype and experiment was 5–7.

Supplementary Materials: The following is available online at http://www.mdpi.com/2223-7747/9/3/316/s1, Table S1: Correlation matrix for response ratio in traits under fluctuating light divided by those under uniform light (Δ = FL/U), in FL experiment 1, Table S2: Correlation matrix for response ratio in traits under fluctuating light divided by those under uniform light (Δ = FL/U), in FL experiment 2, Figure S1: Relationship between correlation coefficients shown in Table S1 (FL 1) vs. correlation coefficients in Table S2, Figure S2: Relationships between the response ratio of dark-adapted F_v/F_m under fluctuating light divided by PLA under uniform light (Δ = FL/U) and A) the response ratio of projected leaf area as well as B) the response ratio of the number of visible leaves, Figure S3: Temperature ($^\circ$C, left, red) and light intensity (μmol m^{-2} s^{-1}, right, blue) measured in climate chamber for U (A) and FL (B,C) over the course of a day (A,B) and for 1 h (C) logged at 30 s interval, Figure S4: Data acquisition from chlorophyll a fluorescence pictures.

Author Contributions: Conceptualization, U.A. and E.K.; Methodology, U.A. and E.K.; Formal Analysis, all authors; Investigation, E.K.; Writing—Original Draft Preparation, E.K. and U.A.; Writing—Review and Editing, all authors; Visualization, all authors; Supervision, U.A.; Project Administration, U.A.; Funding Acquisition, U.A. All authors have read and agreed to the published version of the manuscript.

Funding: This research was funded by an ERA-CAPS grant from the Deutsche Forschungsgemeinschaft (DFG) to U.A (AR 808/4-1).

Acknowledgments: The authors thank Urszula Luzarowska and Roel van Bezouw for providing SNP information.

Conflicts of Interest: The authors declare no conflict of interest.

References

1. Alter, P.; Dreissen, A.; Luo, F.L.; Matsubara, S. Acclimatory responses of Arabidopsis to fluctuating light environment: Comparison of different sunfleck regimes and accessions. *Photosynth. Res.* **2012**, *113*, 221–237. [CrossRef]

2. Matthews, J.S.A.; Vialet-Chabrand, S.; Lawson, T. Acclimation to fluctuating light impacts the rapidity of response and diurnal rhythm of stomatal conductance. *Plant Physiol.* **2018**, *176*, 939–1951. [CrossRef]

3. Schneider, T.; Bolger, A.; Zeier, J.; Preiskowski, S.; Benes, V.; Trenkamp, S.; Usadel, B.; Farré, E.M.; Matsubara, S. Fluctuating light interacts with time of day and leaf development stage to reprogram gene expression. *Plant Physiol.* **2019**, *179*, 1632–1657. [CrossRef]

4. Rooijen, R.V.; Aarts, M.G.M.; Harbinson, J. Natural genetic variation for acclimation of photosynthetic light use efficiency to growth irradiance in Arabidopsis. *Plant Physiol.* **2015**, *167*, 1412–1429. [CrossRef]

5. Van Rooijen, R.; Kruijer, W.; Boesten, R.; Van Eeuwijk, F.A.; Harbinson, J.; Aarts, M.G.M. Natural variation of YELLOW SEEDLING1 affects photosynthetic acclimation of Arabidopsis thaliana. *Nat. Commun.* **2017**, *8*. [CrossRef]

6. Cruz, J.A.; Savage, L.J.; Zegarac, R.; Hall, C.C.; Satoh-Cruz, M.; Davis, G.A.; Kovac, W.K.; Chen, J.; Kramer, D.M. Dynamic Environmental Photosynthetic Imaging Reveals Emergent Phenotypes. *Cell Syst.* **2016**, *2*, 365–377. [CrossRef] [PubMed]

7. Vialet-Chabrand, S.; Matthews, J.S.A.; Simkin, A.J.; Raines, C.A.; Lawson, T. Importance of fluctuations in light on plant photosynthetic acclimation. *Plant Physiol.* **2017**, *173*, 2163–2179. [CrossRef] [PubMed]

8. Kubásek, J.; Urban, O.; Šantrůček, J. C4 plants use fluctuating light less efficiently than do C3 plants: A study of growth, photosynthesis and carbon isotope discrimination. *Physiol. Plant.* **2013**, *149*, 528–539. [CrossRef] [PubMed]

9. Leakey, A.D.B.; Press, M.C.; Scholes, J.D.; Watling, J.R. Relative enhancement of photosynthesis and growth at elevated CO2 is greater under sunflecks than uniform irradiance in a tropical rain forest tree seedling. *Plant, Cell Environ.* **2002**, *25*, 1701–1714. [CrossRef]

10. Vaseghi, M.J.; Chibani, K.; Telman, W.; Liebthal, M.F.; Gerken, M.; Schnitzer, H.; Mueller, S.M.; Dietz, K.J. The chloroplast 2-cysteine peroxiredoxin functions as thioredoxin oxidase in redox regulation of chloroplast metabolism. *Elife* **2018**, *7*, 1–28. [CrossRef]

11. Kaiser, E.; Matsubara, S.; Harbinson, J.; Heuvelink, E.; Marcelis, L.F.M. Acclimation of photosynthesis to lightflecks in tomato leaves: Interaction with progressive shading in a growing canopy. *Physiol. Plant.* **2018**, *162*. [CrossRef] [PubMed]

12. Kaiser, E.; Galvis, V.C.; Armbruster, U. Efficient photosynthesis in dynamic light environments: A chloroplast's perspective. *Biochem. J.* **2019**, *476*, 2725–2741. [CrossRef] [PubMed]

13. Pearcy, R.W.; Krall, J.P.; Sassenrath-Cole, G.F. Photosynthesis in fluctuating light environments. In *Photosynthesis and the Environment*; Baker, N.R., Ed.; Kluwer Academic: Dordrecht, The Netherlands, 1996; pp. 321–346.

14. Zhu, X.G.; Ort, D.R.; Whitmarsh, J.; Long, S.P. The slow reversibility of photosystem II thermal energy dissipation on transfer from high to low light may cause large losses in carbon gain by crop canopies: A theoretical analysis. *J. Exp. Bot.* **2004**, *55*, 1167–1175. [CrossRef] [PubMed]

15. Suorsa, M.; Järvi, S.; Grieco, M.; Nurmi, M.; Pietrzykowska, M.; Rantala, M.; Kangasjärvi, S.; Paakkarinen, V.; Tikkanen, M.; Jansson, S.; et al. PROTON GRADIENT REGULATION5 is essential for proper acclimation of Arabidopsis photosystem I to naturally and artificially fluctuating light conditions. *Plant Cell* **2012**, *24*, 2934–2948. [CrossRef] [PubMed]

16. Armbruster, U.; Carrillo, L.R.; Venema, K.; Pavlovic, L.; Schmidtmann, E.; Kornfeld, A.; Jahns, P.; Berry, J.A.; Kramer, D.M.; Jonikas, M.C. Ion antiport accelerates photosynthetic acclimation in fluctuating light environments. *Nat. Commun.* **2014**, *5*, 1–8. [CrossRef] [PubMed]

17. Takahashi, S.; Badger, M.R. Photoprotection in plants: A new light on photosystem II damage. *Trends Plant Sci.* **2011**, *16*, 53–60. [CrossRef]

18. Kanazawa, A.; Kramer, D.M. In vivo modulation of nonphotochemical exciton quenching (NPQ) by regulation of the chloroplast ATP synthase. *PNAS* **2002**, *99*, 12789–12794. [CrossRef]

19. Armbruster, U.; Leonelli, L.; Galvis, V.C.; Strand, D.; Quinn, E.H.; Jonikas, M.C.; Niyogi, K.K. Regulation and levels of the thylakoid K+/H+ antiporter KEA3 shape the dynamic response of photosynthesis in fluctuating light. *Plant Cell Physiol.* **2016**, *57*, 1557–1567. [CrossRef]

20. Caliandro, R.; Nagel, K.A.; Kastenholz, B.; Bassi, R.; Li, Z.; Niyogi, K.K.; Pogson, B.J.; Schurr, U.; Matsubara, S. Effects of altered α- and β-branch carotenoid biosynthesis on photoprotection and whole-plant acclimation of Arabidopsis to photo-oxidative stress. *Plant, Cell Environ.* **2013**, *36*, 438–453. [CrossRef]

21. Kaiser, E.; Morales, A.; Harbinson, J.; Kromdijk, J.; Heuvelink, E.; Marcelis, L.F.M. Dynamic photosynthesis in different environmental conditions. *J. Exp. Bot.* **2015**, *66*. [CrossRef]

22. Kaiser, E.; Morales, A.; Harbinson, J. Fluctuating light takes crop photosynthesis on a rollercoaster ride. *Plant Physiol.* **2018**, *176*. [CrossRef]

23. Armbruster, U.; Correa Galvis, V.; Kunz, H.H.; Strand, D.D. The regulation of the chloroplast proton motive force plays a key role for photosynthesis in fluctuating light. *Curr. Opin. Plant Biol.* **2017**, *37*, 56–62. [CrossRef] [PubMed]

24. Körner, C. Paradigm shift in plant growth control. *Curr. Opin. Plant Biol.* **2015**, *25*, 107–114. [CrossRef] [PubMed]

25. Driever, S.M.; Lawson, T.; Andralojc, P.J.; Raines, C.A.; Parry, M.A.J. Natural variation in photosynthetic capacity, growth, and yield in 64 field-grown wheat genotypes. *J. Exp. Bot.* **2014**, *65*, 4959–4973. [CrossRef] [PubMed]

26. Carmo-Silva, E.; Andralojc, P.J.; Scales, J.C.; Driever, S.M.; Mead, A.; Lawson, T.; Raines, C.A.; Parry, M.A.J. Phenotyping of field-grown wheat in the UK highlights contribution of light response of photosynthesis and flag leaf longevity to grain yield. *J. Exp. Bot.* **2017**, *68*, 3473–3486. [CrossRef]

27. Qu, M.; Zheng, G.; Hamdani, S.; Essemine, J.; Song, Q.; Wang, H.; Chu, C.; Sirault, X.; Zhu, X.G. Leaf photosynthetic parameters related to biomass accumulation in a global rice diversity survey. *Plant Physiol.* **2017**, *175*, 248–258. [CrossRef]

28. Murchie, E.H.; Kefauver, S.; Araus, J.L.; Muller, O.; Rascher, U.; Flood, P.J.; Lawson, T. Measuring the dynamic photosynthome. *Ann. Bot.* **2018**, *122*, 207–220. [CrossRef]

29. van Bezouw, R.F.H.M.; Keurentjes, J.J.B.; Harbinson, J.; Aarts, M.G.M. Converging phenomics and genomics to study natural variation in plant photosynthetic efficiency. *Plant J.* **2019**, *97*, 112–133.

30. Fiorani, F.; Schurr, U. Future Scenarios for Plant Phenotyping. *Annu. Rev. Plant Biol.* **2013**, *64*, 267–291. [CrossRef]

31. Dutta, S.; Cruz, J.A.; Jiao, Y.; Chen, J.; Kramer, D.M.; Osteryoung, K.W. Non-invasive, whole-plant imaging of chloroplast movement and chlorophyll fluorescence reveals photosynthetic phenotypes independent of chloroplast photorelocation defects in chloroplast division mutants. *Plant J.* **2015**, *84*, 428–442. [CrossRef]

32. Yendrek, C.R.; Tomaz, T.; Montes, C.M.; Cao, Y.; Morse, A.M.; Brown, P.J.; McIntyre, L.M.; Leakey, A.D.B.; Ainsworth, E.A. High-throughput phenotyping of maize leaf physiological and biochemical traits using hyperspectral reflectance. *Plant Physiol.* **2017**, *173*, 614–626. [CrossRef] [PubMed]

33. Hancock, A.M.; Brachi, B.; Faure, N.; Horton, M.W.; Jarymowycz, L.B.; Sperone, F.G.; Toomajian, C.; Roux, F.; Bergelson, J. Adaptation to climate across the Arabidopsis thaliana genome. *Science (80-.)* **2011**, *334*, 83–86. [CrossRef] [PubMed]

34. Lee, H.Y.; Wah, S.C.; Hong, Y.-N. Photoinactivation of Photosystem II in leaves of Capsicum annuum. *Photosynth. Res.* **1999**, *105*, 377–384.

35. Consortium, 1001 Genomes 1,135 Genomes Reveal the Global Pattern of Polymorphism in Arabidopsis thaliana. *Cell* **2016**, *166*, 481–491. [CrossRef] [PubMed]

36. Köhl, K.; Tohge, T.; Schöttler, M.A. Performance of Arabidopsis thaliana under different light qualities: Comparison of light-emitting diodes to fluorescent lamp. *Funct. Plant Biol.* **2017**, *44*, 727–738. [CrossRef]

37. Felsenstein, J. PHYLIP (Phylogeny Inference Package), version 3.698. 2019.

38. Horton, M.W.; Hancock, A.M.; Huang, Y.S.; Toomajian, C.; Atwell, S.; Auton, A.; Muliyati, N.W.; Platt, A.; Sperone, F.G.; Vilhjálmsson, B.J.; et al. Genome-wide patterns of genetic variation in worldwide Arabidopsis thaliana accessions from the RegMap panel. *Nat. Genet.* **2012**, *44*, 212–216. [CrossRef] [PubMed]

39. Benjamini, Y.; Hochberg, Y. Controlling the false discovery rate: A practical and powerful approach to multiple testing. *J. R. Stat. Soc. Ser. B* **1995**, *57*, 289–300. [CrossRef]

Article

Evolution of Photorespiratory Glycolate Oxidase among Archaeplastida

Ramona Kern [1,†], Fabio Facchinelli [2], Charles Delwiche [3], Andreas P. M. Weber [2], Hermann Bauwe [1] and Martin Hagemann [1,4,*]

[1]   Plant Physiology, Institute of Biological Sciences, University of Rostock, Albert-Einstein-Str. 3, D-18051 Rostock, Germany; ramona.kern@uni-rostock.de (R.K.); hermann.bauwe@uni-rostock.de (H.B.)

[2]   Institute of Plant Biochemistry, Cluster of Excellence on Plant Science (CEPLAS), Heinrich-Heine-University Düsseldorf, 40225 Düsseldorf, Germany; fabio.facchinelli@uni-duesseldorf.de (F.F.); andreas.weber@uni-duesseldorf.de (A.P.M.W.)

[3]   Department of Cell Biology and Molecular Genetics, University of Maryland, College Park, MD 20742, USA; delwiche@umd.edu

[4]   Department Life, Light & Matter, University of Rostock, 18051 Rostock, Germany

*   Correspondence: martin.hagemann@uni-rostock.de; Tel.: +49-(0)381-4986110; Fax: +49-(0)381-4986112

†   Current address: University of Rostock, Institute of Biological Sciences, Applied Ecology & Phycology, Albert-Einstein-Str. 3, D-18059 Rostock, Germany.

Received: 29 November 2019; Accepted: 11 January 2020; Published: 15 January 2020

Abstract: Photorespiration has been shown to be essential for all oxygenic phototrophs in the present-day oxygen-containing atmosphere. The strong similarity of the photorespiratory cycle in cyanobacteria and plants led to the hypothesis that oxygenic photosynthesis and photorespiration co-evolved in cyanobacteria, and then entered the eukaryotic algal lineages up to land plants via endosymbiosis. However, the evolutionary origin of the photorespiratory enzyme glycolate oxidase (GOX) is controversial, which challenges the common origin hypothesis. Here, we tested this hypothesis using phylogenetic and biochemical approaches with broad taxon sampling. Phylogenetic analysis supported the view that a cyanobacterial GOX-like protein of the 2-hydroxy-acid oxidase family most likely served as an ancestor for GOX in all eukaryotes. Furthermore, our results strongly indicate that GOX was recruited to the photorespiratory metabolism at the origin of Archaeplastida, because we verified that Glaucophyta, Rhodophyta, and Streptophyta all express GOX enzymes with preference for the substrate glycolate. Moreover, an "ancestral" protein synthetically derived from the node separating all prokaryotic from eukaryotic GOX-like proteins also preferred glycolate over L-lactate. These results support the notion that a cyanobacterial ancestral protein laid the foundation for the evolution of photorespiratory GOX enzymes in modern eukaryotic phototrophs.

Keywords: Glycolate oxidase; photorespiration; evolution; Archaeplastida; Cyanobacteria

---

1. Introduction

Oxygenic photosynthesis is among the most important biological processes on Earth, because it produces the vast majority of organic carbon and nearly all of the atmospheric oxygen. This process is thought to have evolved approximately 2.5 billion years ago in cyanobacteria, and was later conveyed into a eukaryotic host cell via endosymbiosis, giving rise to plastids [1–3]. The phototrophic eukaryotes Glaucophyta, Rhodophyta (red algae), and Viridiplantae (green algae and land plants) form the monophyletic group of the Archaeplastida, which all harbor primary plastids that evolved from a common cyanobacterial ancestor [4] (for an alternative hypothesis of multiple origins and convergent evolution of primary plastids, see [5]). Other eukaryotic algal groups contain plastids, which evolved via secondary- or higher-order endosymbiosis of a green or red algal ancestor [4]. The green lineage

(Viridiplantae) comprises two major clades, the "classical" green algae (Chlorophyta), and their sister lineage, the charophyte green algae, from which land plants evolved (Streptophyta) [6,7].

All organisms performing oxygenic photosynthesis fix inorganic carbon as CO_2 by Rubisco (Ribulose 1,5-bisphosphate carboxylase). In addition to the carboxylation of ribulose 1,5-bisphosphate (RuBP) with CO_2, Rubisco also catalyzes the oxygenation of RuBP, being the first step in a process termed "photorespiration", leading to the appearance of the potent enzyme inhibitor 2-phosphoglycolate (2PG) [8,9]. Despite the evolution of different types of Rubisco with varying substrate affinities and specificity factors, the oxygenase reaction cannot be avoided under the present atmospheric conditions, containing 21% O_2 and 0.04% CO_2 [10]. As 2PG is toxic for the plant's metabolism [11–13], it is rapidly metabolized to glycolate, which is efficiently metabolized in the photorespiratory pathway [14]. However, this salvage process is energetically costly and recovers only 75% of the organic carbon, while 25% is lost as CO_2. It has been estimated that in a crop, C3 plant photorespiration might decrease the yield by approximately 30% under present day atmospheric conditions [15]. Thus, photorespiration is one key target in molecular breeding attempts to improve crop productivity [16,17].

Initially, it was assumed that photorespiration evolved rather late when plants started to colonize the continents and became exposed to the high O_2-containing atmosphere with low CO_2 (e.g., [18]). In contrast, we hypothesized that photorespiration and oxygenic photosynthesis coevolved in ancient cyanobacteria [14,19]. This hypothesis is based on the discovery of the essential function of the photorespiratory 2PG metabolism under ambient air conditions in the model cyanobacterium *Synechocystis* sp. PCC 6803 [19]. However, the cyanobacterial photorespiratory 2PG metabolism differs from that of plants. First, it is not compartmentalized in the prokaryotic cell. Second, glycolate oxidation is performed by glycolate dehydrogenases among cyanobacteria, and not by glycolate oxidase (GOX), as in plants (Figure 1). Third, in addition to a plant-like 2PG metabolism, *Synechocystis* can also metabolize glyoxylate via the bacterial glycerate pathway and by its complete decarboxylation [19]. The subsequent phylogenetic analysis also revealed that not all enzymes of the plant 2PG cycle originated from the cyanobacterial ancestor. In particular, the photorespiratory enzymes that operate in the mitochondrion of plants are more closely related to enzymes of proteobacteria, the ancestors of this organelle [20,21]. We hence hypothesized that these enzymes, which originally also existed in the cyanobacterial endosymbiont, were replaced by the already established mitochondrial enzymes of a proteobacterial origin [22].

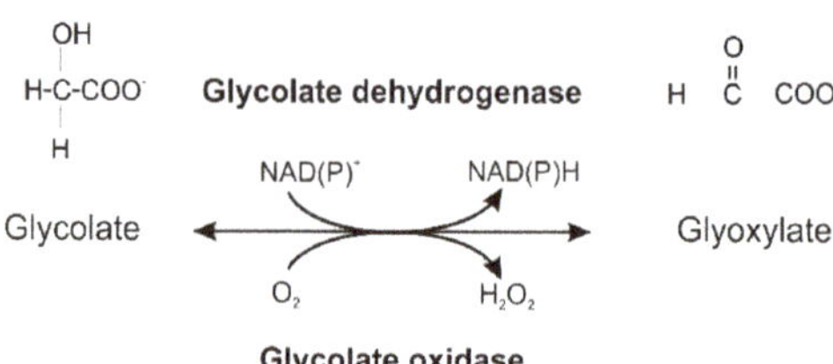

Figure 1. Glycolate oxidation can be catalyzed either by a glycolate dehydrogenase (top) or a glycolate oxidase (bottom).

However, in contrast to plants and similar to cyanobacteria, chlorophyte algae, such as *Chlamydomonas reinhardtii*, also use glycolate dehydrogenase for glycolate oxidation, not GOX [23]. The presence of glycolate dehydrogenases in green algae and the later discovery of similar enzymes in cyanobacteria correlated with the existence of an efficient inorganic carbon concentrating mechanism (CCM), which suppresses photorespiration to a great extent in these organisms. Therefore, it was initially assumed that cyanobacteria and chlorophytes with a lower photorespiratory flux due to the CCM prefer glycolate dehydrogenases, which have a higher affinity to glycolate than GOX [24,25]. Accordingly, it was proposed that plant-like photorespiration, including the peroxisomal glycolate oxidation via GOX, only evolved late among streptophytic green algae, and represented an adaption to a low CO_2 and high O_2 concentration, which was indeed necessary for the later terrestrialization of the plant kingdom [18].

Recently, numerous new genome sequences became publicly available, which permitted a broader analysis of the distribution of GOXes and glycolate dehydrogenases among different algal lineages. Those searches also revealed that some cyanobacterial genomes possess a GOX-like protein [20]. The subsequent biochemical analysis of the GOX-like protein from the diazotrophic cyanobacterium *Nostoc* sp. PCC 7120 showed that this protein is rather a L-lactate oxidase (LOX). LOX and GOX (here summarized as GOX-like proteins) belong to the group of 2-hydroxy-acid oxidases, display highly similar primary and tertiary structures (see Figure S1), and share the same catalytic mechanism [26–29]. Phylogenetic analysis using GOX-like proteins from heterotrophic bacteria, cyanobacteria, eukaryotic algae, and plants implied that the cyanobacterial GOX-like protein is the common ancestor of all plant GOX proteins, which was consistent with the hypothetical transfer of genes for photorespiratory enzymes into the plant genome via primary endosymbiosis from the previously assumed ancestral N_2-fixing cyanobacterium [30]. However, more recent phylogenetic analyses point at the likely uptake of a non-N_2-fixing, early-branching unicellular cyanobacterium as a primary endosymbiont [31]. Another study included GOX-like proteins from non-photosynthetic eukaryotes in the phylogenetic analysis, and postulated that plant and animal GOX proteins share a common non-cyanobacterial ancestry [32].

Hence, to solve this controversy and better understand the evolutionary origin of the photorespiratory GOX, we reanalyzed GOX-like proteins using biochemical and phylogenetic approaches. Our results support the view that GOX became part of the photorespiratory metabolism early in the evolution of Archaeplastida, and that a cyanobacterial GOX-like protein most likely served as the ancestor for GOX in all eukaryotic lineages.

2. Results

2.1. Data Mining

To evaluate the GOX phylogeny, we considered a broad spectrum of phyla, including non-photosynthetic species. The well-characterized At-GOX2 from *Arabidopsis thaliana* (At3g14415) [30] was used in BLASTP [33] searches against defined taxonomic groups (e.g., cyanobacteria or proteobacteria) to find related GOX-like proteins. Proteins from 5 to 14 species of each taxonomic group were selected, which showed the best BLAST hits. In the case of underrepresented taxonomic groups, all of the sequences that could be identified as putative GOX proteins were used for the alignment (Table S1). In addition to the sequences in the databases, we determined the cDNA sequence of the putative GOX from the streptophyte green alga *Spirogyra pratensis* (sequence is shown in Figure S2; accession number AVP27295.1). Furthermore, the obviously miss-annotated GOX sequence of *Cyanophora paradoxa* was corrected (see below). The complete gene with a corrected exon/intron structure and the complete protein coding sequence is shown in Supplementary Material 1 (Figure S3; accession number AVP27296.1).

2.2. Monophyly of Eukaryotic and Cyanobacterial GOX-Like Proteins

To reanalyze the phylogenetic origin of the plant and animal GOX-like proteins, we took sequences from a broad spectrum of phyla into consideration, for example, we also included related proteins from the chromalveolate taxa (Table S1). In total, 111 GOX-like proteins from 11 groups, including Archaeplastida and Metazoa, were incorporated into the phylogenetic analysis. GOX-like proteins from fungi were excluded, as these proteins show an accelerated evolution preventing their comparison [32]. We also restricted the analysis to one putative GOX isoform each from algae, plants, and animals, as the previous study [32] showed that the diversification of GOX into different biochemical subgroups occurred from one of the ancestral proteins within these groups. The alignment was constructed with ProbCons [34]. Additionally, we also constructed alignments using MUSCLE [35], compared both results, and changed the alignment if necessary in order to obtain the best scores. Using the final alignment (Supplementary Material 2), we reconstructed a protein tree using Bayesian interference.

The midpoint-rooted Bayesian tree (Figure 2) is well supported by the Bayesian posterior probabilities (BPP). The GOX-like proteins from all of the eukaryotes form a monophyletic group (BPP = 0.96). The proteins from Chlorophyta cluster in that group, however, they build an outgroup to other eukaryotic GOX-like proteins and are distinct from that of other Archaeplastida. The divergence of the GOX proteins of streptophytes is consistent with the fact that the chlorophyte GOX-like proteins act as LOX enzymes (see below). The GOX-like sequences from the Metazoa cluster together with sequences of chromalveolate taxa, including non-photosynthetic and photosynthetic groups; this clade is the sister clade to all Archaeplastida, except Chlorophyta (BPP = 1). Interestingly, LOX proteins from cyanobacteria cluster as sisters to all eukaryotes, showing a close relationship between cyanobacterial and eukaryotic GOX-like proteins. It is also noteworthy that the GOX-like proteins of chromalveolates do not cluster within red algae, as would be expected from the secondary origin of their plastids, but rather form a clade with Metazoa, although the posterior probabilities for the critical branches placing red algae are relatively low. The cyanobacterial and eukaryotic clades of GOX-like proteins are separated from the proteins of Actinobacteria (also including non-sulfur bacteria), as well as from the clade Proteobacteria (also including Verrucomicrobia; BPP = 1). The outermost clade is built by Firmicutes and Archaea. A similar clustering of GOX-like proteins was obtained using the maximum likelihood (ML) algorithm (Figure S4). However, in contrast to the Bayesian tree, some branches are not well supported in the ML tree, but all cyanobacterial and eukaryotic GOX-like proteins again form one monophyletic, statistically well-supported clade.

Similar to the previous study by Esser et al. [32], we also rooted the tree by assuming the monophyly of the eukaryotic clade (Figure S5B). This rooting resulted in a monophyletic group of all Eubacteria and Archaea, which cluster in a sister group relationship to Eukaryotes, including plants, algae, and animals. However, the clustering of the bacterial and archaeal GOX-like proteins in this rooted tree is not congruent to the clustering of taxonomic groups, which is observed when analyzing, for example, signature sequences in proteins [36]. In both trees, Archaea and Firmicutes cluster as sister groups. Alternatively, we placed the root of the tree between Firmicutes or Archaea and the other clusters of GOX-like proteins (Figure S5C,D). Both of these trees again showed a sister group topology of sequences from eukaryotes and cyanobacteria, as found before with the midpoint-rooted tree.

Overall, our phylogenetic analyses show that eukaryotic GOX-like proteins are more closely related to those of cyanobacteria than to those of any other prokaryotic group, including Proteobacteria and Archaea. As it is known that phylogenetic analysis can be biased when homologous proteins evolved into enzymes with varied metabolic functions, we set out a comprehensive biochemical evaluation of GOX-like proteins among Archaeplastida.

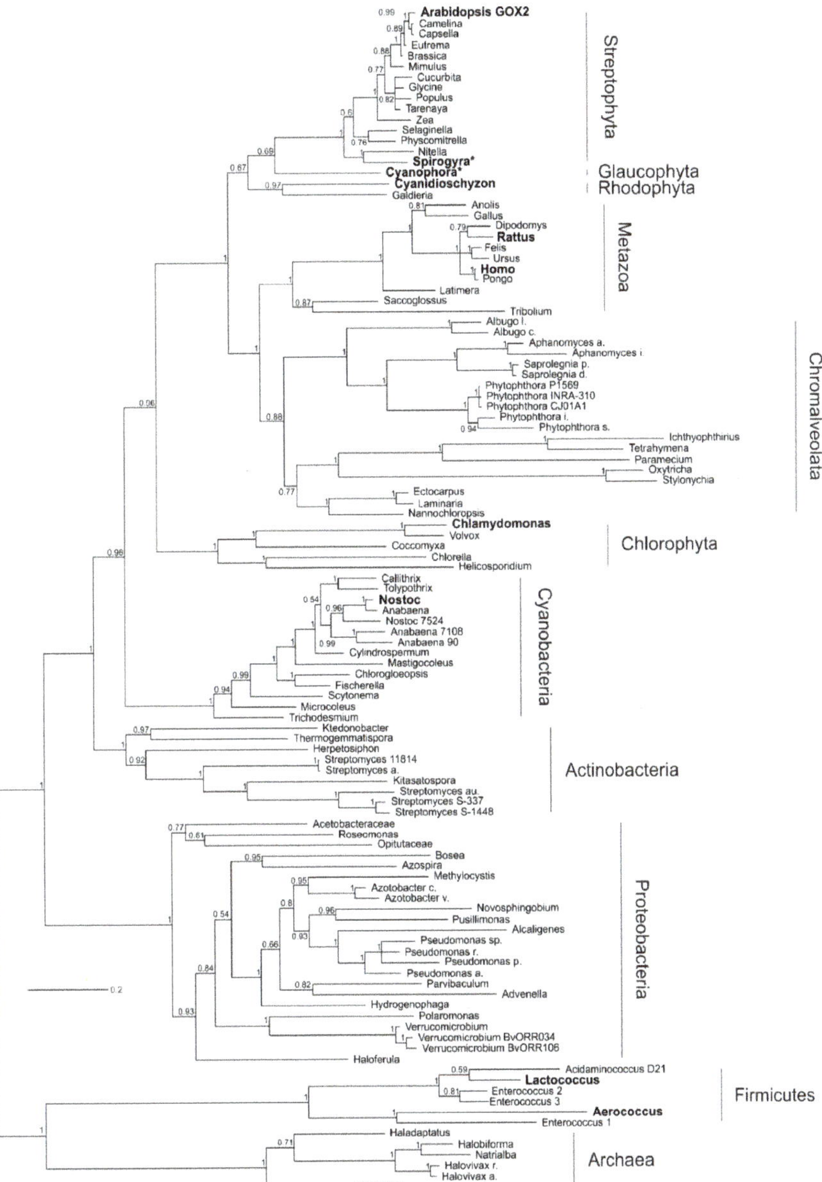

Figure 2. Phylogenetic tree of glycolate oxidase (GOX)-like proteins. The tree is based on GOX-like proteins from all groups in the tree of life and was implemented using Bayesian interference. The monophyletic group of GOX-like proteins from Eukaryotes builds the sister clade to cyanobacteria, pointing to a cyanobacterial origin of all eukaryotic GOX proteins, including heterotrophic organisms like animals. GOX-like proteins that have been biochemically verified are printed in bold (https://www.brenda-enzymes.org/index.php; EC 1.1.3.15; [37]). Proteins with a star were analyzed in this study. The numbers at the nodes show the posterior probability. The full species names and accession numbers are listed in Table S1. GOX—glycolate oxidase; LOX—lactate oxidase.

2.3. Archaeplastida Except Chlorophyta Possess a GOX Protein

It is known that GOX-like proteins show varying (multi-)substrate preferences. For example, many of these enzymes preferentially oxidize L-lactate, that is, they rather represent LOX than GOX enzymes. Therefore, we analyzed the substrate preference of GOX-like proteins from all groups of Archaeplastida. To this end, we overexpressed selected cDNAs in *E. coli* to obtain recombinant proteins for biochemical characterization. In addition to the previously obtained data for GOX and LOX proteins from Cyanobacteria, Rhodophyta, Chlorophyta, and land plants [30,38], the His- or Strep-tagged proteins from the early splitting-off glaucophyte alga *Cyanophora paradoxa* and the streptophyte (sister clade of land plants) alga *Spirogyra pratensis* were biochemically characterized using substrate concentrations of glycolate and L-lactate ranging from 0.1 to 200 mM. We found that both recombinant proteins can oxidize glycolate and, to a lesser extent, L-lactate (Figure 3 and Table 1), which is consistent with earlier reports of clear GOX activity in crude extracts from these algae [24,39,40]. Compared to At-GOX2 from *Arabidopsis*, both enzymes show a higher affinity (at least 10 times) to glycolate than to L-lactate as a substrate. Although the Vmax values are only slightly higher for glycolate than for L-lactate, the catalytic efficiency (k_{cat}/K_m) is 20 to 33 time higher with glycolate as substrate, which is similar to AtGOX2 (Table 1 and Table S2). Based on this preference for glycolate as substrate, the enzymes were named Cp-GOX and Sp-GOX, respectively. Compared with other GOX proteins, Cp-GOX shows the lowest K_m and the second highest k_{cat}/K_m value for glycolate (Table 1 and Table S2). This finding could point to an early evolution of glycolate oxidation activity in the ancestor of cyanobacteria. To verify this hypothesis, we modeled, synthesized, and characterized an ancestral GOX protein.

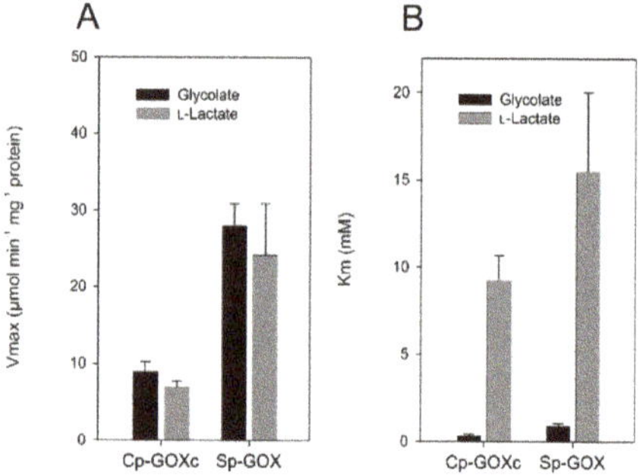

Figure 3. Biochemical characterization of GOX proteins from algae. The nonlinear regression was fitted to the Michaelis–Menten kinetic using Sigma Plot 13.0. The resulting parameters for the maximal enzymatic reaction rate (V_{max}) (**A**) and the substrate affinity expressed as K_m (**B**) were calculated for the substrates glycolate and L-lactate. The standard deviation was calculated from at least two independent biological replicates. Cp-GOXc—*Cyanophora paradoxa*; Sp-GOX—*Spirogyra pratensis*.

2.4. Ancestral GOX-Like Protein Sequence with Active Site Identical to Plant GOX

In contrast to the proteins of Archaeplastida, except Chlorophyta, cyanobacteria (Table 1) and all other analyzed prokaryotes possess LOX, a GOX-like protein that prefers L-lactate over glycolate [27,30,41]. To get an idea about the substrate preference of an ancestral protein, we reconstructed a protein that could correspond to the hypothetical common ancestor of GOX-like proteins from Archaeplastida. The primary structure of this "ancestral" protein was derived from a reduced phylogenetic tree based on Bayesian interference, including 37 sequences from five different taxonomic groups (Figure 4). Also, for the reduced dataset, the monophyly of cyanobacterial and eukaryotic GOX-like proteins is supported by a high posterior probability (BPP = 1). The ancestral protein, named N3-GOX, used for the biochemical analysis referred to the most probable amino acid sequence calculated via the ML algorithm. An amino acid sequence comparison of N3-GOX with

At-GOX2 from *Arabidopsis* revealed that the three amino acid residues in the active site (Table 2), which were previously proven to determine the specificity for the substrate glycolate [30], were identical in these enzymes, pointing to a higher GOX rather than LOX activity of the ancestral protein. Comparing the entire sequence, 68% of the amino acid residues are identical in these two proteins. In contrast, the cyanobacterial No-LOX (from *Nostoc* sp. PCC7120 [30]) shares only 56% identical positions with N3-GOX, and shows different active-site amino acid residues (Table 3).

Table 1. Summary of the K_m and V_{max} values of plant, algal, cyanobacterial, and ancestral GOX-like proteins. The table displays mean values and standard deviation of the V_{max} and K_m values with the substrates glycolate or L-lactate from measurements with at least three biological replicates. The enzymes are ordered as inferred from the phylogenetic tree displayed in Figure 2, that is, starting with the hypothetical ancestral form N3-GOX, followed by the cyanobacterial (No-LOX) and chlorophytic LOX (Cr-LOX) enzymes, and then the GOX proteins from Archaeplastida, namely: glaucophytic (Cp-GOXc), rhodophytic (Cm-GOX), and streptophytic (Sp-GOX) toward plant (At-GOX2) enzymes.

Enzyme	Organism	Reference	L-Lactate		Glycolate	
			V_{max} (μmol min^{-1} mg^{-1})	K_m (mM)	V_{max} (μmol min^{-1} mg^{-1})	K_m (mM)
N3-GOX	Synthetic ancestral protein	This study	0.17 ± 0.01	13.73 ± 1.93	0.87 ± 0.05	11.8 ± 0.61
No-LOX [x]	*Nostoc* sp. PCC 7120	Hackenberg et al.[x]	12.73 ± 1.55	0.04 ± 0.01	0.05 ± 0.02	0.23 ± 0.05
Cr-LOX [x]	*Chlamydomonas reinhardtii*	Hackenberg et al.[x]	10.59 ± 0.46	0.08 ± 0.03	0.19 ± 0.06	1.24 ± 0.06
Cp-GOXc	*Cyanophora paradoxa*	This study	6.94 ± 0.76	9.27 ± 1.44	8.98 ± 1.25	0.38 ± 0.05
Cm-GOX [y]	*Cyanidioschyzon merolae*	Rademacher et al.[y]	3.27 ± 0.35	14.92 ± 2.99	1.6 ± 0.29	0.9 ± 0.23
Sp-GOX	*Spirogyra pratensis*	This study	24.21 ± 6.73	15.52 ± 4.55	28.09 ± 2.78	0.94 ± 0.12
At-GOX2 [x]	*Arabidopsis thaliana*	Hackenberg et al.[x]	0.74 ± 0.04	0.36 ± 0.18	35.64 ± 11.16	1.91 ± 0.64

[x] [30]; [y] [38].

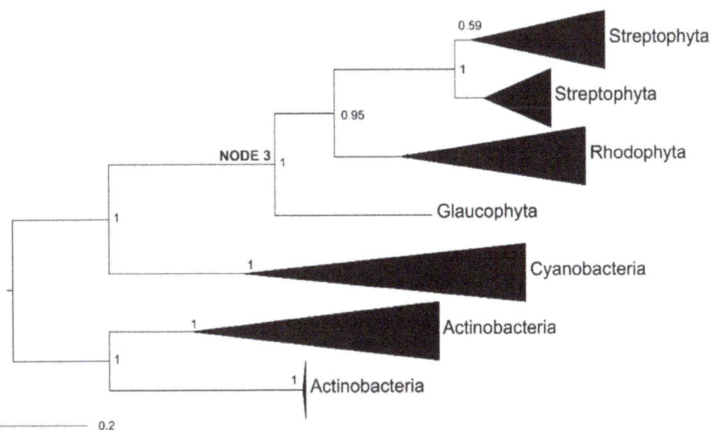

Figure 4. Phylogenetic tree of GOX-like proteins used for "ancestral" protein sequence reconstruction based on Bayesian interference. The numbers at nodes show the posterior probability. The full species names of all of the summarized groups can be found in Table S1. The derived amino acid for Node 3 is the ancestral sequence of eukaryotic GOX-like proteins.

Table 2. Amino acid positions of GOX-like proteins responsible for L-lactate or glycolate specificity. The proteins from the cyanobacterium *Nostoc* and the green algae *Chlamydomonas* exhibit amino acids shown to determine for L-lactate specificity, whereas the corresponding three amino acids of the ancestral (N3-GOX), glaucophytic (Cp-GOXc), rhodophytic (Cm-GOX), streptophytic (Sp-GOX) and plant (At-GOX2) protein, determine for glycolate specificity [30]. M—methionine; L—leucine; F—phenylalanine; T—threonine; W—tryptophan; V—valine.

	Amino Acid Position in No-LOX		
	82	**112**	**212**
	L-Lactate oxidases		
No-LOX	M	L	F
Cr-LOX	M	V	F
	Glycolate oxidases		
N3-GOX	T	W	V
Cp-GOXc	T	W	V
Cm-GOX	T	W	V
Sp-GOX	T	W	V
At-GOX2	T	W	V

Table 3. Sequence similarities (in %) between the ancestral oxidase (N3-GOX), the plant glycolate oxidase (At-GOX2), and the cyanobacterial L-lactate oxidase (No-LOX).

	N3-GOX	**No-LOX**	**At-GOX2**
N3-GOX	1	0.56	0.68
No-LOX	0.56	1	0.46
At-GOX2	0.68	0.46	1

2.5. Activity of Ancestral GOX Proteins Point to Early Evolution of Preferential Glycolate Oxidation

To analyze the enzymatic activity of the ancestral N3-GOX protein, which represents a proxy for the common ancestor of GOX-like proteins from Archaeplastida, the gene was synthesized, cloned, and expressed in *E. coli*. The subsequent biochemical analysis was done as described for the extant GOX-like proteins, using glycolate and L-lactate concentrations ranging from 0.2 to 100 mM and 0.5 to 200 mM, respectively. The "ancestral" N3-GOX showed a preference for glycolate as a substrate, as reflected in a higher V_{max} and k_{cat}/K_m value for glycolate (Figure 5 and Table S2). However, compared with the extant proteins At-GOX2 and Sp-GOX from the streptophyte clade, N3-GOX shows a 30 to 35 times lower V_{max} (0.87 µmol min^{-1} mg^{-1} protein) and a slightly higher k_{cat}/K_m with glycolate as a substrate (Figure 5, Table 1, and Table S2). Although the K_m value for glycolate is lower, the difference is not significant, which is in contrast to most of the other GOX-like proteins [30,38]. While the proteins from *Cyanidioschyzon* and *Spirogyra* show a similar affinity for L-lactate, the affinity for glycolate is at least 10 times lower in the "ancestral" N3-GOX compared with the extant GOX and LOX proteins (Table 1).

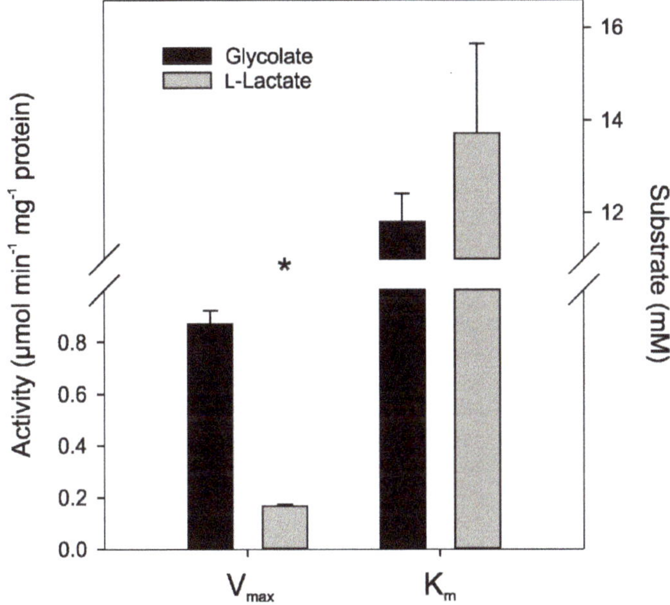

Figure 5. Biochemical characterization of the synthetic "ancestral" N3-GOX protein. The K_m and V_{max} values for glycolate and L-lactate were calculated by non-linear regression implemented in Sigma Plot 13.0 using the Michaelis–Menten kinetic model. For each substrate, three biological replicates were used. Asterisk indicates the significant difference ($p < 0.05$) determined with the two-tailed Student's *t*-test.

3. Discussion

GOX-like proteins belong to the (S)-2-hydroxy-acid oxidase protein family. Besides the important role of GOX in the photorespiration of plants, where it oxidizes glycolate to glyoxylate (Figure 1; e.g., [42]), GOX-like enzymes are also present in animals and heterotrophic prokaryotes, where they can also metabolize other hydroxy acids such as L-lactate, and are additionally involved in fatty acid metabolism or in the detoxification of critical intermediates [43,44]. Despite the differences in their substrate preference [30,32,43,45], all of these enzymes are clearly members of a single gene family and share similar homo-tetrameric quaternary structures (see Figure S1) [46,47].

Their wide-spread occurrence in almost all bacterial and eukaryotic groups makes the emergence of the eukaryotic GOX-like proteins from a bacterial ancestor very likely. Indeed, our phylogenetic analysis (Figure 2) indicates that all eukaryotic GOX-like proteins evolved from an ancestral protein of ancient cyanobacteria. This scenario includes the LOX proteins of extant cyanobacteria, which build the sister group to the clade of eukaryotic GOX-like proteins. This sister group topology is preserved when alternative roots are used for tree building (see Figure S5A,C,D). Thus, the phylogenetic reconstruction supports the view that all eukaryotic GOX-like proteins most likely evolved from prokaryotic ancestors, likely acquired from ancient cyanobacteria. This view, also supported by the close biochemical and protein similarities, is more parsimonious than the assumption of an independent evolution of these proteins among eukaryotes, although it is surprising in the sense that it implies gene transfer from a cyanobacterium to a very early eukaryote. As expected, the topology of the tree shown in Figure 2 shares many features with a previously reported reconstruction [32]. However, the inclusion of important additional taxa and different rooting have significant new implications for our understanding of the evolution of GOX-like proteins.

Placing the root of the tree for GOX-like proteins in between eukaryotic and prokaryotic proteins, as was done by Esser et al. [32], unsurprisingly results in the separation of eukaryotic from all prokaryotic GOX-like proteins (see Figure S5B), and an appearance as if the eukaryotic and prokaryotic proteins would have evolved independently. However, given the evidence that Eukarya are closely related to, and may in fact be derived from Archaea [48,49], the interpretation of the eukaryotic GOX-like proteins as being the native eukaryotic form is difficult to reconcile with the phylogenetic reconstructions. If eukaryotic GOX-like proteins indeed originated from the archaeal ancestor, then related GOX-like proteins should be found in at least some Archaea. However, only a few GOX-like sequences have been found among Archaea, and all of them are restricted to Euryarchaeota. In our midpoint-rooted tree (Figure 2 and Figure S5A), the GOX-like proteins from Archaea cluster with the Firmicutes as the sister group to all other sequences. An analysis of the protein signatures has placed the Firmicutes as an outgroup to most other bacteria [36], although other analyses place them closer to cyanobacteria [50]. Furthermore, the rare occurrence of GOX-like proteins among Archaea and their sister-group relation with Firmicutes suggests that related proteins were present in the last universal common ancestor and lost from other Archaea, or, alternatively, have undergone a horizontal gene transfer (HGT) event. In fact, HGT is a prominent force in prokaryotic and archaeal genome evolution [51]. Placing the GOX-like proteins of Archaea as the outgroup, all bacterial and eukaryotic sequences are found in one clade, where cyanobacterial and eukaryotic proteins are found as sister groups (see Figure S5C).

Hence, our phylogenetic tree strongly implies an origin of GOX in all eukaryotes from a GOX-like protein of the 2-hydroxy-acid oxidase family of ancient cyanobacteria. Such a relation is easily understandable for eukaryotic phototrophs, which evolved because of the endosymbiotic uptake of a cyanobacterial cell that eventually formed the plastids [1]. However, the origin of the GOX-like proteins of chromalveolate taxa and Metazoa from cyanobacteria is more difficult to understand. It is widely accepted that animal (and fungal) lineages separated from the phototrophic eukaryotes before the endosymbiotic engulfment of plastids. Hence, the sister group relationship of animal GOX-like proteins with cyanobacterial proteins is not easily explainable. One possibility is that the placement of this clade is incorrect because of a phylogenetic reconstruction artifact; however, the clade is quite distant from Proteobacteria (mitochondria) and Archaea, the other expected placements for eukaryotic genes obtained from prokaryotes. It is interesting to note that a few other animal proteins appear to be of a cyanobacterial origin. For example, the animal alanine–glyoxylate aminotransferase, which is, like GOX, located in the peroxisome, appears to be derived from cyanobacteria [52]. Moreover, the animal aldehyde dehydrogenases and cytochrome P450 enzymes have their closest relatives with cyanobacterial orthologs [53]. Finally, the photorespiratory glycerate 3-kinase (GLYK) is shared by cyanobacteria, fungi, and all eukaryotic phototrophs [20,54,55]. A simple and parsimonious explanation for the unexpected cyanobacterial protein origins is an early HGT between an ancestral cyanobacterium and the common ancestor of eukaryotes. The occurrence of not only one, but several cyanobacterial proteins in the non-phototrophic eukaryotes could further indicate an earlier endosymbiotic event between a eukaryotic cell and a cyanobacterium, which, however, did not result in a stable establishment of a plastid. Thus, these cyanobacterial genes could be seen as relics from earlier transfers, possibly even ones that somehow prepared the eukaryotic host for the final successful plastid incorporation as suggested by the "shopping bag" hypothesis [56], that is, multiple endosymbiotic gene transfer.

Our results support the inference that the photorespiratory GOX in all phototrophic prokaryotes originated from an ancestral cyanobacterial protein [30]. We verified the presence of a biochemically active GOX in glaucophyte algae, red algae [38], and streptophyte green algae (Table 1). Thus, the consistent utilization of a photorespiratory GOX among all groups of Archaeplastida points to an early evolution of the plant-like photorespiratory cycle in the common ancestor of phototrophic eukaryotes. This notion is also supported by the biochemical analysis of the reconstructed ancestral N3-GOX, which displays some preference of glycolate over L-lactate as a substrate. Previously, it was assumed that the photorespiratory glycolate oxidation via GOX only appeared in Streptophyta, because chlorophytes such as *Chlamydomonas* perform this reaction by a mitochondrion localized glycolate dehydrogenase [57–59]. The genome sequence of *Chlamydomonas* revealed that, in addition to glycolate dehydrogenase, this chlorophyte also possesses a GOX-like protein. However, biochemical analyses showed that, in contrast to the GOX of all other Archaeplastida, the *Chlamydomonas* enzyme must be assigned as LOX, as it clearly prefers the substrate L-lactate over glycolate, similar to the GOX-like proteins from N_2-fixing cyanobacteria such as *Nostoc* (Table 1) [30].

The close similarity of cyanobacterial and chlorophyte LOX proteins is also reflected in our phylogenetic tree, where the chlorophytic LOX proteins form an extra clade at the base of all eukaryotic GOX-like sequences, clearly separated from the GOX clade of all remaining Archaeplastida. There are two possible scenarios how LOX evolved among chlorophytes. The most likely explanation is that the LOX in Chlorophyta reflects a functional reversal from an early GOX; this may reflect a clade that diverged before the ancestral cyanobacterial 2-hydroxy-acid oxidase diversified among Archaeplastida, but may also reflect a phylogenetic artifact stemming from differential selection, depending on the substrate. As mentioned before, all GOX-like enzymes can use L-lactate and glycolate to some degree, and it is to be expected that over the course of evolutionary history, they were optimized to fit the specific metabolic requirements of the respective organism as GOX or LOX. A second possible explanation is that an ancient cyanobacterial LOX received by the endosymbiotic uptake of the plastid ancestor was retained as a second gene in the chlorophyte genome, while it was lost from other algal genomes, or has been obtained by a second HGT from cyanobacteria, as suggested by Esser et al. [32]. Regardless of which scenario actually occurred, LOX in chlorophytes also originated from a protein closely related to those of cyanobacteria.

Furthermore, the sequence and biochemical analyses of the GOX-like proteins among Archaeplastida and cyanobacteria revealed that the experimentally verified amino acid residues that determine whether glycolate or L-lactate is the preferred substrate [30] also distinguish the cyanobacterial and chlorophyte LOX from GOX, such as the Cp-GOX from *Cyanophora*, the Cm-GOX from *Cyanidioschyzon*, and the streptophytic Sp-GOX from *Spirogyra* (Table 2). Interestingly, the hypothetical ancestral N3-GOX also possesses the glycolate-preferring amino acid signature, corresponding to its dominant enzymatic activity as GOX. The amino acid residues responsible for the binding of the flavin mononucleotide (FMN) cofactor are also highly conserved in all GOX and LOX proteins, analyzed here or previously [30,38]. Interestingly, we found two clear trends among the biochemically verified GOX enzymes among Archaeplastida (Table 1). The GOX of the early branching Glaucophyta showed the highest affinity for glycolate among phototrophic eukaryotes, whereas streptophyte GOX proteins have a five-times lower affinity for glycolate. The inverse trend is observed regarding the V_{max} values, which are the highest among Streptophytes and the lowest for the enzyme from Glaucophyta. These findings are consistent with the hypothesis that Streptophytes, especially C3 plants, are characterized by much higher photorespiratory fluxes compared with algae, in which photorespiration is often inhibited by the presence of a CCM. Thus, a highly active peroxisomal GOX allows for the rapid degradation of glycolate and its recycling to 3-phosphoglycerate [24,60]. The relatively low V_{max} of the N3-GOX may also result from incorrectly predicted amino acid residues, which can lead to protein misfolding [61].

4. Materials and Methods

4.1. Phylogenetic Analysis

The selected amino acid sequences of all GOX-like proteins were aligned using ProbCons [34], which produces the best alignment scores [62]. An alternative alignment was constructed using MUSCLE [35]. Both alignments, especially their non-ambiguous parts, were compared via the program SuiteMSA [63]. To make sure that the incorrect aligned position was not considered for the phylogenetic analyses, we used the alignment curation program GBlocks [64]. Nearly identical sequences (after Gblocks curation) or sequences with a different evolutionary rate (calculated via Tajima's rate test included in MEGA 5.1; [65]) were excluded from further analyses. We also excluded all sequences from single species, which did not cluster into the clade of their taxonomic group to minimize HGT influences. ProtTest 3 [66] was used to find the best fitting substitution model for the given data set. The model of Le and Gascuel [67], including a proportion of invariant sites and a gamma model of rate heterogeneity, was found to be the most appropriate model of evolution for the alignment. For the tree reconstruction, we used the Bayesian approach implemented in MrBayes [68], running 2,000,000 generations and disregarding the first 25% of samples as burn-in. The likelihoods and tree topologies of the independent runs were analyzed and compared. In addition to the Bayesian approach, we also performed a maximum likelihood analysis using RAxML version 8 [69], with a bootstrap test of the statistical support from 1000 replicates.

4.2. Ancestral Sequence Reconstruction

For the reconstruction of the ancestral eukaryotic GOX protein sequence, we first reduced the number of sequences used for the phylogeny (shown in Figure 2) to 36 species, which are situated around the nodes of the eukaryotic GOX-like proteins. The phylogenetic analysis of the reduced dataset was performed as described above, and is shown in Figure 4. Because of the long evolutionary distance of the analyzed proteins, ancestral reconstruction based on amino acid inference were performed using CODEML included in PAML [70]. In addition, we used the FASTML program including the marginal and joint method of ancestral sequence reconstruction [71]. We also performed a Maximum Likelihood and Maximum Parsimony indel reconstruction, which provided the same results. We focused on the last common ancestor of all eukaryotes (Node 3). The deduced amino acid sequences of the Node 3 ancestral GOX-like protein was translated into a DNA sequence applying the codon-usage of *E. coli* (sequences are displayed in the Figure S6). The codon-optimized gene for the ancestral GOX protein was synthesized (Thermo Fisher, Darmstadt, Germany). For cloning into the expression vector pASG-IBA43+, the restriction sites *Nhe*I and *Bam*HI were added at the beginning and the end of the ancestral gene, respectively.

4.3. Sequencing of the cDNA Encoding the GOX from Spirogyra Pratensis

The complete sequence of the *Spirogyra* GOX was extracted from an existing Expressed sequence tag (EST) library [72], which was screened via BLASTX using the *Arabidopsis* At-GOX2 (At3g14415) sequence. The full-length cDNA sequence from *Spirogyra* (Figure S2) was compared to known GOX and LOX proteins, and checked for completeness. The sequence of *Spirogyra* GOX (Sp-GOX) was deposited in GenBank under the accession number AVP27295.1.

4.4. Estimation and Synthesis of the cDNA Encoding the GOX from Cyanophora Paradoxa

The GOX sequence from *C. paradoxa* was identified by BLAST searches with plant and algal GOX proteins. The originally annotated *Cyanophora* protein of the genomic Contig54585 consists of 316 amino acids, which is considerably shorter than the GOX protein from *Arabidopsis* (367 amino acids). Initially, we used this cDNA to synthesize the *Cyanophora gox* gene; however, we failed to obtain an enzymatic active protein, most likely because of the missing C-terminus. Subsequent BLAST searches including EST sequences identified two different EST sequences (EG947183.1 and ES232585.1), which partially overlap with the *gox* gene in the *Cyanophora* genome database (Contig54585; http://cyanophora.rutgers.edu/cyanophora/home.php). The newly obtained cDNA for the *Cyanophora gox* gene (348 amino acids long, sequence called Cp-GOXb) was also used for gene synthesis (Thermo Fisher), but we again failed to produce an enzymatic active protein. Finally, we re-sequenced the entire *gox* gene from *Cyanophora* gDNA using new primer sets (CpGOXb_275_fw, CpGOXb_369_fw, and CpGOXb_655_rv; Table S3). The new sequence revealed a missing 15-nucleotides-long insertion in the originally annotated genomic version of this *gox* gene. The final cDNA sequence gave rise to the corrected protein, which was named Cp-GOXc. The entire gene was synthesized by PCR using the DNA of the previously obtained expression vector pASG-IBA43plus-*cp-goxb* as a template and the phosphorylated primers CpGOX_RV_5P and CpGOX_FW_5P (Table S3). The previously 15 missing bases coding for five amino acids were included via the used primer. After the ligation of the final PCR product, we obtained the expression vector pASG-IBA43plus-*cp-goxc* harboring the correct cDNA sequence of the *gox* gene from *Cyanophora*. A scheme including all steps leading to the corrected *gox* gene from *Cyanophora* is shown in Figure S7.

4.5. Cloning, Heterologous Expression and Purification of Recombinant GOX Proteins

The obtained *cp-gox* gene was subsequently cloned without a stop codon into the expression vector pASG-IBA43plus. The resulting strep-tagged fusion proteins were generated in *E. coli* strain BL21 GOLD and purified using the Strep-tactin matrix, according to the supplier's protocol (IBA Bio technology, Göttingen, Germany). The *gox* gene of *Spirogyra* and the gene encoding the ancestral GOX protein were also cloned into pASG-IBA43plus. However, the N-terminal His-tag was used for purification with Ni-NTA Sepharose, according to the supplier's protocol (Invitrogen, Karlsruhe, Germany).

For the production of Sp-GOX and Cp-GOX, the recombinant *E. coli* strain was grown in an LB medium to an OD_{750} of 0.6. The gene expression was initiated with the addition of 200 μg/L anhydrotetracycline for 4 to 16 h at 20 to 30 °C. A soluble protein of the ancestral N3-GOX could be only obtained when cultivating *E. coli* cultures at 18 °C for 16 h after induction with anhydrotetracycline. *E. coli* cells were harvested by centrifugation and re-suspended in buffer A (20 mM Tris-HCl, 500 mM NaCl, 1 mM dithiothreitol (DTT), and 0.1 mM Flavin mononucleotide (FMN)) in case of His-tag fusion proteins or buffer W (100 mM Tris pH 8.0, 150 mM NaCl, 1 mM EDTA, 1 mM DTT, and 0.1 mM FMN) in case of Strep-tag proteins. The protein extraction was done by ultrasonic treatments (4 times 30 s, 90 W) on ice. After centrifugation (20,000 g, 20 min, 4 °C, Sorvall SS34 rotor), the cell-free protein extract was directly loaded onto Ni-NTA or Strep-tactin columns. The His-tagged proteins were washed using buffer A supplemented with 40 to 80 mM imidazole, whereas His-tagged GOX was eluted with buffer A supplemented with 200 mM imidazole. For Strep-tagged proteins, the washing steps were performed using the buffer W, and the proteins were eluted by buffer W, supplemented with 2.5 mM desthiobiotin. Subsequently, the eluted proteins were desalted using PD-10 columns (GE Healthcare, Schwerte, Germeny), and finally dissolved in 20 mM Tris/HCl pH 8.0 containing 1 mM DTT and 0.1 mM FMN. The purity of the eluted proteins was checked by SDS-PAGE and staining with Coomassie Brilliant Blue (Figure S8).

4.6. Enzyme Activity Assays

The GOX or LOX activity was measured by the detection of O_2 consumption in the presence of different concentrations of L-lactate and glycolate using Hansatech oxygen electrodes (Oxygraph), as described by Hackenberg et al. [30]. One unit of enzyme activity was defined as the consumption of 1 μmol O_2 in 1 min at 30 °C. The protein concentrations were estimated according to the literature [73].

5. Conclusions

The monophyly of eukaryotic and cyanobacterial GOX-like proteins points to a cyanobacterial origin of all eukaryotic GOX-like proteins, including the photorespiratory GOX of phototrophic eukaryotes. Biochemically, these proteins operate as GOX rather than LOX in the Archaeplastida. The slight preference for glycolate of the synthetic ancestral N3-GOX suggests that glycolate oxidation could have been the ancestral function, and during the course of evolution, this protein has changed its preference to glycolate or L-lactate and other related substrates according to the requirements of the specific host.

The likely scenario for the evolution of GOX among eukaryotes and particularly of the photorespiratory GOX of Archaeplastida is summarized in Figure 6. A bifunctional GOX-like protein (represented by N3-GOX) was transferred from an ancient cyanobacterium via HGT into eukaryotes before the split of animal, fungal, and plant lineages occurred. Most extant cyanobacteria lost this enzyme, and use a glycolate dehydrogenase for photorespiratory glycolate oxidation, whereas N_2-fixing cyanobacteria optimized the ancestral GOX-like protein to operate as LOX, which helps consume oxygen, protecting nitrogenase [30]. Among the Metazoa, the ancestral cyanobacterial protein evolved into GOX-like enzymes with varying substrate specificities after gene duplication and biochemical specialization, as suggested by Esser et al. [32]. After the engulfment of an ancient cyanobacterium as a plastid ancestor, possibly two gene copies for GOX-like proteins co-existed in the proto-alga, but only one of these copies was retained. Among the Archaeplastida, the ancestral GOX-like protein early on evolved to the main photorespiratory glycolate oxidizing enzyme and became localized in peroxisomes. The evolution of peroxisomes in eukaryotes is still a matter of discussion, but it has been shown that the proteome of peroxisomes is variable, pointing to an evolutionary optimization of peroxisome functions in time by protein acquisitions and losses [74]. The increasing photorespiratory flux due to the increasing oxygen concentration in the environment was matched by increasing the V_{max} of glycolate oxidation.

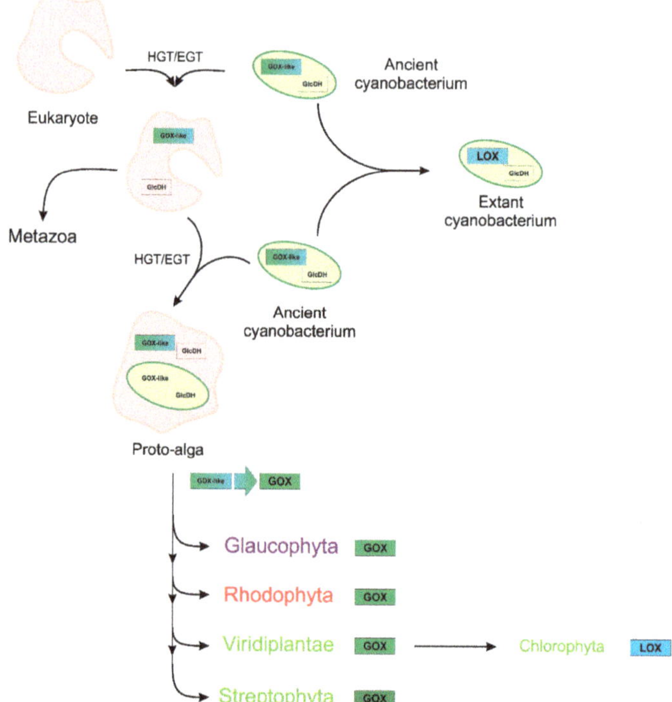

Figure 6. Hypothetical evolutionary scenario of GOX-like proteins of the 2-hydroxy-acid oxidase family among Eukaryotes. A bifunctional GOX-like protein and a glycolate dehydrogenase (GlcDH) existed in ancient cyanobacteria. In most extant cyanobacteria the GOX-like gene was lost, and they use GlcDH for photorespiratory glycolate oxidation. The cyanobacterial gene of the GOX-like protein was initially transferred to the eukaryotic genome via horizontal (endosymbiotic) gene transfer (HGT/EGT). After the engulfment of an ancient cyanobacterium as a plastid ancestor, probably two gene copies for GOX-like proteins existed. Subsequently, one of these copies was lost during plastid establishment. The early evolution of the glycolate specificity of the GOX-like protein most likely took place in the proto-alga before the split-off of Archaeplastida lineages. Only among cyanobacteria and chlorophytes, did GOX-like proteins evolve into L-lactate oxidase (LOX).

Hence, the early acquisition of a peroxisomal GOX-like protein by the hypothetical host cell [75–77] can be regarded as further exaptation to integrate the photosynthetic plastid ancestor with an already active host glycolate metabolism in the oxygen-containing atmosphere [22].

Supplementary Materials: The following are available online at http://www.mdpi.com/2223-7747/9/1/106/s1, Supplementary material 1: Figure S1: GOX-like proteins from eukaryotes and prokaryotes with different substrate preferences share a common tertiary and quaternary structure; Figure S2: Sequence of the cDNA and protein of the GOX from *Spirogyra pratensis* used in this study; Figure S3: Corrected sequence of the Cp-GOX coding gene in *Cyanophora paradoxa*; Figure S4: Collapsed phylogenetic tree of GOX-like proteins based on 111 amino acid sequences from all organismic groups in the tree of life; Figure S5: Alternative rooting of phylogenetic trees of GOX- and LOX-like sequences; Figure S6: Sequence of the protein and the codon-optimized DNA of the ancestral GOX-like protein (N3-GOX) used in this study; Figure S7: Schematic display of the work to obtain the correct complete *cp-goxc* gene from *Cyanophora paradoxa*; Figure S8: Purification of recombinant GOX proteins from *Spirogyra* (Sp-GOX) and *Cyanophora* (Cp-GOXc) and the ancestral GOX protein (N3-GOX); Table S1: Species names and accession numbers for phylogeny; Table S2: Catalytic efficiency of recombinant GOX and GOX-like proteins; Table S3: Sequences of the used primer sequences Supplementary Material 2: Alignment of GOX-like proteins, accession numbers: Sp-GOX: AVP27295.1 and Cp-GOX: AVP27296.

Author Contributions: Conceptualization, R.K. M.H., and H.B.; methodology and investigation, R.K. and F.F.; validation and interpretation, R.K. and M.H.; formal analysis, R.K., M.H., C.D., and A.P.M.W.; resources, A.P.M.W., H.B., and C.D.; writing (original draft), R.K. and M.H.; writing (review and editing), all co-authors; supervision, M.H., A.P.M.W., and H.B. All authors have read and agreed to the published version of the manuscript.

Funding: This research was funded by the DFG (Deutsche Forschungsgemeinschaft), in the frame of the Forschergruppe FOR 1186-Promics.

Acknowledgments: We thank Endymion D. Cooper (University of Maryland, USA) for his help obtaining the Sp-GOX cDNA sequence. We appreciate the technical assistance of Manja Henneberg and the discussions with Lukas Krebes.

Conflicts of Interest: The authors declare no conflict of interest. The funders had no role in the design of the study; in the collection, analyses, or interpretation of data; in the writing of the manuscript; or in the decision to publish the results.

References

1. Hohmann-Marriott, M.F.; Blankenship, R.E. Evolution of photosynthesis. *Annu. Rev. Plant Biol.* **2011**, *62*, 515–548. [CrossRef]

2. Soo, R.M.; Hemp, J.; Parks, D.H.; Fischer, W.W.; Hugenholtz, P. On the origins of oxygenic photosynthesis and aerobic respiration in Cyanobacteria. *Science* **2017**, *355*, 1436–1440. [CrossRef]

3. Nowack, E.C.; Weber, A.P. Genomics-informed insights into endosymbiotic organelle evolution in photosynthetic Eukaryotes. *Annu. Rev. Plant Biol.* **2018**, *69*, 51–84. [CrossRef]

4. Keeling, P.J. The endosymbiotic origin, diversification and fate of plastids. *Philos. Trans. R. Soc. B Boil. Sci.* **2010**, *365*, 729–748. [CrossRef] [PubMed]

5. Stiller, J.W.; Reel, D.E.C.; Johnson, J.C. A single origin of plastids revisited: Convergent evolution in organellar genome content. *J. Phycol.* **2003**, *39*, 95–105. [CrossRef]

6. Leliaert, F.; Smith, D.R.; Moreau, H.; Herron, M.D.; Verbruggen, H.; Delwiche, C.F.; De Clerck, O. Phylogeny and Molecular Evolution of the Green Algae. *Crit. Rev. Plant Sci.* **2012**, *31*, 1–46. [CrossRef]

7. Nishiyama, T.; Sakayama, H.; de Vries, J.; Buschmann, H.; Saint-Marcoux, D.; Ullrich, K.K.; Haas, F.B.; Vanderstraeten, L.; Becker, D.; Lang, D.; et al. The *Chara* genome: Secondary complexity and implications for plant terrestrialization. *Cell* **2018**, *174*, 448–464. [CrossRef]

8. Bowes, G.; Ogren, W.; Hageman, R. Phosphoglycolate production catalyzed by ribulose diphosphate carboxylase. *Biochem. Biophys. Res. Commun.* **1971**, *45*, 716–722. [CrossRef]

9. Ogren, W.L. Affixing the O to Rubisco: Discovering the source of photorespiratory glycolate and its regulation. *Photosynth. Res.* **2003**, *76*, 53–63. [CrossRef]

10. Tcherkez, G.G.B.; Farquhar, G.D.; Andrews, T.J. Despite slow catalysis and confused substrate specificity, all ribulose bisphosphate carboxylases may be nearly perfectly optimized. *Proc. Natl. Acad. Sci. USA* **2006**, *103*, 7246–7251. [CrossRef]

11. Kelly, G.J.; Latzko, E. Inhibition of spinach-leaf phosphofructokinase by 2-phosphoglycollate. *FEBS Lett.* **1976**, *68*, 55–58. [CrossRef]

12. Anderson, K.L.; Tayne, T.A.; Ward, D.M. Formation and fate of fermentation products in hot spring cyanobacterial mats. *Appl. Environ. Microbiol.* **1987**, *53*, 2343–2352. [CrossRef] [PubMed]

13. Flügel, F.; Timm, S.; Arrivault, S.; Florian, A.; Stitt, M.; Fernie, A.R.; Bauwe, H. The Photorespiratory Metabolite 2-Phosphoglycolate Regulates Photosynthesis and Starch Accumulation in *Arabidopsis*. *Plant Cell* **2017**, *29*, 2537–2551. [CrossRef] [PubMed]

14. Bauwe, H.; Hagemann, M.; Fernie, A.R. Photorespiration: Players, partners and origin. *Trends Plant Sci.* **2010**, *15*, 330–336. [CrossRef] [PubMed]

15. Walker, B.J.; Vanloocke, A.; Bernacchi, C.J.; Ort, D.R. The Costs of Photorespiration to Food Production Now and in the Future. *Annu. Rev. Plant Boil.* **2016**, *67*, 107–129. [CrossRef]

16. Ort, D.R.; Merchant, S.S.; Alric, J.; Barkan, A.; Blankenship, R.E.; Bock, R.; Croce, R.; Hanson, M.R.; Hibberd, J.M.; Long, S.P.; et al. Redesigning photosynthesis to sustainably meet global food and bioenergy demand. *Proc. Natl. Acad. Sci. USA* **2015**, *112*, 8529–8536. [CrossRef]

17. Hagemann, M.; Bauwe, H. Photorespiration and the potential to improve photosynthesis. *Curr. Opin. Chem. Biol.* **2016**, *35*, 109–116. [CrossRef]

18. Becker, B. Snow ball earth and the split of Streptophyta and Chlorophyta. *Trends Plant Sci.* **2013**, *18*, 180–183. [CrossRef]

19. Eisenhut, M.; Ruth, W.; Haimovich, M.; Bauwe, H.; Kaplan, A.; Hagemann, M. The photorespiratory glycolate metabolism is essential for cyanobacteria and might have been conveyed endosymbiontically to plants. *Proc. Natl. Acad. Sci. USA* **2008**, *105*, 17199–17204. [CrossRef]

20. Kern, R.; Bauwe, H.; Hagemann, M. Evolution of enzymes involved in the photorespiratory 2-phosphoglycolate cycle from cyanobacteria via algae toward plants. *Photosynth. Res.* **2011**, *109*, 103–114. [CrossRef]

21. Kern, R.; Eisenhut, M.; Bauwe, H.; Weber, A.P.M.; Hagemann, M. Does the *Cyanophora paradoxa* genome revise our view on the evolution of photorespiratory enzymes? *Plant Biol.* **2013**, *15*, 759–768. [CrossRef] [PubMed]

22. Hagemann, M.; Kern, R.; Maurino, V.G.; Hanson, D.T.; Weber, A.P.M.; Sage, R.F.; Bauwe, H. Evolution of photorespiration from cyanobacteria to land plants, considering protein phylogenies and acquisition of carbon concentrating mechanisms. *J. Exp. Bot.* **2016**, *67*, 2963–2976. [CrossRef] [PubMed]

23. Nakamura, Y.; Kanakagiri, S.; Van, K.; He, W.; Spalding, M.H. Disruption of the glycolate dehydrogenase gene in the high-CO_2-requiring mutant HCR89 of *Chlamydomonas reinhardtii*. *Can. J. Bot.* **2005**, *83*, 820–833. [CrossRef]

24. Kehlenbeck, P.; Goyal, A.; Tolbert, N.E. Factors affecting development of peroxisomes and glycolate metabolism among algae of different evolutionary lines of the Prasinophyceae. *Plant Physiol.* **1995**, *109*, 1363–1370. [CrossRef] [PubMed]

25. Stabenau, H.; Winkler, U. Glycolate metabolism in green algae. *Physiol. Plant.* **2005**, *123*, 235–245. [CrossRef]

26. Lindqvist, Y.; Brändén, C.I.; Mathews, F.S.; Lederer, F. Spinach glycolate oxidase and yeast flavocytochrome b2 are structurally homologous and evolutionarily related enzymes with distinctly different function and flavin mononucleotide binding. *J. Biol. Chem.* **1991**, *266*, 3198–3207.

27. Maeda-Yorita, K.; Aki, K.; Sagai, H.; Misaki, H.; Massey, V. l-Lactate oxidase and l-lactate monooxygenase, Mechanistic variations on a common structural theme. *Biochimie* **1995**, *77*, 631–642. [CrossRef]

28. Stenberg, K.; Lindqvist, Y. Three-dimensional structures of glycolate oxidase with bound active-site inhibitors. *Protein Sci.* **1997**, *6*, 1009–1015. [CrossRef]

29. Leiros, I.; Wang, E.; Rasmussen, T.; Oksanen, E.; Repo, H.; Petersen, S.B.; Heikinheimo, P.; Hough, E. The 2.1 Å structure of *Aerococcus viridans* l-lactate oxidase (LOX). *Acta Crystallogr. Sect. F Struct. Boil. Cryst. Commun.* **2006**, *62*, 1185–1190. [CrossRef]

30. Hackenberg, C.; Kern, R.; Huge, J.; Stal, L.J.; Tsuji, Y.; Kopka, J.; Shiraiwa, Y.; Bauwe, H.; Hagemann, M. Cyanobacterial lactate oxidases serve as essential partners in N_2 fixation and evolved into photorespiratory glycolate oxidases in plants. *Plant Cell* **2011**, *23*, 2978–2990. [CrossRef]

31. Ponce-Toledo, R.I.; Deschamps, P.; López-García, P.; Zivanovic, Y.; Benzerara, K.; Moreira, D. An early-branching freshwater cyanobacterium at the origin of plastids. *Curr. Biol.* **2017**, *27*, 386–391. [CrossRef] [PubMed]

32. Esser, C.; Kuhn, A.; Groth, G.; Lercher, M.J.; Maurino, V.G. Plant and animal glycolate oxidases have a common eukaryotic ancestor and convergently duplicated to evolve long-chain 2-hydroxy acid oxidases. *Mol. Biol. Evol.* **2014**, *31*, 1089–1101. [CrossRef] [PubMed]

33. Altschul, S.F.; Gish, W.; Miller, W.; Myers, E.W.; Lipman, D.J. Basic local alignment search tool. *J. Mol. Biol.* **1990**, *215*, 403–410. [CrossRef]

34. Do, C.B.; Mahabhashyam, M.S.P.; Brudno, M.; Batzoglou, S. ProbCons, Probabilistic consistency-based multiple sequence alignment. *Genome Res.* **2005**, *15*, 330–340. [CrossRef] [PubMed]

35. Edgar, R.C. MUSCLE, multiple sequence alignment with high accuracy and high throughput. *Nucleic Acids Res.* **2004**, *32*, 1792–1797. [CrossRef] [PubMed]

36. Griffiths, E.; Gupta, R.S. Signature sequences in diverse proteins provide evidence for the late divergence of the order Aquificales. *Int. Microbiol.* **2004**, *7*, 41–52.

37. Placzek, S.; Schomburg, I.; Chang, A.; Jeske, L.; Ulbrich, M.; Tillack, J.; Schomburg, D. BRENDA in 2017, new perspectives and new tools in BRENDA. *Nucleic Acids Res.* **2017**, *45*, 380–388. [CrossRef]

38. Rademacher, N.; Kern, R.; Fujiwara, T.; Mettler-Altmann, T.; Miyagishima, S.-y.; Hagemann, M.; Eisenhut, M.; Weber, A.P.M. Photorespiratory glycolate oxidase is essential for the survival of the red alga *Cyanidioschyzon merolae* under ambient CO_2 conditions. *J. Exp. Bot.* **2016**, *67*, 3165–3175. [CrossRef]

39. Stabenau, H. Microbodies from *Spirogyra*, organelles of a filamentous alga similar to leaf peroxisomes. *Plant Physiol.* **1976**, *58*, 693–695. [CrossRef]

40. Betsche, T.; Schaller, D.; Melkonian, M. Identification and characterization of glycolate oxidase and related enzymes from the endocyanotic alga *Cyanophora paradoxa* and from *Pea* leaves. *Plant Physiol.* **1992**, *98*, 887–893. [CrossRef]

41. Seki, M.; Iida, K.-I.; Saito, M.; Nakayama, H.; Yoshida, S.-I. Hydrogen Peroxide Production in Streptococcus pyogenes: Involvement of Lactate Oxidase and Coupling with Aerobic Utilization of Lactate. *J. Bacteriol.* **2004**, *186*, 2046–2051. [CrossRef] [PubMed]

42. Zelitch, I.; Schultes, N.P.; Peterson, R.B.; Brown, P.; Brutnell, T.P. High glycolate oxidase activity is required for survival of maize in normal air. *Plant Physiol.* **2009**, *149*, 195–204. [CrossRef] [PubMed]

43. Jones, J.M.; Morrell, J.C.; Gould, S.J. Identification and characterization of HAOX1; HAOX2; and HAOX3; three human peroxisomal 2-hydroxy acid oxidases. *J. Biol. Chem.* **2000**, *275*, 12590–12597. [CrossRef]

44. Pellicer, M.T.; Badia, J.; Aguilar, J.; Baldoma, L. *glc* locus of *Escherichia coli*, Characterization of genes encoding the subunits of glycolate oxidase and the *glc* regulator protein. *J. Bacteriol.* **1996**, *178*, 2051–2059. [CrossRef]

45. Gunshore, S.; Brush, E.J.; Hamilton, G.A. Equilibrium constants for the formation of glyoxylate thiohemiacetals and kinetic constants for their oxidation by O_2 catalyzed by L-hydroxy acid oxidase. *Bioorg. Chem.* **1985**, *13*, 1–13. [CrossRef]

46. Dupuis, L.; Caro, J.D.; Brachet, P.; Puigserver, A. Purification and some characteristics of chicken liver L-2-hydroxyacid oxidase A. *FEBS Lett.* **1990**, *266*, 183–186. [CrossRef]

47. Streitenberger, S.A.; López-Más, J.A.; Sánchez-Ferrer, Á.; García-Carmona, F. Use of Dye Affinity Chromatography for the Purification of *Aerococcus viridans* Lactate Oxidase. *Biotechnol. Prog.* **2002**, *18*, 657–659. [CrossRef]

48. Barns, S.M.; Delwiche, C.F.; Palmer, J.D.; Dawson, S.C.; Hershberger, K.L.; Pace, N.R. Phylogenetic perspective on microbial life in hydrothermal ecosystems; past and present. *Ciba Found. Symp.* **1996**, *202*, 24–39.

49. Da Cunha, V.; Gaia, M.; Gadelle, D.; Nasir, A.; Forterre, P. Lokiarchaea are close relatives of Euryarchaeota; not bridging the gap between prokaryotes and eukaryotes. *PLoS Genet.* **2017**, *13*, e1006810. [CrossRef]

50. Schulz, F.; Eloe-Fadrosh, E.A.; Bowers, R.M.; Jarett, J.; Nielsen, T.; Ivanova, N.N.; Kyrpides, N.C.; Woyke, T. Towards a balanced view of the bacterial tree of life. *Microbiome* **2017**, *5*, 140. [CrossRef]

51. Koonin, E.V.; Wolf, Y.I. Genomics of bacteria and archaea, the emerging dynamic view of the prokaryotic world. *Nucleic Acids Res.* **2008**, *36*, 6688–6719. [CrossRef] [PubMed]

52. Gabaldón, T.; Snel, B.; van Zimmeren, F.; Hemrika, W.; Tabak, H.; Huynen, M.A. Origin and evolution of the peroxisomal proteome. *Biol. Direct* **2006**, *1*, 8. [CrossRef] [PubMed]

53. Millard, A.; Scanlan, D.J.; Gallagher, C.; Marsh, A.; Taylor, P.C. Unexpected evolutionary proximity of eukaryotic and cyanobacterial enzymes responsible for biosynthesis of retinoic acid and its oxidation. *Mol. BioSyst.* **2014**, *10*, 380–383. [CrossRef]

54. Boldt, R.; Edner, C.; Kolukisaoglu, U.; Hagemann, M.; Weckwerth, W.; Wienkoop, S.; Morgenthal, K.; Bauwe, H. D-Glycerate 3-kinase; the last unknown enzyme in the photorespiratory cycle in *Arabidopsis*; belongs to a novel kinase family. *Plant Cell* **2005**, *17*, 2413–2420. [CrossRef] [PubMed]

55. Maruyama, S.; Matsuzaki, M.; Misawa, K.; Nozaki, H. Cyanobacterial contribution to the genomes of the plastid-lacking protists. *BMC Evol. Biol.* **2009**, *9*, 197. [CrossRef] [PubMed]

56. Howe, C.; Barbrook, A.; Nisbet, R.; Lockhart, P.; Larkum, A. The origin of plastids. *Philos. Trans. R. Soc. B Boil. Sci.* **2008**, *363*, 2675–2685. [CrossRef]

57. Atteia, A.; Adrait, A.; Brugiere, S.; Tardif, M.; van Lis, R.; Deusch, O.; Dagan, T.; Kuhn, L.; Gontero, B.; Martin, W.; et al. A proteomic survey of *Chlamydomonas reinhardtii* mitochondria sheds new light on the metabolic plasticity of the organelle and on the nature of the alpha-proteobacterial mitochondrial ancestor. *Mol. Biol. Evol.* **2009**, *26*, 1533–1548. [CrossRef]

58. Nelson, E.B.; Tolbert, N.E. Glycolate dehydrogenase in green algae. *Arch. Biochem. Biophys.* **1970**, *141*, 102–110. [CrossRef]

59. Frederic, S.E.; Gruber, P.J.; Tolbert, N.E. Occurrence of glycolate dehydrogenase and glycolate oxidase in green plants - Evolutionary survey. *Plant Physiol.* **1973**, *52*, 318–323. [CrossRef]

60. Hagemann, M.; Fernie, A.R.; Espie, G.S.; Kern, R.; Eisenhut, M.; Reumann, S.; Bauwe, H.; Weber, A.P.M. Evolution of the biochemistry of the photorespiratory C2 cycle. *Plant Biol.* **2013**, *15*, 639–647. [CrossRef]

61. Shih, P.M.; Occhialini, A.; Cameron, J.C.; Andralojc, P.J.; Parry, M.A.J.; Kerfeld, C.A. Biochemical characterization of predicted Precambrian RuBisCO. *Nat. Commun.* **2016**, *7*, 10382. [CrossRef] [PubMed]

62. Pervez, M.T.; Babar, M.E.; Nadeem, A.; Aslam, M.; Awan, A.R.; Aslam, N.; Hussain, T.; Naveed, N.; Qadri, S.; Waheed, U.; et al. Evaluating the accuracy and efficiency of multiple sequence alignment methods. *Evol. Bioinform.* **2014**, *10*, 205–217. [CrossRef] [PubMed]

63. Anderson, C.L.; Strope, C.L.; Moriyama, E.N. SuiteMSA, visual tools for multiple sequence alignment comparison and molecular sequence simulation. *BMC Bioinform.* **2011**, *12*, 184.

64. Castresana, J. Selection of conserved blocks from multiple alignments for their use in phylogenetic analysis. *Mol. Biol. Evol.* **2000**, *17*, 540–552. [CrossRef]

65. Tamura, K.; Peterson, D.; Peterson, N.; Stecher, G.; Nei, M.; Kumar, S. MEGA5, molecular evolutionary genetics analysis using maximum likelihood; evolutionary distance; and maximum parsimony methods. *Mol. Biol. Evol.* **2011**, *28*, 2731–2739. [CrossRef] [PubMed]

66. Darriba, D.; Taboada, G.L.; Doallo, R.; Posada, D. ProtTest 3, fast selection of best-fit models of protein evolution. *Bioinformatics* **2011**, *27*, 1164–1165. [CrossRef] [PubMed]

67. Le, S.Q.; Gascuel, O. An improved general amino acid replacement matrix. *Mol. Biol. Evol.* **2008**, *25*, 1307–1320. [CrossRef] [PubMed]

68. Ronquist, F.; Teslenko, M.; van der Mark, P.; Ayres, D.L.; Darling, A.; Hohna, S.; Larget, B.; Liu, L.; Suchard, M.A.; Huelsenbeck, J.P. MrBayes 3.2, efficient Bayesian phylogenetic inference and model choice across a large model space. *Syst. Biol.* **2012**, *61*, 539–542. [CrossRef]

69. Stamatakis, A. RAxML version 8, a tool for phylogenetic analysis and post-analysis of large phylogenies. *Bioinformatics* **2014**, *30*, 1312–1313. [CrossRef]

70. Yang, Z. PAML 4, phylogenetic analysis by maximum likelihood. *Mol. Biol. Evol.* **2007**, *24*, 1586–1591. [CrossRef]

71. Ashkenazy, H.; Penn, O.; Doron-Faigenboim, A.; Cohen, O.; Cannarozzi, G.; Zomer, O.; Pupko, T. FastML, a web server for probabilistic reconstruction of ancestral sequences. *Nucleic Acids Res.* **2012**, *40*, W580–W584. [CrossRef] [PubMed]

72. Timme, R.E.; Delwiche, C.F. Uncovering the evolutionary origin of plant molecular processes, comparison of *Coleochaete* (Coleochaetales) and *Spirogyra* (Zygnematales) transcriptomes. *BMC Plant Biol.* **2010**, *10*, 96. [CrossRef] [PubMed]

73. Bradford, M.M. A rapid and sensitive method for the quantitation of microgram quantities of protein utilizing the principle of protein-dye binding. *Anal. Biochem.* **1976**, *72*, 248–254. [CrossRef]

74. Gabaldón, T. Evolution of the Peroxisomal Proteome. *Alzheimer's Dis.* **2018**, *89*, 221–233.

75. Gabaldón, T. Evolutionary considerations on the origin of peroxisomes from the endoplasmic reticulum; and their relationships with mitochondria. *Cell Mol. Life Sci.* **2014**, *71*, 2379–2382. [CrossRef]

76. Sugiura, A.; Mattie, S.; Prudent, J.; McBride, H.M. Newly born peroxisomes are a hybrid of mitochondrial and ER-derived pre-peroxisomes. *Nature* **2017**, *542*, 251–254. [CrossRef]

77. Gabaldón, T. Peroxisome diversity and evolution. *Philos. Trans. R. Soc. B Boil. Sci.* **2010**, *365*, 765–773. [CrossRef] [PubMed]

Article

The Development of Crassulacean Acid Metabolism (CAM) Photosynthesis in Cotyledons of the C_4 Species, *Portulaca grandiflora* (Portulacaceae)

Lonnie J. Guralnick *, Kate E. Gilbert, Diana Denio and Nicholas Antico

Department of Biology, Roger Williams University, One Old Ferry Rd., Bristol, RI 02809, USA; kgilbert721@g.rwu.edu (K.E.G.); ddenio335@g.rwu.edu (D.D.); nantico949@g.rwu.edu (N.A.)
* Correspondence: lguralnick@rwu.edu; Tel.: +1-401-254-3581; Fax: +1-401-254-3360

Received: 10 December 2019; Accepted: 27 December 2019; Published: 2 January 2020

Abstract: *Portulaca grandiflora* simultaneously utilizes both the C_4 and Crassulacean acid metabolism (CAM) photosynthetic pathways. Our goal was to determine whether CAM developed and was functional simultaneously with the C_4 pathway in cotyledons of *P. grandiflora*. We studied during development whether CAM would be induced with water stress by monitoring the enzyme activity, leaf structure, JO_2 (rate of O_2 evolution calculated by fluorescence analysis), and the changes in titratable acidity of 10 and 25 days old cotyledons. In the 10 days old cotyledons, C_4 and CAM anatomy were evident within the leaf tissue. The cotyledons showed high titratable acid levels but a small CAM induction. In the 25 days old cotyledons, there was a significant acid fluctuation under 7 days of water stress. The overall enzyme activity was reduced in the 10 days old plants, while in the 25 days old plants CAM activity increased under water-stressed conditions. In addition to CAM, the research showed the presence of glycine decarboxylase in the CAM tissue. Thus, it appears both pathways develop simultaneously in the cotyledons but the CAM pathway, due to anatomical constraints, may be slower to develop than the C_4 pathway. Cotyledons showed the ancestral Atriplicoid leaf anatomy, which leads to the question: Could a CAM cell be the precursor to the C_4 pathway? Further study of this may lead to understanding into the evolution of C_4 photosynthesis in the *Portulaca*.

Keywords: *Portulaca grandiflora*; C_4 photosynthesis; Crassulacean acid metabolism (CAM), evolution; development; PEP carboxylase; Portulacaceae; glycine decarboxylase

---

1. Introduction

CO_2 concentrating mechanisms have evolved in terrestrial plants in response to changing environmental conditions. Two different mechanisms have evolved that involve a similar suite of enzymes utilized in a different fashion to overcome photorespiration and increased water loss. Photorespiration increases when CO_2 becomes limited under high light intensities and high evaporative demand resulting in increased transpirational water loss [1]. The C_4 pathway overcomes these limitations with the CO_2 being initially captured as HCO_3^- by phosphoenolpyruvate carboxylase (PEPCase) and then fixed via the C_3 pathway by Rubisco. C_4 plants have a spatial separation of the C_4 and C_3 pathways occurring within two different cell types in the leaf. The C_4 pathway, located in the palisade mesophyll cells, is radially arranged around the C_3 pathway located in the bundle sheath cells, which surround the vascular tissue. This is typically referred to as Kranz anatomy [1,2]. Research by Voznesenskaya et al. [3,4] has shown that the Kranz anatomy is not essential for terrestrial C_4 photosynthesis to occur but it can occur in a single cell with a spatial separation of the C_4 and C_3 pathways within a single chlorenchyma cell. The C_4 pathway concentrates CO_2 at the site of Rubisco and helps to suppress photorespiration in the bundle sheath cells. The C_4 pathway has been found in approximately 19 plant families and over 8000 species [1] and has evolved independently many times.

Crassulacean acid metabolism (CAM) is a metabolic and anatomical adaptation that is characterized by net nocturnal carbon dioxide uptake with a temporal separation of the C_4 and C_3 pathway resulting in decreased transpiration rates and water loss [5,6]. The CO_2 is similarly fixed (as in C_4 plants) by PEPCase, converted to malate and stored as malic acid in the large central vacuole during the night period. In the subsequent light period, the malate is transported out of the vacuole and then decarboxylated to release CO_2 for utilization by Rubisco in the C_3 cycle. CAM plants typically have a mesophyll anatomy with primarily spongy parenchyma cells with a large central vacuole, which has the ability to store the increasing accumulation of malic acid during the nocturnal period [7]. CAM plants anatomically have very low mesophyll airspace, so when the stomata are closed the CO_2 concentrations inside the leaf can reach very high levels to suppress photorespiration [6,7]. CAM has evolved in at least 34 different plant families including 6 aquatic families and over 20,000 species [1,8].

CAM and C_4 photosynthesis have evolved independently multiple times in many different plant families. One might hypothesize both CAM and C_4 could have evolved in the same plant families numerous times due to the similarity of the enzymes involved in the pathways. The two pathways have evolved in the same family four times (Aizoaceae, Asteraceae, Euphorbiaceae, Portulacaceae). This raises an interesting question about why the C_4 and CAM pathways have only concurrently evolved in these four families. The original circumscription of the Portulacaceae describes the family as containing approximately 29 genera [9]. Guralnick and Jackson [10] have reported the evolution and distribution of C_4 and CAM photosynthesis found in this family. It has been observed that some members are C_3 plants, while others are C_3 plants with attributes of CAM. Others are C_4 plants with some CAM characteristics, and the more advanced species are facultative CAM plants [10–12].

The genus *Portulaca* is known to have the only C_4 photosynthetic members of the family despite previous reports to the contrary [11]. This revision reveals the genus *Portulaca* (Portulacaceae) contains the only known C_4 members that are capable of CAM photosynthesis [13]. *Portulaca spp.* tend to inhabit environments with high light intensities, which periodically become dry. The genus *Portulaca* has succulent stems and leaves and because of the high degree of succulence, there have been reports in the literature that members of the *Portulaca* show a diurnal acid fluctuation characteristic of CAM species. Koch and Kennedy [14,15] showed *Portulaca oleracea* having a diurnal acid fluctuation in both the stems and leaves. They also measured low levels of net nighttime CO_2 uptake. Research done by Guralnick and Jackson [10] showed that *Portulaca mundula*, *P. pilosa*, and *P. oleracea* exhibited high acid levels and diurnal acid fluctuations indicative of CAM photosynthesis. Kraybill and Martin [16] showed both *P. oleracea* and *P. grandiflora* both undergo CAM cycling with little or no nocturnal CO_2 uptake. Mazen [17] indicated that under water stress conditions that *P. oleracea* had increased levels of PEP (phosphoenolpyruvate) carboxylase protein. Further research showed the genus contains a C_3-C_4 intermediate species, *Portulaca cryptopetala*, in which recent work showed *P. cryptopetala* induced CAM under water stress conditions [2,18]. Winter et al. [18] extended the findings to additional species in the *Portulaca* and also consider them to be facultative CAM species.

Portulaca grandiflora is a small herbaceous annual utilizing the C_4 photosynthetic pathway. *P. grandiflora* has small succulent leaves with a Pilosoid-type Kranz leaf anatomy where the C_4 tissue in the succulent leaves surround the large water storage tissue [2,19]. *P. grandiflora* is known to maintain high organic acid levels and shows a large diurnal acid fluctuation when water stressed, typical of CAM species [20]. Research has indicated the increase in CAM of this species occurs in the water storage portion of the leaf and the stem during water stressed conditions [20]. *Portulaca grandiflora* is unique because it has both C_4 and CAM photosynthetic pathways operating simultaneously in the leaf tissue [20]. This situation is unique due to the proposed incompatibility of both pathways to operate in the same leaf [21]. Phylogenetic analysis has indicated the genus *Portulaca* evolved CAM from a C_3 ancestor prior to the appearance of the C_4 pathway [10]. The objective of this study was to study cotyledon leaf tissue to determine if both the CAM and C_4 pathways were developing and operating simultaneously. CAM induction in developing cotyledons was monitored by withholding water for 3 and 7 days. An understanding of the developmental process of these pathways will aid in clarifying the evolutionary origins of the CAM and C_4 pathways in the Portulacaceae.

2. Results

2.1. Titratable Acidity

Titratable acidity levels for 10 days old cotyledons were at approximately 50–60 µeq gFW^{-1} (FW = Fresh Weight; Figure 1). At 10 days, the control groups and water-stressed leaves showed a slight acid fluctuation of 10–20 µeq gFW^{-1} from a.m. to p.m. (Figure 1). There was no significant difference between a.m. and p.m. acid levels. At 20–25 days old, cotyledons, under control conditions, there was no acid fluctuation observed from a.m. to p.m. levels. Under water stress, cotyledons showed a small significant titratable acid fluctuation from the morning to the evening (Figure 1). Continued water stress to 7 days of the 20–25 days old cotyledons induced a large and significant acid fluctuation of 83 µeq gFW^{-1} in both the cotyledons and primary leaves of *P. grandiflora* (Figure 2). The a.m. acid levels had increased to more than double the control cotyledons.

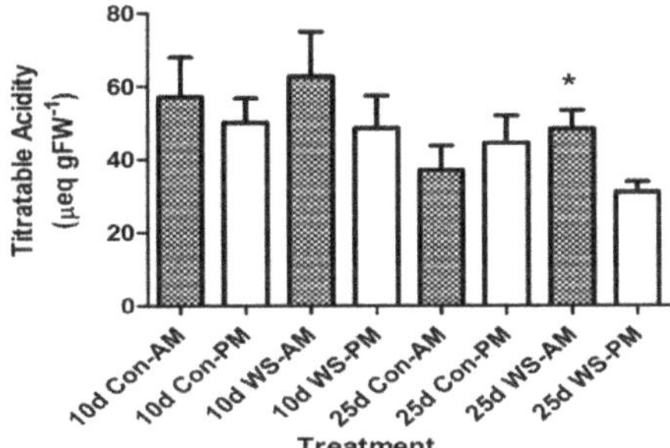

Figure 1. Titratable acidity of *Portulaca grandiflora* in 10 days and 25 days old cotyledons under control and 3 days water-stress conditions. Bars represent the means (SEM). For 10 days old, N = 9–11 leaves per treatment; 25 days old, N = 8–13 leaves per treatment; and * indicates a significant difference between a.m. and p.m. acid levels for 25 days old treatment. Con = Control; WS = Water Stress for all figures.

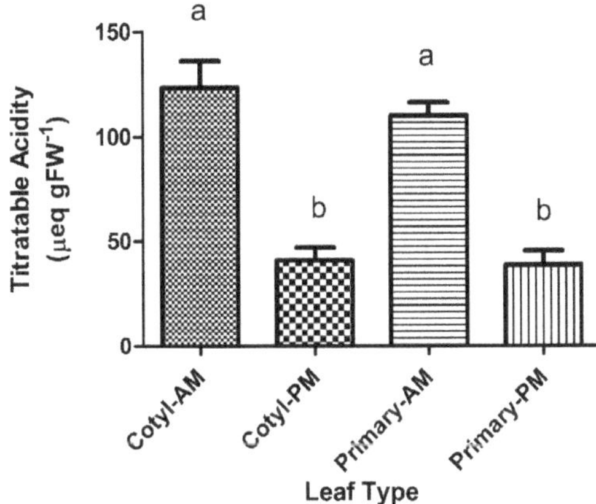

Figure 2. Diurnal titratable acidity levels in cotyledons and primary leaves of *P. grandiflora* after 7 days of water stress. Bars represent the means (SEM). Bars with different letters are significantly different ($p < 0.05$, N = 4).

2.2. Enzyme Activity: PEP Carboxylase and NADP-ME

Ten day old cotyledons showed PEPCase activity higher during the day than at night (Figure 3). Water stress lowered the activity of PEPCase in the 10 days old cotyledons during the day and night. There was a significant difference in daytime activity between the control and water stress at 10 days. At 25 days old, the activity remained high during the day but lower at midnight (Figure 3). In the 25 days old cotyledons, water stress had less of an effect on the overall activity of PEPCase. The PEPCase activity during the day increased in the water-stress plants compared to the control plants (Figure 3). There was no significant differences in PEPCase activity in the 25 days old cotyledons (Figure 3).

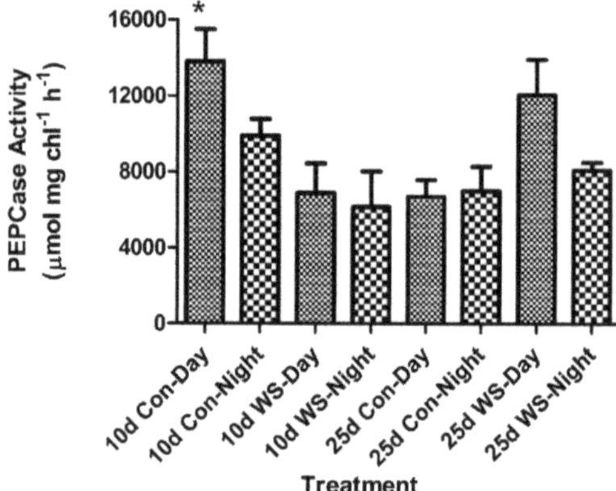

Figure 3. PEP carboxylase activity of *P. grandiflora* in 10 days and 25 days old cotyledons under control and 3 days water-stress conditions. Bars represent the means (SEM); * = 10 days Con-Day significantly different from 10 days WS-Day ($p < 0.05$, N = 8–12); 25 days old cotyledons N = 21–25 control plants, N = 8–11 water-stress plants.

The decarboxylase, NADP-Malic Enzyme (ME), showed high daytime activity in both the 10 days and 25 days old control cotyledons (Figure 4). Water stress significantly lowered the NADP-ME activity at 10 days, while at 25 days there was no significant change in the level of activity between the control and water-stressed plants.

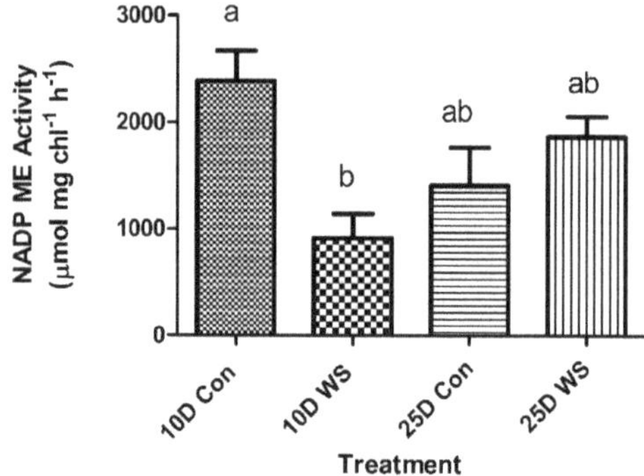

Figure 4. NADP-ME activity of *P. grandiflora* in 10 days and 25 days old cotyledons under control and 3 days water-stress conditions. Bars represent the means (SEM). N = 6–13. Bars with different letters are significantly different ($p < 0.05$).

2.3. Electron Transport Rate/JO₂

The JO_2 rates showed no differences between control and water-stressed plants in 10 days old and 25 days old cotyledons (Figure 5). The JO_2 rates were lower in 25 days old cotyledons than in the 10 days old cotyledons.

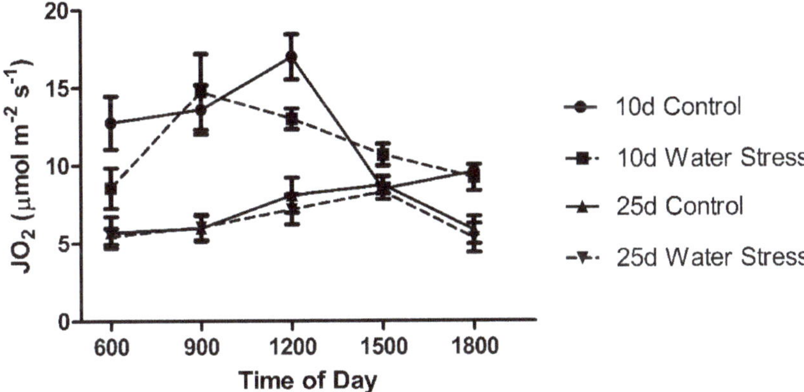

Figure 5. JO_2 rates of leaf samples for 10 days and 25 days control and water-stress treatments. Error bars represent one standard error of the mean. N = 12–14 leaves per time point.

The gross rate of photosynthesis (JO_2) showed a peak at midday for the well-watered 10 days old cotyledons while the stressed cotyledons showed a peak earlier in the morning (Figure 5). In the 25 days old cotyledons, the JO_2 calculated rates were relatively constant over the course of the light period. There was no difference between the control and water-stressed plants in the JO_2 rates (Figure 5).

2.4. Leaf Anatomy

Portulaca grandiflora cotyledons were dissected and photographed to display the internal gross anatomy of the tissue (Figure 6). The sections illustrate the Atriplicoid Kranz anatomy in the cotyledon (Figure 6A), and its linear arrangement compared to a mature leaf where the Kranz anatomy shows a radial arrangement (Figure 6B). A low-magnification view of a cotyledon leaf tissue showed a well-developed Atriplicoid Kranz anatomy arranged in a linear fashion (Figure 7). Mesophyll cells were well developed with chloroplast along the cell wall (Figure 7A). The bundle sheath cells were also well developed with chloroplasts located along the periphery of the inner cell wall close to the vascular bundles (Figure 7A). The CAM/hydrodermal tissue showed fewer chloroplasts than the C_4 tissue (Figure 7A). At 25 days, the Kranz anatomy was very well developed and the vascular tissue was more prominent. Under water-stress conditions, the CAM/hypodermal tissue at 10 days showed signs of water loss compared to the control cotyledon tissue. The cells appeared to have some shriveling and a more irregular shape. At 25 days, there were less noticeable changes under water-stress conditions in the succulent mesophyll and hypodermal tissue (Figure 7).

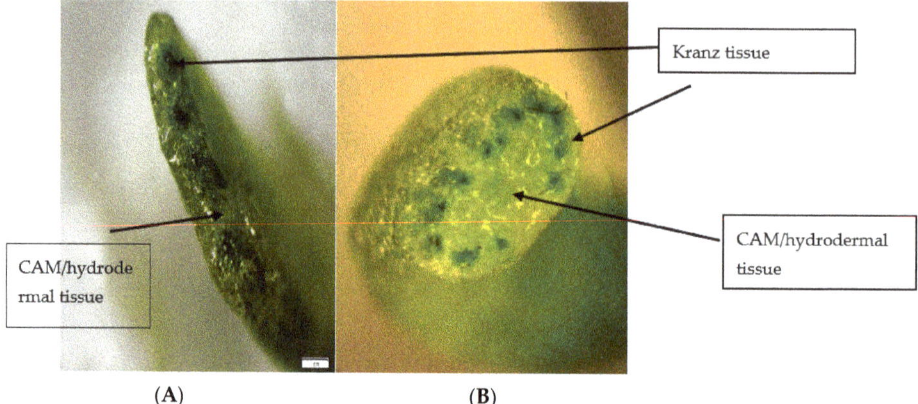

Figure 6. Cross section of 1.7× magnification of (**A**) ~20 days old cotyledon and (**B**) mature leaf of *P. grandiflora*. The dark green bundles within the leaf tissue are the C_4 Kranz anatomy with high levels of chlorophyll. The lighter areas are the CAM/hydrodermal tissue.

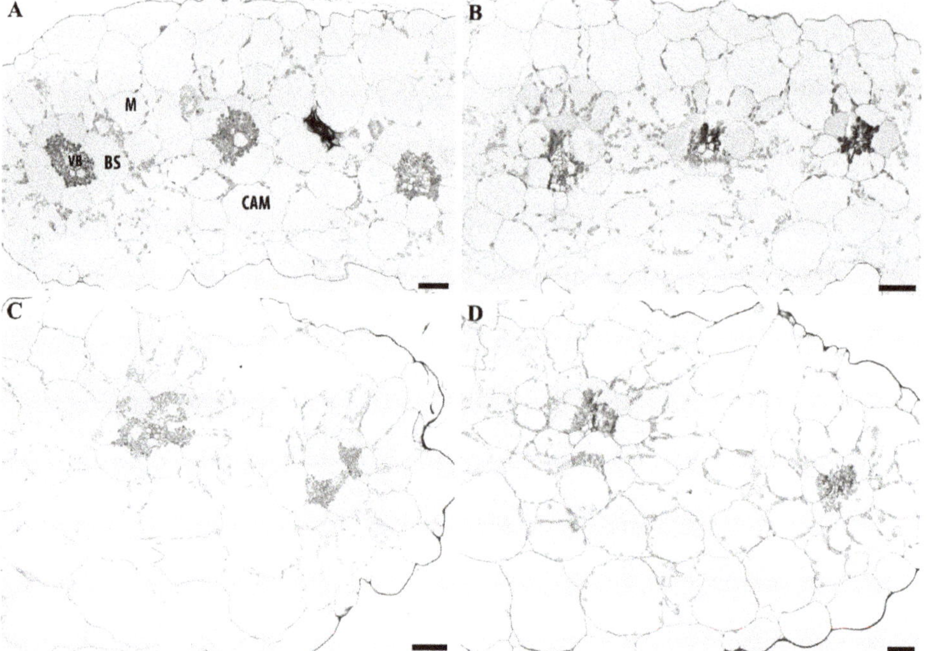

Figure 7. Light microscope images of cotyledons of *P. grandiflora* at low magnification, (**A**) 10 days control, (**B**) 10 days water stress, (**C**) 25 days control, (**D**) 25 days water stress. Bars = 50 μm, VB = vascular bundle, BS = bundle sheath, M = mesophyll, CAM = hypodermal tissue.

The sections illustrate the Atriplicoid Kranz anatomy in its linear arrangement compared to a mature leaf where the Kranz anatomy shows a radial arrangement (Figure 6B). We performed tissue prints of cotyledons to determine the presence of glycine decarboxylase (GDC, a mitochondrial marker for the photorespiratory pathway). GDC was found to be present in the CAM/hydrodermal tissue of *P. grandiflora* cotyledons (Figure 8B). In mature leaves of *P. grandiflora*, GDC was also located in the inner CAM/hydrodermal tissue (data not shown). For comparison, *Portulacaria afra*, a facultative CAM species, showed the presence of GDC throughout the spongy parenchyma mesophyll tissue (8b).

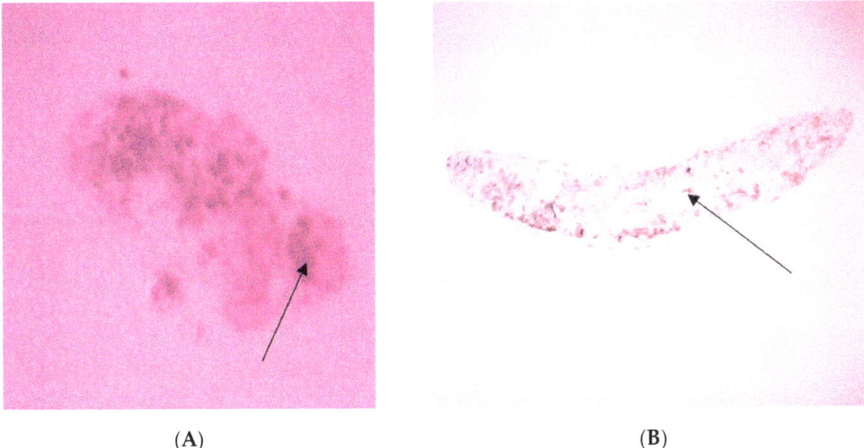

(A)   (B)

Figure 8. Glycine decarboxylase (GDC) protein presence by tissue printing. (**A**) Cotyledon leaf samples were sectioned and photographed, then printed onto nitrocellulose, incubated with GDC antibody, and visulaized. The arrow points to the bundle sheath tissue which showed more GDC protein. The GDC protein was also found present in the surroounding succulent CAM tissue. (**B**) *Portulacaria afra*, a facultative CAM species, leaf tissue print with the arrow indicating GDC antibody found throughout the spongy parenchyma tissue.

3. Discussion

Guralnick and Jackson [10] previously suggested the CAM pathway appeared evolutionarily first in the leaf and the C_4 photosynthetic pathway overlaid the CAM tissue. These conclusions were supported by research which showed the CAM PEPCase gene was more primitive to the C_4 derived PEPCase gene [22]. Sage [21] described the leaf anatomy found in the *Portulaca* as being unique because of the apparent incompatibility of the photosynthetic pathways. The genus *Portulaca* shows nearly all members display C_4 photosynthesis with attributes of the CAM pathway in the same leaf tissue and are considered facultative CAM species [18]. Previous research showed the CAM pathway aids in the retention of water for maintenance of the C_4 pathway and the recycling of carbon to the mesophyll and bundle sheath cells [20]. This cooperation between the tissues may be the result of the close proximity of the tissues to each other because previous work has proposed that the tissues are functioning independently of each other [20].

3.1. C_4 Development

Research has been performed on the physiological and structural development of C_4 photosynthesis in cotyledons of the *Portulaca* but little of CAM development in these C_4 species [23]. The research presented here indicated it was apparent that the C_4 pathway was well developed in 10 days old cotyledons based on the anatomy and physiology. The leaf anatomy of the cotyledons showed a well-developed Kranz C_4 anatomy as found previously in the genus *Portulaca* [23]. *P. grandiflora* cotyledons displayed an Atriplicoid-type anatomy where the vascular bundles are in one plane of the leaf. The Atriplicoid Kranz anatomy showed CAM tissues around the periphery of the leaf surrounding the Kranz bundles. This differs from the Pilosoid Kranz anatomy found in mature leaves of *P. grandiflora* with the C_4 tissue in a ring surrounding the CAM water storage tissue [2,20].

The enzymes of the C_4 pathway were fully functional at 10 days, as shown by a high activity of PEPCase and NADP-ME. This was comparable to the conclusions reached by Dengler et al. [24] for *Atirplex rosea*, which showed accumulation of PEPCase was detected 2–4 days after leaf development and expansion. The expression was limited to mesophyll tissue adjacent to the bundle sheath tissue. The JO_2 studies in this study showed high rates of photosynthesis indicative of C_4 activity and were comparable to the rates previously measured in mature leaf tissue [20]. The C_4 pathway is the primary CO_2 acquiring pathway in this genus and it was predictable the C_4 pathway developed quickly for seedling establishment.

3.2. CAM Development

The development of CAM, which can be measured by titratable acidity levels and diurnal acid fluctuations, showed a slower development in cotyledons from that of the C_4 pathway. The study of CAM development showed comparable titratable acidity levels but slightly lower and comparable to mature leaf tissues under control conditions [20]. One can measure significant total titratable acidity levels in the cotyledons at 10 and 25 days. The acid levels found in the 10 days old cotyledons of *P. grandiflora* are similar but larger than those found in well-watered cotyledons of *Mesembryanthemum crystallinum*, a facultative CAM species [25], and higher than the acid levels in the CAM-cycling species, *Lewisia cotyledon* [26]. The acid levels of *P. grandiflora* were lower than those found in a number of columnar cactus seedlings at 1 day and 7 days old [27]. The acid levels were lower than those for *Opuntia elatior* at 23 days old cotyledons [28]. However, these cactus species primarily utilize the CAM pathway for CO_2 uptake, even in cotyledons. The cotyledon anatomy in cactus has succulent anatomy with a spongy parenchyma tissue with little airspace, which is more conducive to CAM photosynthesis [29,30]. The results in the present study indicated *P. grandiflora* cotyledons showed similar total titratable acid levels comparable to other CAM species.

We investigated the potential for CAM activity by water stressing the cotyledons for three and seven days. Our results differed with the age of the cotyledons and duration of the drought. Ten days old cotyledons showed small acid fluctuations indicative of CAM activity after three days of water stress. The 25 days old water-stressed cotyledons showed a small acid fluctuation compared to control cotyledons which showed no acid fluctuation. Continuation of water stress for seven days induced a very large and significant CAM acid fluctuation in both cotyledons and primary leaf tissue. The acid fluctuation measured in the cotyledons was comparable to acid levels in mature leaf tissue. This indicated an induction of CAM similar to previous research on *P. grandiflora* [20] and was indicative of CAM activity.

In addition, enzyme activity, as measured by nocturnal PEPCase activity, was quite high and comparable to C_4 rates of PEPCase measured during the day. Water stress only had a slight effect on the overall activity of PEPCase. NADP-ME was affected more by water stressed conditions and this result had been observed previously in *P. grandiflora* [20]. Water stress in other facultative CAM plants, such as *Portulacaria afra*, show the decarboxylase enzymes are more affected than PEPCase [31]. The enzyme activities of PEPCase and NADP-ME reported here in cotyledons of *P. grandiflora* were much higher than those reported for cotyledons of *Salsola* spp. [32] and for mature leaves of *P. grandiflora* [20] and may be related to the lower chlorophyll concentrations.

The physiological and anatomical changes differed with the age. At 10 days, water stressing the cotyledons induced a small acid fluctuation and it lowered both the day and night activity of PEPCase. NADP-ME activity also decreased in 10 days old cotyledons. This indicated a decrease in C_4 and CAM activity. It appears that NADP-ME is more sensitive to water stress than Rubisco and PEPCase. Anatomically, the mesophyll CAM tissue showed some shrinkage, which may be due to a redistribution of water from the CAM cells to the C_4 metabolic cells. This has been observed in mature leaves of *P. grandiflora* during a 10 days drought [20]. The sensitivity of the enzymes due to drought at 10 days may be due more in part to structural development of the CAM tissue and water storage at this stage. The enzymatic activity appeared earlier prior to full anatomical development of the CAM tissue. This is similar to the development of C_4 photosynthesis in *A. rosea*, which showed differential enzyme expression of the C_4 photosynthetic enzymes by 4 days [24].

By 20–25 days old, the response to water stress and CAM activity was similar to the 10 days old cotyledons. There was a similar induction of an acid fluctuation in water-stressed plants. Seven days of water stress induced a much larger acid fluctuation indicative of CAM activity and greater than observed after three days of water stress. This acid fluctuation was similar to acid fluctuations in other CAM-cycling species [10,20]. The results indicated an induction of CAM activity similar to control plants of *Lewisia cotyledon* [26]. Induction of water stress did not lower PEPCase activity in 25 days old cotyledons compared to 10 days old cotyledons. NADP-ME did not show a decrease in activity due to water stress and showed a slight increase in activity, which may be due to increased CAM activity of the leaf. The enzyme activity indicated a maintenance of the C_4 pathway in the 25 days old cotyledons. This was supported by the JO_2 data, which showed no changes in activity during three days of drought. Structurally, the CAM/hydrodermal tissue was more mature in the 25 days old cotyledons and did not appear to show as much water loss as the 10 days old cotyledons. This may explain the ability of older cotyledons to maintain C_4 activity when compared to 10 days old cotyledons. The continuation of drought showed a much stronger induction of CAM and was very similar to the response of mature leaves [20]. The CAM tissue in mature leaves recycles CO_2, redistributes water to the C_4 tissue to aid in the survival of the plant, and may play a similar role in cotyledons [20].

Mature leaves of *P. grandiflora* have a Pilosoid-type anatomy with a ring of bundles surrounding the water storage tissue [2,20]. *P. grandiflora* cotyledons have an Atriplicoid-type anatomy where the vascular bundles are in one plane of the leaf. This arrangement is considered ancestral to the Pilosoid arrangement with the water storage tissue found on both the abaxial and adaxial sides of the leaf [33]. Voznesenskaya et al. [2] reported on the diversity of structure in the *Portulaca* and found the arrangement of an inner water storage, the derived condition found in *P. grandiflora*, *P. pilosa*, and *P. amilis* [2], to be due to the ecological constraints of being found in more semi-arid regions than other *Portulaca* spp. However, they did not report any CAM activity in the cotyledons of the *Portulaca spp.* in their study. Further, we report the presence of glycine decarboxylase, a mitochondrial marker for the C_2 photorespiratory pathway in the CAM tissue of cotyledons. Previous research has shown the GDC to be localized in the mitochondria of bundle sheath cells in cotyledons [2]. The presence of GDC adds another dimension to another function of CAM in *P. grandiflora*. Previous work indicated CAM tissue might transfer water from the tissue to the Kranz anatomy to help maintain C_4 photosynthetic activity under water stress. Additionally, decarboxylation of the acid may produce CO_2 for the adjacent mesophyll and bundle sheath tissue. We now suggest the C_2 pathway may provide an additional source of CO_2 for the C_4 pathway. It was shown the photorespiratory pathway may elevate CO_2 levels three-fold in the leaf in *Flaveria pubescens*, (a C_3-C_4 intermediate) [34]. The photorespiratory pathway in the CAM tissue of *P. grandiflora* may provide an additional source of CO_2 for the C_4 pathway. Since CAM tissue has reduced airspace, the Atriplicoid arrangement may reduce diffusion of water out of the leaf and maintain CO_2 levels around the Kranz anatomy. This may aid in survival of the seedlings. Further work will be needed to investigate the role of photorespiration in the different Kranz types CAM tissue in *P. grandiflora*.

This is the first report of CAM activity in the cotyledons and the results indicated CAM can be induced in cotyledons of *P. grandiflora*. The development of CAM occurred after the development of the C_4 photosynthetic pathway and may be due to more of a constraint of leaf anatomy of the CAM tissue in cotyledons. The retention of CAM in the cotyledons of *P. grandiflora* further supports the idea that CAM was an ancient pathway in the genus *Portulaca* including the retention of the C_2 pathway. Since the cotyledons have a different type of Kranz anatomy and water storage tissue than the mature leaves in *P. grandiflora*, this leads to questions on the evolution of C_4 and CAM within the genus *Portulaca*. Research on the origins of CAM in *Portulaca* using PEPCase have indicated the CAM-specific gene was similar in sequence in other species utilizing CAM [22]. The question that arises: Could C_4 photosynthesis have evolved from the CAM cells? Christin et al. [22] have shown the C_4 specific PEPCase gene appears to have evolved from a non-photosynthetic form rather than the CAM form. They also suggest the other enzymes required for C_4 photosynthesis may not have required new genes but utilization of genes already present in the cell. Since CAM cells in the *Portulaca* retain all features of C_3 and CAM photosynthesis including the C_2 pathway, it maybe that novel cells types of the mesophyll and bundle sheath may have differentiated from CAM-like cells. During the slow evolution of C_4 photosynthesis as suggested by Christin et al. [22], a change in the regulatory sequences of PEPCase and GDC would be required as the cells evolved. Current research does not support the idea but questions remain on the change from a CAM-type leaf structure evolved into a C_4 photosynthetic leaf in *Portulaca*.

4. Materials and Methods

4.1. Plant Material

Seeds of *Portulaca grandiflora* were germinated in flats (Figure 9A,B). All plants grown were irrigated with half-strength Hoagland's solution prior to sampling. The plants were grown under natural light conditions, supplemented with artificial light for a light intensity of 300–400 μmol m^{-2} s^{-1}.

(**A**) (**B**)

Figure 9. (**A**) Ten-day-old cotyledons growing flat. (**B**) Approximately 20–25 days old *P. grandiflora* seedling with cotyledons and primary leaves.

The day/night temperature in the glasshouse was 27 °C/17 °C. Seedlings were irrigated daily to maintain high water potential, and water stressed for three days prior to sampling. Experiments were performed when the leaf tissues were 10–14 days old and 22–26 days old post-germination. Leaf samples were taken during the course of the day/night cycle and frozen until assayed for titratable acidity and enzyme activity.

4.2. Titratable Acidity

Cotyledons were harvested in the morning and evening, and frozen (−80 °C) until assayed. Leaf samples were weighed, ground in glass-distilled water, and titrated with 0.01 N KOH to a pH 7 endpoint as described by Guralnick et al. [20]. Results were expressed as μeq gFW^{-1}.

4.3. Enzyme Activity

Frozen leaf tissue collected midday and during the middle of the night period was utilized for phosphoenolpyruvate carboxylase (PEPCase) and NADP-malic enzyme (ME) activity. Assays of crude extracts were done in triplicate under well-watered and water-stress conditions. The samples were assayed spectrophotometrically by following the oxidation of NADH (for PEPCase) or the reduction of $NADP^+$ at 340 nm as previously described by Guralnick and Ting [31]. Results were expressed as μmol mg chl^{-1} h^{-1}.

4.4. Leaf Anatomy

Cotyledons were harvested and cut into sections. They were fixed in 1.5% glutaraldehyde in sodium cacadylate buffer followed by fixation in osmium. Sections were dehydrated in a series to 80% ethanol. The leaf material for transverse cross-section was fixed with fluoroacetic acid and then embedded in JB4 plastic resin (2-hydroxyethyl methacrylate). Sections were cut with a microtome and stained with 1% aniline B.

4.5. Electron Transport and JO_2

The electron transport rate was determined using the pulse amplitude modulated fluorometer (OSP1 Fluorometer), according to Guralnick et al. [20]. Samples were taken over the course of the light period beginning at 6:00 in the greenhouse. Intact leaf tissue of well-watered and water-stressed cotyledons were measured [20]. The true rate of O_2 evolution (JO_2) from Photosystem II activity was calculated, according to the method of Lal and Edwards [35].

4.6. Tissue Printing

Cotyledons were cut with a razor and then photographed. The sections were pressed onto nitroceullusoe membranes. Protein localization of Glycine decarboxylase (GDC-H) antibodies was performed according to the method of Guralnick et al. [20]. Rabbit anti-GDC-H antibodies were purchased from Agrisera and diluted to a 1:5000 concentration. The presence of GDC was visualized using a secondary antibody linked to alkaline phosphatase and then photographed. GDC was a mitochondrial marker for the C_2 photorespiratory pathway.

5. Conclusions

Research has shown the cotyledons of *Portulaca grandiflora* could be induced to perform CAM similar to mature leaves [20]. The CAM pathway appears to develop later due to anatomical but not physiological aspects of CAM. In the species, *P. grandiflora*, both pathways are present and appear to be operating independently in the cotyledons but CAM still aids the C_4 pathway under water-stress conditions. Additionally, the C_2 pathway may raise CO_2 concentrations inside the cotyledons, aiding in raising the internal CO_2 concentrations. More research will be needed to understand the evolution of CAM and C_4 photosynthesis in the genus *Portulaca*.

Author Contributions: The authors on this manuscript contributed to many portions as undergraduate researchers. Conceptualization, L.J.G.; methodology, L.J.G., K.E.G., D.D., and N.A.; data collection, L.J.G., K.E.G., D.D., and N.A.; data analysis, L.J.G., K.E.G., D.D., and N.A.; writing, original draft preparation, L.J.G. and D.D.; writing, review, L.J.G., K.E.G., and N.A.; supervision and project administration, L.J.G.; funding acquisition, L.J.G., D.D., and N.A. All authors have read and agreed to the published version of the manuscript.

Funding: The research received no external funding.

Acknowledgments: The research was supported in part from an Internal RI INBRE fellowship (IDeA Networks of Biomedical Research Excellence) and a Center for Economic and Environmental Development (CEED) fund grant awarded to Diana Denio and Nicolas Antico. Part of this research was presented at the CAM 2010 conference held at the Smithsonian Tropical Research Institute in Panama City, Panama, by Diana Denio. The authors would like to thank Tammy Sage and Rowan Sage for their help with the light microscopy. The authors would like to thank Tim Pelletier and Colleen McKnight-Torres for their support with equipment and supplies for this study.

Conflicts of Interest: The authors declare no conflict of interest.

References

1. Sage, R.F.; Monson, R.K. Preface. In *C₄ Plant Biology*; Academic: Philip Drive Norwell, MA, USA, 1999; pp. xiii–xv.
2. Voznesenskaya, E.V.; Koteyeva, N.K.; Edwards, G.E.; Ocampo, G. Revealing diversity in structural and biochemical forms of C_4 photosynthesis and a C_3-C_4 intermediate in genus *Portulaca* L. (Portulacaceae). *J. Exp. Bot.* **2010**, *61*, 3647–3662. [CrossRef] [PubMed]
3. Voznesenskaya, E.V.; Franceschi, V.R.; Kiirats, O.; Freitag, H.; Edwards, G.E. Kranz anatomy is not essential for terrestial C4 plant photosynthesis. *Nature* **2001**, *414*, 543–546. [CrossRef] [PubMed]
4. Voznesenskaya, E.V.; Franceschi, V.R.; Kiirats, O.; Artyusheva, E.G.; Freitag, H.; Edwards, G.E. Proof of C_4 photosynthesis without Kranz anatomy in *Bienertia cycloptera* (Chenopodiaceae). *Plant J.* **2002**, *31*, 649–662. [CrossRef] [PubMed]
5. Osmond, C.B. Crassulacean acid metabolism: A curiosity in context. *Ann. Rev. Plant Physiol.* **1978**, *29*, 379–414. [CrossRef]
6. Ting, I.P. Crassulacean acid metabolism. *Ann. Rev. Plant Physiol.* **1985**, *36*, 595–622. [CrossRef]
7. Gibson, A. Anatomy of Succulence. In *Crassulacean Acid Metabolism: Proceedings of the Fifth Symposium in Botany*; Ting, I.P., Gibbs, M., Eds.; Waverly Press: Baltimore, MD, USA, 1982; pp. 1–17.
8. Winter, K.; Smith, J.A.C. An Introduction to Crassulacean acid metabolism. In *Crassulacean Acid Metabolism: Biochemical Principles and Ecological Diversity*; Springer: Berlin/Heidelberg, Germany, 1996; pp. 1–13.
9. Eggli, U.; Ford-Werntz, D. Portulacaceae. In *Illustrated Handbook of Succulent Plants: Dicotyledons*; Springer: Berlin/Heidelberg, Germany, 2002; p. 370.
10. Guralnick, L.J.; Jackson, M.D. The occurrence and phylogenetics of Crassulacean acid metabolism activity in the Portulacaceae. *Int. J. Plant Sci.* **2001**, *162*, 257–262. [CrossRef]
11. Guralnick, L.J.; Cline, A.; Smith, M.; Sage, R. Evolutionary Physiology: The extent of C_4 and CAM photosynthesis in the Genera *Anacampseros* and *Grahamia* of the Portulacaceae. *J. Exp. Bot.* **2008**, *59*, 1735–1742. [CrossRef]
12. Hershokivitz, M.A.; Zimmer, E.A. On the evolutionary origins of the cacti. *Taxon* **1997**, *46*, 217–232. [CrossRef]
13. Lara, M.V.; Andreo, C.S. Photosynthesis in nontypical C_4 species. In *Handbook of Photosyntheis*, 2nd ed.; Pessarakli, M., Ed.; Taylor & Francis: London, UK, 2005; pp. 392–421.
14. Koch, K.E.; Kennedy, R.A. Characteristics of Crassulacean acid metabolism in the succulent C_4 dicot, *Portulaca oleracea* L. *Plant Physiol.* **1980**, *65*, 193–197. [CrossRef]
15. Koch, K.E.; Kennedy, R.A. Crassulacean acid metabolism in the succulent C_4 dicot, *Portulaca oleracea* L. under natural environmental conditions. *Plant Physiol.* **1982**, *69*, 757–761. [CrossRef]
16. Kraybill, A.A.; Martin, C.E. Crassulacean acid metabolism in three of the C_4 genus *Portulaca*. *Int. J. Plant Sci.* **1996**, *57*, 103–109. [CrossRef]
17. Mazen, A.M.A. Changes in the levels of phosphoenolpyruvate carboxylase with induction of Crassulacean acid metabolism (CAM)-like behaviour in the C_4 plant *Portulaca oleracea*. *Physiol. Plant.* **1996**, *98*, 111–116. [CrossRef]
18. Winter, K.; Sage, R.F.; Edwards, E.J.; Virgo, A.; Holtum, J.A.M. Facultative Crassulacean acid metabolism in a C_3-C_4 intermediate. *J. Exp. Bot.* **2019**. [CrossRef] [PubMed]
19. Ku, S.B.; Shieh, Y.J.; Reger, B.J.; Black, C.C. Photosynthetic characteristics of *Portulaca* grandiflora, a succulent C_4 dicot. *Plant Physiol.* **1981**, *68*, 1073–1080. [CrossRef]
20. Guralnick, L.J.; Edwards, G.E.; Ku, M.S.B.; Hockema, B.; Franceschi, V.R. Photosynthetic and anatomical characteristics in the C_4-crassulacean acid metabolism-cycling plant, *Portulaca grandiflora*. *Funct. Plant Biol.* **2002**, *29*, 763–773. [CrossRef]
21. Sage, R.F. Are CAM and C_4 photosynthesis incompatible? *Funct. Plant Biol.* **2002**, *29*, 775–785. [CrossRef]
22. Chrisitin, P.A.; Arakaki, M.; Osborne, C.P.; Brautigam, A.; Sage, R.F.; Hibberd, J.M.; Kelly, S.; Covshoff, S.; Wong, G.K.S.; Hancock, L.; et al. Shared origins of key enzyme during the evolution of C_4 photosynthesis and CAM metabolism. *J. Exp. Bot.* **2014**, *65*, 3609–3621. [CrossRef]

23. Voznesenskaya, E.V.; Koteyeva, N.K.; Edwards, G.E.; Ocampo, G. Unique photsynthetic phenotypes in *Portulaca* (Portulacceae): C_3-C_4 intermediates and NAD-ME C_4 species with Pilosoid-type Kranz anatomy. *J. Exp. Bot.* **2017**, *68*, 225–239. [CrossRef]
24. Dengler, N.G.; Dengler, R.E.; Donnelly, P.M.; Filosa, M.F. Expression of the C_4 pattern of photosynthetic enzyme accumulation during leaf development in *Atriplex rosea* (Chenopodiaceae). *Am. J. Bot.* **1995**, *82*, 319–327. [CrossRef]
25. Von Willert, D.J.; EUer, B.M.; Werger, M.J.A.; Brinckmann, E.; Ihlenfeldt, H.-D. *Life Strategies of Succulents in Deserts. With Special Reference to the Namib Desert. Cambridge Studies in Ecology;* von Willert, D.J., Matyssek, R., Herppich, W.B., Eds.; Cambridge University Press: Cambridge, UK, 1992.
26. Guralnick, L.J.; Marsh, C.; Asp, R.; Karjala, A. Physiological and anatomical aspects of CAM-cycling in *Lewisia cotyledon* var cotyledon (Portulacaceae). *Madrono* **2001**, *48*, 131–137.
27. Hernandez-Gonzales, O.; Villarreal, O.B. Crassulacean acid metabolism photosynthesis in columnar cactus seedling during ontogeny: The effect of light on nocturnal acidity accumulation and chlorophyll fluorescence. *Am. J. Bot.* **2007**, *94*, 1344–1351. [CrossRef] [PubMed]
28. Winter, K.; Garcia, M.; Holtum, J.A.M. Drought-stress-induced up-regulation of CAM in seedlings of a tropical cactus, *Opuntia elatior*, operating predominantly in the C_3 mode. *J. Exp. Bot.* **2011**, *62*, 4037–4042. [CrossRef] [PubMed]
29. Ayala-Cordero, G.; Terrazas, T.; Lopez-Mata, L.; Trejo, C. Morpho-anatomical changes and photosynthetic metabolism of *Stenocereus beneckei* seedlings under soil water deficit. *J. Exp. Bot.* **2006**, *57*, 3165–3174. [CrossRef] [PubMed]
30. Secorum, A.C.; de Souza, L.A. Morphology and anatomy of *Rhipsalis cereuscula*, *Rhipsalis floccosa* subsp. *Hohenauensis* and *Lepismium cruciforme* (Cactaceae) seedlings. *Rev. Mex. Biodivers.* **2011**, *82*, 131–143.
31. Guralnick, L.J.; Ting, I.P. Seasonal response to drought and rewatering in *Portulacaria afra* (L.) Jacq. *Oecologia* **1987**, *70*, 85–91. [CrossRef] [PubMed]
32. Pyankov, V.I.; Voznesenskaya, E.V.; Kuz'min, A.N.; Ku, M.S.B.; Ganko, E.; Franceschi, V.R.; Black, C.C.; Edwards, G.E. Occurrence of C_3 and C_4 photosynthesis in coltyledons of leaves of *Salsola* species (Chenopodiaceae). *Photosynth. Res.* **2000**, *63*, 69–84. [CrossRef]
33. Artyushera, E.G.; Edwards, G.E.; P'yankov, V.I. Photosynthesizing tissue development in C_4 cotyledons of two Salsola species (Chenopodiaceae). *Russ. J. Plant Physiol.* **2003**, *50*, 4–18. [CrossRef]
34. Keerberg, O.; Parnik, T.; Ivanova, H.; Bassuner, B.; Bauwe, H. C_2 photosynthesis generates about 3-fold elevated leaf CO_2 levels in the C_3–C_4 intermediate species *Flaveria pubescens*. *J. Exp. Bot.* **2014**, *65*, 3649–3656. [CrossRef]
35. Lal, A.; Edwards, G.E. Analysis of inhibition of photosynthesis wunder waer stress in the C_4 species *Amaranthus cruentus* and *Zea mays*: Electron transport, CO_2 fixation, and carboxylation capacity. *Australian J. Exp. Bot.* **1996**, *23*, 403–412.

Article

Enzymatic Properties of Recombinant Phospho-Mimetic Photorespiratory Glycolate Oxidases from *Arabidopsis thaliana* and *Zea mays*

Mathieu Jossier, Yanpei Liu, Sophie Massot and Michael Hodges *

Institute of Plant Sciences Paris-Saclay, CNRS, Université Paris-Sud, INRA, Université d'Evry, Université Paris-Diderot, Université Paris-Saclay, 91405 Orsay CEDEX, France; mathieu.jossier@ips2.universite-paris-saclay.fr (M.J.); yanpei.liu@ips2.universite-paris-saclay.fr (Y.L.); sophie.massot@ips2.universite-paris-saclay.fr (S.M.)
* Correspondence: michael.hodges@cnrs.fr

Received: 22 November 2019; Accepted: 21 December 2019; Published: 24 December 2019

Abstract: In photosynthetic organisms, the photorespiratory cycle is an essential pathway leading to the recycling of 2-phosphoglycolate, produced by the oxygenase activity of ribulose-1,5-bisphosphate carboxylase/oxygenase, to 3-phosphoglycerate. Although photorespiration is a widely studied process, its regulation remains poorly understood. In this context, phosphoproteomics studies have detected six phosphorylation sites associated with photorespiratory glycolate oxidases from *Arabidopsis thaliana* (*At*GOX1 and *At*GOX2). Phosphorylation sites at T4, T158, S212 and T265 were selected and studied using Arabidopsis and maize recombinant glycolate oxidase (GOX) proteins mutated to produce either phospho-dead or phospho-mimetic enzymes in order to compare their kinetic parameters. Phospho-mimetic mutations (T4D, T158D and T265D) led to a severe inhibition of GOX activity without altering the K_M glycolate. In two cases (T4D and T158D), this was associated with the loss of the cofactor, flavin mononucleotide. Phospho-dead versions exhibited different modifications according to the phospho-site and/or the GOX mutated. Indeed, all T4V and T265A enzymes had kinetic parameters similar to wild-type GOX and all T158V proteins showed low activities while S212A and S212D mutations had no effect on *At*GOX1 activity and *At*GOX2/*Zm*GO1 activities were 50% reduced. Taken together, our results suggest that GOX phosphorylation has the potential to modulate GOX activity.

Keywords: *Arabidopsis thaliana*; glycolate oxidase; photorespiration; protein phosphorylation; *Zea mays*

---

1. Introduction

Photorespiration begins with the fixation of O_2 to ribulose-1,5-bisphosphate by ribulose-1,5-bisphosphate carboxylase/oxygenase (RuBisCO) leading to the formation of one molecule of 2-phosphoglycolate (2PG) and one molecule of 3-phosphoglycerate (3PGA). 2PG is then metabolized to produce 3PGA by the photorespiratory cycle which occurs in four subcellular compartments (chloroplasts, peroxisomes, mitochondria and cytosol), and involves eight core enzymes and several transporters [1]. Photorespiration has a negative impact on plant yield since it limits photosynthetic CO_2 assimilation due to competition at the RuBisCO active site, and it releases assimilated carbon and nitrogen as CO_2 and ammonium that have to be either reassimilated at an energetic cost or lost. This has led to efforts to minimize the negative effects of photorespiration to improve plant yield by producing plants containing a chloroplastic bypass to metabolize photorespiratory glycolate (the most recent examples being [2,3]). However, photorespiratory glycolate produced in the chloroplast from toxic 2PG [4] is normally metabolized in peroxisomes by glycolate oxidase (GOX), a flavin mononucleotide (FMN) containing enzyme that catalyzes the transformation of glycolate to glyoxylate with the production of hydrogen peroxide [5]. This enzyme evolved from a bacterial lactate oxidase and it is a member of the α-hydroxy-acid oxidase superfamily [6]. In *Arabidopsis*, there are five GOX-related genes: *At3g14420*, *At3g14415*, *At4g18360*, *At3g14130* and *At3g14150* (encoding *At*GOX1, *At*GOX2, *At*GOX3, *At*HAOX1 and *At*HAOX2, respectively). According to transcriptomic analyses, *At3g14415* (*At*GOX2) and *At3g14420* (*At*GOX1) are highly expressed in leaves and they represent the major GOX isoforms in *Arabidopsis thaliana* [7]. *At*GOX3 is mainly expressed in senescing leaves and roots, where it has been proposed to also function as a lactate oxidase and thus play a role in lactate metabolism [8]. *At*HAOX1 and *At*HAOX2 are expressed in seeds and encode proteins preferring medium- and long-chain hydroxyl acids as substrates [6]. Knock-out mutants of each Arabidopsis *GOX* gene do not exhibit a photorespiratory growth phenotype although they were all more sensitive to *Pseudomonas syringae* and to ozone [9,10]. A photorespiratory phenotype in air that was reversed by elevated CO_2 (3000 ppm) was observed however in an Arabidopsis artificial miRNA GOX line (*amiRgox1/2*) with both *At*GOX1 and *At*GOX2 knocked-down and only 5%–10% of wild-type Arabidopsis leaf GOX activity [7]. The transfer of *amiRgox1/2* plants from high CO_2 to ambient air led to a 700-fold accumulation of glycolate and a reduced carbon allocation to sugars, organic acids and amino acids that induced an early senescence of older leaves [7]. Even though *At*GOX1 and *At*GOX2 showed a redundant photorespiratory function, *At*GOX1 appeared to have a more predominant role in photorespiration since it attenuated the phenotype of a *cat2* mutant [11]. The importance of GOX has been observed also in tobacco, rice and even maize (a C4-plant), where a reduction of GOX activity in RNAi, antisense or mutant lines led to delayed growth in air associated with a decrease of net CO_2 assimilation rate [12–15].

Since photorespiration interacts with several metabolic processes including photosynthesis, nitrogen metabolism, respiration, C1 metabolism as well as H_2O_2 production by GOX [16], it might be expected that the photorespiratory cycle would be coordinated with these processes and modulated according to metabolic needs and perhaps environmental cues. Even though photorespiration has been widely studied over the last few decades, regulation of this C2-cycle and its enzymes is still poorly understood. It has been proposed that serine acts as a signal to regulate the expression of photorespiratory genes [17]. Several peroxisomal photorespiratory enzymes have been associated with putative post-translational modifications such as ubiquitination, nitration, persulfidation, and acetylation (for a review see [18]). Indeed, the activity of pea GOX was found to be inhibited by S-nitrosylation [19] as was Arabidopsis glycine decarboxylase [20]. The glycerate kinase of maize was shown to be redox-regulated by thioredoxin f; however, this was not the case for the Arabidopsis enzyme [21]. Arabidopsis mitochondrial glycine decarboxylase L-protein activity was found to be redox regulated by thioredoxin [22,23]. Protein phosphorylation could be another post-translational mechanism involved in the regulation of the photorespiratory cycle as phosphoproteomics studies have identified a number of phosphopeptides associated with all but one of the core photorespiratory enzymes [24]. Concerning photorespiratory *At*GOX1 and *At*GOX2, several phosphorylation sites (T4, T155, T158, S212, T265 and T355) have been reported (see Table 1).

In this study, the consequences of T4, T158, S212 and T265 phosphorylation on *At*GOX1 and *At*GOX2 enzymatic activities and kinetic properties were analysed using purified recombinant wild-type, phospho-dead and phosphorylation-mimetic GOX proteins. This was also carried out to test the equivalent residues (T5, T159, S213 and T266) of maize photorespiratory GOX (*Zm*GO1) [15]. Our results are discussed in terms of the possible consequences of each GOX phosphorylation on enzyme structure and function and in response to environmental stresses and future avenues to explore to better understand photorespiratory GOX phosphorylation are proposed.

2. Results

2.1. Phosphopeptides and Phosphorylated Residues of Arabidopsis GOX1 and GOX2

In the context of a possible regulation of photorespiratory GOX by protein phosphorylation, a first step was to retrieve potential phosphorylated residues from phosphopeptides associated with *At*GOX1 and *At*GOX2 from published phosphoproteomics studies and the PhosPhAt 4.0 database (http://phosphat.mpimp-golm.mpg.de/) (see Table 1). In this way, six phosphopeptides were found with six different phosphorylated residues; T4, T155, T158, S212, T265 phosphopeptides associated with both *At*GOX1 and *At*GOX2 and T355 that was in a peptide associated only with *At*GOX1 (Table 1).

Table 1. Phosphopeptides associated with *At*GOX1 and *At*GOX2.

Gene Name, Locus	Phosphopeptide	Peptide Location	Sample Type, Age	Treatment	References
	MEI(pT$_4$)NVTEYDAIAK	1–14/367	Leaves	Oxygen depletion	PhosPhAt 4.0 [1]
	AIAL(pT$_{155}$)VDTPRL	151-161/367	Seedlings, 11 days	Sucrose depletion	[25]
AtGOX1, *At3g14420* *orAtGOX2,* *At3g14415*	AIALTVD(pT$_{158}$)PRL	151-161/367	Seedling, 2 weeks	ABA and dehydration	[26]
			Seedlings, 10 days	Continuous light for 24 h	[27]
			Rosette	Varying O$_2$/CO$_2$ conditions	[28]
	TL(pS$_{212}$)WK	210-214/367	Seedlings, 2 weeks	ABA and dehydration	[26]
	QLDYVPA(pT$_{265}$)ISALEEVVK	258-273/367	Cauline leaves, 2 months		[29]
AtGOX1 *At3g14420*	NHI(pT$_{355}$)TEWDTPR	352-362/367	Seedlings, 10 days	Continuous light for 24 h	[27]

The phosphorylated residue in each peptide is shown in bold with its position numbered according to *At*GOX1. Peptide location shows the position of each phosphopeptide with respect to the number of residues in *At*GOX1.
[1] PhosPhAt 4.0: http://phosphat.mpimp-golm.mpg.de/.

All of these phosphorylated amino acids were conserved among *At*GOX1, *At*GOX2 and *At*GOX3 proteins while T4 and T158 were replaced by a valine in *At*HAOX1 and *At*HOAX2 (Figure S1). To analyse the conservation of these phosphorylated residues amongst (S)-2-hydroxy-acid oxidases (EC 1.1.3.15) of bacteria, cyanobacteria, algae, human and plants, we compared amino acid sequences of lactate oxidase (LOX) from *Aerococcus viridians*, *Lactococcus lactis*, *Nostoc* and *Chlamydomonas reinhardtii*, GOX from *Zea mays*, *Arabidopsis thaliana*, *Brassica napus*, *Vitis vinifera*, *Nicotiana benthamiana*, *Spinacia oleracea*, *Populus alba* and *Homo sapiens* and HAOX from *Arabidopsis thaliana* and *Homo sapiens* (Figure S2). The residues corresponding to T155, S212 and T265 of Arabidopsis GOX were conserved as either a serine or a threonine in all species (except for T265 in *Chlamydomonas reinhardtii* LOX, S212 in *Lactococcus* LOX and HAOX2 of *Homo sapiens*, T265 in *Chlamydomonas reinhardtii* LOX), whereas T4 was conserved only in the plant GOX proteins while T158 was conserved in all GOX proteins, human HAOX2, *Aerococcus viridians* LOX and *Lactococcus lactis* LOX (Figure S2). Based on these observations, it was decided to investigate further the phosphoregulation of photorespiratory GOX by characterising phospho-mimetic recombinant *At*GOX1 and *At*GOX2 as well as the C4-plant enzyme, *Zm*GO1 that had been shown to be important for growth in air (400 ppm CO$_2$) [15]. Among the six phosphorylation sites, the highly conserved plant GOX residues, T4, T158, S212 and T265 of *At*GOX1 and *At*GOX2 were chosen for this study that corresponded to T5, T159, S213, T266 of *Zm*GO1 (Figure S2).

2.2. Phospho-Mimetic AtGOX1, AtGOX2 and ZmGO1 Exhibit Altered Glycolate Oxidase Activities

To investigate the potential regulation of *At*GOX1 and *At*GOX2 activity by protein phosphorylation and to see if this was conserved in maize photorespiratory GO1, the kinetic parameters of purified recombinant N-terminal His-tagged GOX proteins (see Figure S3) was undertaken. All selected phosphorylated residues were replaced by an aspartate to mimic a constitutive phosphorylation.

To produce phospho-dead GOX proteins, S212/213 and T265/266 (Arabidopsis GOX/ZmGO1 numberings) were replaced by an alanine while T4/5 and T158/159 were changed to a valine to also mimic the sequence of *At*HOAX1 and *At*HOAX2 (and *Homo sapiens* HAOX2 for the T4 position) (Figure S2) so as to evaluate the role of these amino acids in substrate specificity. Using glycolate as a substrate, *At*GOX1, *At*GOX2 and *Zm*GO1 presented rather similar K_M glycolate (210 μM, 279 μM and 126 μM) and k_{cat} (11.12 s^{-1}, 10.93 s^{-1} and 14.45 s^{-1}) values (Table 2). The T4/5V mutations did not have any consequences on either K_M glycolate or k_{cat} while the T4/5D phospho-mimetic mutations drastically decreased (by 10–20 fold) the k_{cat} of the three recombinant GOX enzymes without significantly altering the Km glycolate (Table 2). Similar results were observed for the T265/266 phospho-site since its mutation to alanine did not have any effect on the calculated kinetic parameters while k_{cat} was strongly decreased for $AtGOX1_{T265D}$, $AtGOX2_{T265D}$ and $ZmGO1_{T266D}$ proteins by 99%, 81% and 95% (Table 2). $AtGOX1_{T158V}$, $AtGOX2_{T158V}$ and $ZmGO1_{T159V}$ also showed altered kinetic parameters with an improved K_M glycolate (2–3 fold lower) but a 4–5 fold decreased k_{cat} compared to their wild-type GOX counterparts (Table 2). When this residue was mutated to mimic a phosphorylated GOX, the resulting recombinant proteins ($AtGOX1_{T158D}$, $AtGOX2_{T158D}$ and $ZmGO1_{T159D}$) were inactive (Table 2). Perhaps surprisingly, the mutated S212/213 phospho-site gave a differential affect amongst the three GOX proteins studied. While $AtGOX1_{S212A}$ and $AtGOX1_{S212D}$ did not show any differences in their kinetic parameters compared to $AtGOX1_{WT}$, both $AtGOX2_{S212A}$ and $AtGOX2_{S212D}$ as well as $ZmGO1_{S213A}$ and $ZmGO1_{S213D}$ exhibited an approximately 2-fold decrease of their k_{cat} with no change in K_M glycolate (Table 2).

Table 2. Effect of phospho-site mutations on glycolate-dependent kinetic parameters of recombinant *At*GOX1, *At*GOX2 and *Zm*GO1.

	K_M	k_{cat}		K_M	k_{cat}		K_M	k_{cat}
*At*GOX1	(μM)	(s^{-1})	*At*GOX2	(μM)	(s^{-1})	*Zm*GO1	(μM)	(s^{-1})
WT	210 ± 90	11.12 ± 3.08	WT	279 ± 30	10.93 ± 3.48	WT	126 ± 34	11.45 ± 1.86
T4V	123 ± 35	9.59 ± 1.57	T4V	251 ± 70	12.09 ± 1.50	T5V	135 ± 56	8.67 ± 2.32
T4D	151 ± 44	**0.61 ± 0.46 ***	T4D	390 ± 215	**1.40 ± 0.42 ***	T5D	157 ± 31	**0.53 ± 0.22 ***
T158V	**100 ± 19 ***	**2.36 ± 0.47 ***	T158V	**133 ± 83 ***	**2.49 ± 0.60 ***	T159V	**46 ± 20 ***	**2.40 ± 0.35 ***
T158D	no activity		T158D	no activity		T159D	no activity	
S212A	257 ± 31	12.13 ± 2.15	S212A	249 ± 14	**6.10 ± 0.64 ***	S213A	89 ± 13	**7.19 ± 0.46 ***
S212D	277 ± 24	11.46 ± 2.31	S212D	325 ± 13	**5.23 ± 0.41 ***	S213D	129 ± 28	**6.39 ± 1.45 ***
T265A	276 ± 112	14.02 ± 2.65	T265A	237 ± 67	11.33 ± 3.27	T266A	91 ± 18	11.75 ± 2.94
T265D	**531 ± 50 ***	**0.14 ± 0.06 ***	T265D	388 ± 159	**2.06 ± 1.09 ***	T266D	173 ± 65	**0.62 ± 0.77 ***

Mean values ± SD from three independent biological replicates. Statistical significance was determined by a Student's *t*-test. Values in bold and marked by an asterisk were significantly different compared to the corresponding WT protein ($p < 0.05$).

2.3. Phospho-site Mutations at T4/5 and T158/159 Have a Limited Effect on Substrate Specificity

*At*HOAX1 and *At*HAOX2 proteins preferentially use long-chain hydroxy-acids as substrates although *At*HOAX1 can also quite efficiently use both lactate and glycolate [6]. Since the two Arabidopsis HAOX enzymes had a valine at the T4 and T158 positions, it was decided to test whether the mutation of these residues could give rise to substrate-specific effects. To achieve this, activity measurements were repeated with either L-lactate or 2-hydroxyoctanoate using T4/5 and T158/159 phospho-site mutated GOX proteins and their kinetic parameters were calculated and compared (Table 3). We chose to test 2-hydroxyoctanoate because it was found to be a good substrate for both Arabidopsis HAOX enzymes [6]. First, however, the substrate specificity of GOX$_{WT}$ proteins were compared and as previously reported, the Km L-lactate was higher than the Km glycolate [8] with an approximate 8-fold higher value for *At*GOX1 and *At*GOX2 while only a 4-fold difference was seen for *Zm*GO1, and surprisingly, the calculated k_{cat} values for the lactate oxidase reaction were not significantly different when compared to the glycolate oxidase activities (Table 2; Table 3). When using 2-hydroxyoctanoate as a substrate, again a higher Km was observed compared to glycolate for *At*GOX1 and *At*GOX2 (3.6-fold and 1.8-fold, respectively), while the Km 2-hydroxyoctanoate for *Zm*GO1 remained unchanged (Tables 2 and 3). However, the k_{cat} for the 2-hydroxyoctanoate reaction was halved for all recombinant GOX_{WT} proteins when compared to glycolate oxidase activity (Tables 2 and 3). Taken together, these results showed that our recombinant GOX_{WT} proteins were more efficient (based on k_{cat}/Km) using glycolate as a substrate, although *Zm*GO1 appeared to be less selective.

Table 3. Effect of selected phospho-site mutations on L-lactate and 2-hydroxy-octanoate dependent kinetic parameters of recombinant *At*GOX1, *At*GOX2 and *Zm*GO1.

Enzyme	L-lactate		2-hydroxy-octanoate	
	K_M	k_{cat}	K_M	k_{cat}
	(μM)	(s^{-1})	(μM)	(s^{-1})
*At*GOX1$_{WT}$	1664 ± 293	9.09 ± 1.37	757 ± 82	5.95 ± 0.58
*At*GOX1$_{T4V}$	1844 ± 911	9.01 ± 3.59	**350 ± 221** *	4.66 ± 1.56
*At*GOX1$_{T4D}$	**2377 ± 270** *	**0.30 ± 0.07** *	1255 ± 818	**0.39 ± 0.17** *
*At*GOX1$_{T158V}$	**549 ± 160** *	**2.30 ± 0.56** *	**191 ± 51** *	**2.89 ± 0.89** *
*At*GOX1$_{T158D}$	no activity		no activity	
*At*GOX2 $_{WT}$	2094 ± 791	6.81 ± 1.54	487 ± 99	3.78 ± 1.26
*At*GOX2 $_{T4V}$	2119 ± 453	6.65 ± 1.17	**951 ± 91** *	4.05 ± 2.94
*At*GOX2 $_{T4D}$	2019 ± 169	**0.67 ± 0.17** *	**1721 ± 677** *	**0.59 ± 0.30** *
*At*GOX2 $_{T158V}$	**439 ± 171** *	**1.71 ± 0.53** *	501 ± 116	2.46 ± 0.83
*At*GOX2 $_{T158D}$	no activity		no activity	
*Zm*GO1 $_{WT}$	495 ± 75	10.85 ± 0.08	136 ± 19	5.89 ± 0.14
*Zm*GO1 $_{T4V}$	488 ± 233	11.99 ± 1.68	168 ± 59	5.88 ± 1.59
*Zm*GO1 $_{T4D}$	719 ± 233	**0.57 ± 0.26** *	**414 ± 84** *	**0.53 ± 0.40** *
*Zm*GO1 $_{T158V}$	**161 ± 29** *	**3.55 ± 0.53** *	**48 ± 36** *	**3.22 ± 0.39** *
*Zm*GO1 $_{T158D}$	no activity		no activity	

Mean values ± SD from three independent biological replicates. Statistical significance was determined by a Student's *t*-test. Values in bold and marked by an asterisk were significantly different compared to the corresponding WT protein ($p < 0.05$).

The effect of the selected (T4/5 and T158/159) phospho-site mutations on the calculated kinetic parameters of the lactase oxidase and 2-hydroxyoctanoate oxidase reactions was compared. It was found that T158V/D and T159V/D mutations led to similar effects on K_M and k_{cat} (T158/159V) and activity (T158/159D) using either L-lactate or 2-hydroxyoctanoate when compared to glycolate (Tables 2 and 3). For the T4V/D and T5V/D mutations, various consequences were observed (Table 3). As seen when testing the glycolate oxidase activity, the k_{cat} of *At*GOX1$_{T4D}$, *At*GOX2$_{T4D}$ and *Zm*GO1$_{T5D}$ was strongly reduced compared to GOX_{WT} proteins when either L-lactate or 2-hydroxyoctanoate was used (Tables 2 and 3). In these conditions, some significant changes were observed also for certain K_M values but this depended on the GOX protein since only the K_M L-lactate of *At*GOX1$_{T4D}$ was slightly increased (1.4-fold) while the K_M 2-hydroxyoctanoate of *At*GOX2$_{T4D}$ and *Zm*GO1$_{T5D}$ was increased by 3.5-fold and 3-fold, respectively, when compared to their corresponding GOX_{WT} proteins (Table 3). When the T4/T5 phospho-site was replaced by a valine, only *At*GOX1$_{T4V}$ and *At*GOX2$_{T4V}$ differed from their *At*GOX$_{WT}$ counterparts with respect to Km 2-hydroxyoctanoate since a 2-fold decrease for *At*GOX1$_{T4V}$ and a 2-fold increase for *At*GOX2$_{T4V}$ was observed (Table 3). Therefore, in general, the T4/5 and T158/159 mutations led to similar effects on the calculated kinetic parameters and enzyme activities when either glycolate, L-lactate or 2-hydroxyoctanoate was used as substrate. The T4/5 and T158/159 mutations to valine did not appear to alter substrate specificity since enzymatic efficiency using either L-lactate or 2-hydroxyoctanoate was not improved except in the cases of *At*GOX1$_{T4V}$, *At*GOX1T$_{158V}$ and *Zm*GO1$_{T159V}$ where k_{cat}/Km ratios for the 2-hydroxyoctanoate reaction appeared to be higher.

2.4. Phospho-Mimetic T4/5D and T158/159D Recombinant Proteins Lack FMN

GOX activity requires FMN as a cofactor [30] and therefore a change in FMN content could explain the altered activities of our phospho-mimetic recombinant GOX proteins (Tables 2 and 3). By measuring the ratio of the absorbance between protein (at 280 nm) and FMN (at 450 nm) ($A_{280nm/450nm}$), it was possible to determine the presence of FMN in the purified GOX proteins used to measure enzymatic activities. An example of absorption spectra of WT and T158/159 mutated GOX proteins highlights the typical spectra associated with FMN that was missing in the $GOX_{T158/159D}$ forms thereby indicating an absence of cofactor (Figure 1). Similar absorption spectra were carried out for each recombinant protein and used to calculate the $A_{280nm/450nm}$ ratios given in Table 4. It can be seen that all FNR-containing GOX_{WT} proteins had a similar low $A_{280nm/450nm}$ ratio of between 8 and 9. On the other hand, *At*GOX1$_{T158D}$, *At*GOX2$_{T158D}$ and *Zm*GO1$_{T159D}$ exhibited higher $A_{280nm/450nm}$ values of 23, 20.5 and 19.5, respectively (Table 4) since they all lacked a significant typical FMN absorption signature (Figure 1), thus indicating an absence (or an extremely reduced amount) of cofactor.

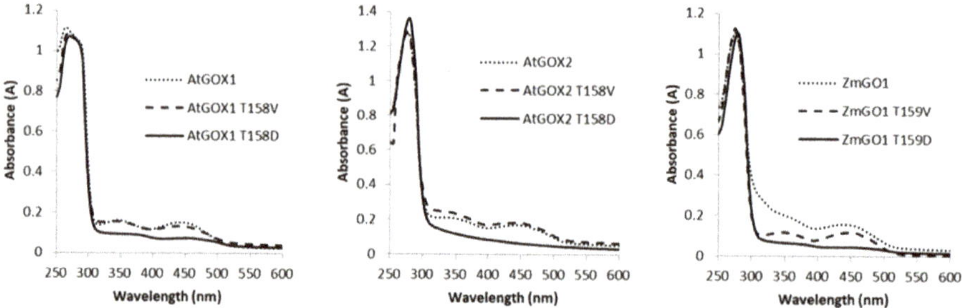

Figure 1. Absorption spectra of recombinant *At*GOX1, *At*GOX2 and *Zm*GO1 and their T158/T159 phospho-site mutated forms.

In this way, *At*GOX1$_{T4D}$, *At*GOX2$_{T4D}$ and *Zm*GO1$_{T5D}$ were seen also to lack FMN since they had high $A_{280nm/450nm}$ ratios of 28.3, 36.5 and 25.7, respectively (Table 4). Therefore, there appeared to be a good correlation between FMN content and GOX activity for these mutated forms. However, this was not always the case. *At*GOX1$_{T158V}$, *At*GOX2$_{T158V}$ and *Zm*GO1$_{T159V}$ as well as *At*GOX1$_{T265D}$, *At*GOX2$_{T265D}$ and *Zm*GO1$_{T266D}$ appeared to have a normal FMN content (Table 4) even though they exhibited a strong decrease of k_{cat} (Table 2). This was also seen for *At*GOX2$_{S212A}$, *At*GOX2$_{S212D}$, *Zm*GO1$_{S213A}$ and *Zm*GO1$_{S231D}$ which showed normal $A_{280nm/450nm}$ ratios (Table 4) but reduced k_{cat} values (Table 2).

Table 4. $A_{280/450nm}$ ratios of recombinant *At*GOX1, *At*GOX2 and *Zm*GO1 proteins.

*At*GOX1	Ratio 280/450 nm	*At*GOX2	Ratio 280/450 nm	*Zm*GO1	Ratio 280/450 nm
WT	8.7 ± 2.1	WT	8.3 ± 0.7	WT	8.1 ± 1.8
T4V	10.7 ± 4.1	T4V	6.9 ± 2.2	T5V	6.7 ± 3.0
T4D	**28.3 ± 6.1 ***	T4D	**36.5 ± 7.3 ***	T5D	**25.7 ± 6.2 ***
T158V	8.7 ± 2.3	T158V	8.2 ± 1.7	T159V	7.9 ± 1.1
T158D	**23.0 ± 8.3 ***	T158D	**20.5 ± 9.3 ***	T159D	**19.5 ± 7.1 ***
S212A	6.9 ± 3.3	S212A	12.4 ± 5.1	S213A	6.6 ± 3.5
S212D	6.6 ± 4.0	S212D	10.4 ± 2.2	S213D	6.1 ± 1.7
T265A	9 ± 3.3	T265A	9.6 ± 2.4	T266A	7.0 ± 0.1
T265D	8.5 ± 2.5	T265D	11.8 ± 4.5	T266D	21.2 ± 12.1

Mean values ± SD from three independent experiments. Statistical significance was determined by a Student's *t*-test. Values in bold and marked by an asterisk were significantly different compared to the corresponding WT protein ($p < 0.05$).

3. Discussion

Regulation of the photorespiratory cycle is still poorly understood even if phosphoproteomics studies have indicated that all photorespiratory enzymes except glycerate kinase can be phosphorylated [24]. In this context, GOX phosphorylation could be important since several phospho-sites have been reported and the phosphorylated residues have been conserved during the evolution of land plants (Figure S2). In this study, we explored the role of GOX phosphorylation by analyzing the kinetic parameters of both phospho-dead (to control the importance of the original residue) and phospho-mimetic recombinant Arabidopsis (C3-plant) and maize (C4-plant) photorespiratory GOX proteins.

3.1. Phospho-Mimetic GOX and Inhibition of Enzyme Activity

Phospho-mimetic GOX proteins exhibited different degrees of inhibition of their glycolate oxidase activity, except in one case, AtGOX1$_{T212D}$, where the recombinant protein maintained GOX1$_{WT}$ activity (Table 2). This inhibition was seen to correlate with reduced amounts (or the absence) of FMN (Figure 1, Table 4) for GOX$_{T4/5D}$ and GOX$_{T158/158D}$ but this was not observed for GOX$_{S212/213D}$ and GOXT$_{265/266D}$. When the phospho-mimetic mutation affected FMN content, the degree of inhibition was not constant since the mutation of T4/T5 to aspartate dramatically decreased GOX activity without changing the K_M glycolate, while T158D and T159D mutations resulted in inactive recombinant GOX proteins (Table 2). These modifications in both Arabidopsis and maize GOX kinetics parameters could be attributed to the lack of FMN in these mutated proteins since its presence is essential for GOX activity (Table 4) [30]. Recently, the 3D structure of apo-GOX (lacking FMN) and holo-GOX from *Nicotiana benthamiana* revealed that loop4 (residues 157–165; 156–164 of *At*GOX) and loop6 (residues 253–265; 252–264 of *At*GOX) together with a loop situated between residues 28–33 (27–32 for *At*GOX) formed a lid preventing the loss of FMN [30] (Figure 2A, Figure S1). Moreover, this lid had a strong interaction with a neighbouring GOX subunit of the tetrameric *Nb*GOX, with α-helix4 (next to loop4) forming H-bonds with E3, T5 and N6 (E2, T4 and N5 of *At*GOX) allowing a cooperative mechanism in FMN binding between GOX subunits of the same tetramer [30] (Figure 2B, Figure S1). Structural models of *Arabidopsis thaliana* GOX1 and GOX2 as well as *Zm*GO1 based on the structure of *Spinacia oleracea* GOX (PDB 1AL7) indicated that T4 formed H-bonds with R163, E165 and K169 of α-helix4 of a neighbouring GOX subunit.

When T4 was replaced by an aspartate in these models, only an H-bond with D163 of α-helix4 was present. Moreover, T158 is located close to a residue implicated in FMN-binding (T155) [31,32], the latter being also potentially phosphorylated [25]. Thus, T158 (present in loop4) and T4 (at the N-terminus) are located in essential domains for FMN binding; therefore, their mutation to aspartate, and probably their phosphorylation, may disturb FMN binding to holoGOX resulting in a less active or inactive enzyme. Furthermore, structural models indicated that when T158/159 was replaced by an aspartate, a new H-bond formed with Y129. This tyrosine residue normally formed an H-bond with FMN and when mutated to a phenylalanine (spinach GOX$_{Y129F}$) the resulting protein had only 3.5% of GOX$_{WT}$ activity and a k_{cat} of only 0.74 s^{-1} instead of 20 s^{-1}; however, FMN content was not affected [33]. Therefore, it is possible that phosphorylation of T158/159 could alter GOX activity even if FMN was retained in the GOX protein.

The mutation of T265/266 to aspartate decreased the glycolate-dependent k_{cat} of *At*GOX1/2$_{T265D}$ and *Zm*GO1$_{T266D}$ compared to the wild-type recombinant protein but only changed the K_M glycolate of *At*GOX1$_{T265D}$ (Table 2). As mentioned earlier, contrary to GOX$_{T4D/T5D}$ and GOX$_{T158D/T159D}$, the FMN content of phospho-mimetic GOX$_{T265D/T266D}$ was comparable to recombinant GOX$_{WT}$ protein (Table 4) even though T265/T266 is located just at the end of loop6 (residues 253–265; 252–264 of *At*GOX1/2) which is part of the "lid" structure involved in FMN loss in the pH sensor model [30]. Again, based on GOX structural models, T265 formed an H-bond with D285 of β-sheet7; however, when replaced by an aspartate this bond was broken and a new H-bond was formed with N253 of loop6 that is next to the active site H254 involved in proton abstraction during catalysis. We propose that GOX$_{T265/T266D}$ inhibited glycolate oxidase activity by bringing about a conformational change that interfered with catalysis via the displacement of H254.

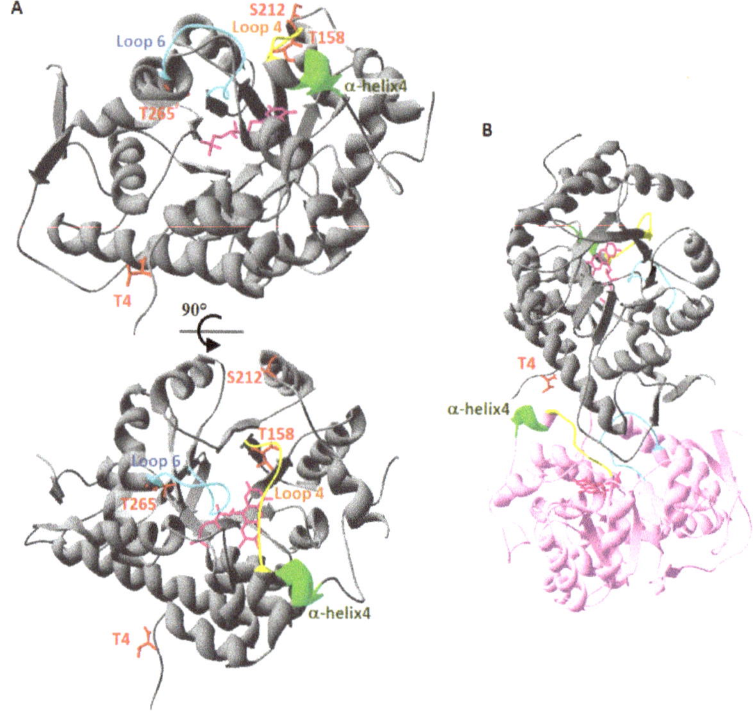

Figure 2. Localisation of phosphorylated residues and important structures of *At*GOX1. A structural model of *At*GOX1 based on the 3D-structure of spinach GOX [31]. (**A**) *At*GOX1 monomer showing phosphorylated residues T4, T158, S212 and T265 (red), loops 4 (yellow) and 6 (blue), α-helix4 (green) and the flavin mononucleotide (FMN) (pink). The bottom structure is rotated by 90° with respect to the top structure. (**B**) View of two *At*GOX monomers (in grey and in pink) and the proximity between T4 (red) of a GOX subunit with α-helix4 (green) of a neighbouring GOX subunit.

3.2. Phospho-Site Mutations Suggest Differences between Photorespiratory GOX Proteins

Our studies to decipher photorespiratory GOX regulation by protein phosphorylation appeared to indicate that our different GOX enzymes (*At*GOX1 versus *At*GOX2 and *At*GOX1/2 versus *Zm*GO1) did not always respond in a similar manner thus suggesting subtle differences between them even though they have a conserved photorespiratory role in planta. Both mutations at S212/S213 led to a similar decrease in the glycolate-dependent k_{cat} of *At*GOX2 and *Zm*GO1 while the k_{cat} of *At*GOX1 remained unaltered (Table 2). On the other hand, these mutations did not modify the K_M glycolate of the different proteins (Table 2). Therefore, while *At*GOX1 and *At*GOX2 have been shown to have a redundant photorespiratory function [7] and similar kinetic properties ([34], Table 2), these observations suggest that they could have subtle regulatory and structural differences which could potentially explain reported differences in planta [11]. Furthermore, since S212/S213 was found to be important for *At*GOX2 and *Zm*GO1 activity but not for *At*GOX1 activity, we propose that the reduced k_{cat} observed with GOX$_{S212/213D}$ does not reflect the actual consequences of a phosphorylated GOX at S212. It is indeed difficult to explain why *At*GOX1 did not behave in a similar way to the other two GOX proteins since S212 is located in a loop between α-helixD and α-helixE that contains no known residues involved in glycolate oxidase activity. Our structural models showed that S212 was not involved in any H-bond formation; however, when replaced by an aspartate it could form a new H-bond with D54 in a loop between β-sheetA and β-sheetB of a neighbouring GOX subunit, but this was the case for all three proteins.

*At*GOX1/2 T4 and *Zm*GO1 T5 mutations to valine were done to suppress any phosphorylation but also to mimic the sequence found in both *At*HAOX1 and *At*HAOX2, two enzymes that use more efficiently long-chain hydroxyl-acids [6], and *Hs*HAOX2. As previously shown [8], *At*GOX1 and *At*GOX2 were less efficient using L-lactate as a substrate with both enzymes showing an 8-fold increase in K_M although the k_{cat} values were not significantly different when compared to glycolate (Tables 2 and 3). However, contrary to the literature [8], both *At*GOX1 and *At*GOX2 were able use 2-hydroxy-octanoate as a substrate but again they were less efficient with lower k_{cat} and increased K_M values when compared to glycolate (Tables 2 and 3). Interestingly, the C4-plant enzyme *Zm*GO1 had a similar k_{cat} using either L-lactate or glycolate as a substrate while displaying only a 4-fold increase of K_M and even though the k_{cat} with 2-hydroxy-octanoate decreased in a similar manner to *At*GOX proteins, its K_M 2-hydroxy-octanoate was unaltered when compared to its K_M glycolate (Tables 2 and 3). Thus, even if the kinetics parameters of *Zm*GO1, *At*GOX1 and *At*GOX2 were similar for the glycolate oxidase reaction ([34], Table 2), *Zm*GO1 appeared to be able to use more efficiently both L-lactate and 2-hydroxy-octanoate (as seen from a comparison of k_{cat}/K_M ratios).

Differences between GOX proteins with respect to 2-hydroxy-octanoate were also observed when comparing the effect of T4/5 and T158/159 mutations to valine. Indeed *At*GOX1$_{T4V}$, *At*GOX2$_{T4V}$ and ZmGO1T5V each exhibited a different alteration of K_M 2-hydroxy-octanoate compared to their GOX$_{WT}$ counterparts (Table 3). This amino acid being important for GOX quaternary structure [30], its replacement by a valine may modify 2-hydroxy-octanoate specificity towards that of *At*HAOX1 and *At*HAOX2 due to possible subtle differences in 3D structure.

T158V and T159V mutations induced the same modifications of K_M and k_{cat} when using either of the three substrates tested thus indicating that the mutation affected the global enzymatic mechanism and it did not appear to be involved in determining substrate specificity (Tables 2 and 3). Finally, *At*GOX1/2$_{T158V}$ and *Zm*GO1$_{T159v}$ proteins were poorly active whatever the substrate used however for each protein the K_M appeared to be lower than their GOX$_{WT}$ protein counterparts (Tables 2 and 3). This inferred that T158 was involved in substrate binding as well as enzymatic activity and that when mutated to valine (as in Arabidopsis HAOX proteins) this somehow improved substrate binding. In conclusion to this part of the discussion, we can say that V4/5 and V157/158 containing GOX proteins were not transformed into HAOX enzymes as they were unable to use 2-hydroxy-octanoate more efficiently.

3.3. Discovering Conditions Inducing GOX Phosphorylation and Identifying GOX Kinases

To date, the only evidence of GOX protein phosphorylation has come from phosphoproteomics studies using mass spectroscopy analyses. From Table 1 it can be seen that only the peptide containing phosphorylated T158 has been reported more than once and although the accuracy of the methods in this domain have increased over the years, there are still a number of technical constraints and peptide identification is prone to error. It is therefore important to confirm GOX phosphorylation using other methods and/or confirm by mass spectroscopy using less complex samples instead of total protein extracts. Furthermore, it is necessary to identify conditions that bring about GOX phosphorylation so as to better understand its function. The T4 phospho-site of *At*GOX1 and *At*GOX2 was identified in leaves of plants subjected to oxygen depletion (5% of O_2 for 3 h) (Table 1). In such low O_2 conditions, photorespiration would be less important and GOX activity could be reduced via T4/T5 phosphorylation. Although the T158 phospho-site has been seen in several different phosphoproteome experiments, no significant differences in the quantity of this phosphopeptide were reported whatever the conditions tested (Table 1), even when comparing dark versus light and low versus high CO_2 concentrations that are expected to modulate photorespiratory activity [28]. Thus, T158 phosphorylation may have a role that is not linked to photorespiration. Several studies have implicated GOX in response to pathogen attack [9,35,36], and therefore regulation of GOX activity, together with catalase, may be a way to modulate H_2O_2 production as part of plant defence signalling and this could involve protein phosphorylation. Arabidopsis phospho-site S212 was identified in a phosphoproteomic study of the triple kinase mutant *snrk2.2/2.3/2.6* subjected either to ABA or dehydration treatments [26]. SnRK2.2, SnRK2.3 and SnRK2.6 (also known as OST1) are three kinases activated by ABA [37] and we have recently shown that another photorespiratory enzyme (SHMT1) responded to ABA and altered stomatal movements in response to salt stress [38]. Thus, GOX could be a target of leaf ABA signalling either in mesophyll cells or stomata. Indeed, the quantity of the TL(pS)WK phosphopeptide was shown to increase in response to either ABA or dehydration and this increase was absent in *snrk2.2/2.3/2.6* seedlings [26].

Therefore, phosphoproteomics studies have had very limited success in providing insights into the eventual role of GOX phosphorylation. Since this post translational modification is a rapid and reversible response to environmental stimuli, experiments should be conducted to identify when GOX is differentially phosphorylated. With respect to GOX photorespiratory function, different conditions expected to modulate photorespiratory flux should be examined such as day/night cycle, variable CO_2/O_2, and stresses like heat, drought, salt and high light as well as pathogen attack. To simplify the detection of GOX phosphopeptides by mass spectroscopy, GOX should be specifically immunoprecipitated from soluble protein extracts at different times during the stress treatments or during the day/night cycle. In this way, it should be possible to identify conditions where GOX phosphorylation is modulated.

Of course, our in vitro data using recombinant phospho-mimetic GOX proteins only give an indication of what GOX phosphorylation might actually be doing with respect to enzyme activity. Indeed, GOX was produced in *Escherichia coli* as a recombinant phospho-mimetic protein and therefore we do not know whether this led to a non-physiological alteration of GOX structure (and activity). Since the modification was constitutive and present as soon as the protein was synthesized, it does not reflect a reversible phosphorylation of a protein. For instance, is the absence of FMN in $GOX_{T4/5D}$ and $GOX_{T158/159D}$ due to structural changes induced by the aspartate that inhibits FMN entry into the apo-protein to form the halo-protein, and therefore would phosphorylation lead to changes that favour FMN removal from the halo-protein? Furthermore, phospho-mimetic mutations do not always lead to the modifications in protein function brought about by an actual phosphorylation. Thus, to assess the real role of GOX phosphorylation, identification of the protein kinases (and also the phosphatases) involved would be crucial. Based on the SUBA database (http://suba.plantenergy. uwa.edu.au/), 33 protein kinases are predicted to be addressed to the peroxisome. Using machine learning methods to identify proteins carrying plant peroxisomal PST1 targeting signals, 11 protein kinases were identified as being addressed to the peroxisome [39]. More recently, a list of about 200 confirmed Arabidopsis peroxisomal proteins was compiled and four kinases and eight phosphatases or phosphatase subunits were identified [40]. A strategy to identify *At*GOX1/2 kinases could be to retrieve knock-out mutant lines for predicted and/or verified peroxisomal protein kinases and compare *At*GOX1/2 phosphorylation status by mass spectroscopy after *At*GOX1/2 immunoprecipitation from soluble proteins extracted from wild-type and mutant plants treated to previously identified conditions that induce *At*GOX1/2 phosphorylation.

In conclusion, photorespiratory GOX has the potential to be regulated by protein phosphorylation at several distinct sites. In general, phospho-mimetic recombinant GOX proteins exhibited reduced activities and this could be explained by predicted changes in structural interactions affecting key residues involved in FMN binding and catalysis. Phospho-mimetic GOX did not show any alteration in substrate specificity although C4-plant *Zm*GO1 did appear to have a relaxed substrate specificity when compared to C3-plant *At*GOX1 and *At*GOX2.

4. Materials and Methods

4.1. Plasmid Constructions and Site-Directed Mutagenesis

To produce recombinant proteins, previously made pET28a-*At*GOX1, pET28a-*At*GOX2 and pET28a-*Zm*GO1 expression plasmids [34] were used as templates to introduce point mutations using specific primers pairs (Table S1) and the QuikChange® II XL site-directed mutagenesis kit (Agilent®, Les Ulis, France), according to the manufacturer's instructions. This strategy generated T4V, T4D, T158V, T158D, S212A, S212D, T265A, T265D (*At*GOX1 and *At*GOX2) and T5V, T5D, T159V, T159D, S213A, S213D, T266A, T266D (*Zm*GO1) mutated proteins. All constructions were subsequently verified by DNA sequencing using T7 and T7-term primers (Table S1).

4.2. Purification of Recombinant GOX Proteins, SDS-PAGE and Determination of FMN Content

GOX proteins were purified as described by Dellero et al. [34]. The purity of each recombinant GOX protein was checked by SDS-PAGE (10% acrylamide) stained with Coomassie Brilliant Blue [41]. Recombinant GOX proteins used to measure FMN content were purified without FMN in buffers and FMN levels were evaluated by spectrophotometry using the $A_{280}/A_{450\,nm}$ ratio.

4.3. GOX Activity Measurements

Enzyme activities were measured using 5 µg of purified recombinant GOX in 50 mM Tris-HCl, 0.1 mM FMN, pH 8.0 and different glycolate (0.05 to 10 mM), L-lactate (0.3 to 10 mM) and 2-hydroxy-octanoate (0.15 to 5 mM) concentrations by an enzyme-coupled reaction at 30 °C. H_2O_2 produced by GOX activity was quantified in the presence of 0.4 mM *o*-dianisidine and 2 U horseradish peroxidase by measuring the ΔA_{440nm} using a Varian Cary 50 spectrophotometer. K_M and k_{cat} values were calculated using SigmaPlot 13.0 software, based on the curve fitting Michaelis–Menten equation: $v_0 = V_{max}[S]/(K_M + [S])$.

4.4. Structural Analysis

*At*GOX1, *At*GOX2 and *Zm*GO1 3D-structures were designed based on the spinach GOX (UniProtKB: P05414) 3D-structure (PDB: 1al7.1) using the SWISS-MODEL server (https://swissmodel.expasy.org/) modelling service. Manipulation of 3D-structures was realized using Deepview (Swiss Pdb-viewer) (https://spdbv.vital-it.ch/) and included the compute H-bond formation and mutation tools.

Supplementary Materials: The following are available online at http://www.mdpi.com/2223-7747/9/1/27/s1, Figure S1: *At*GOX1, *At*GOX2, *At*GOX3, *At*HAOX1 and *At*HAOX2 protein sequence alignments showing the conservation of phosphorylated residues and other important residues and structures, Figure S2: Amino acid sequence alignments of *Lactococcus lactis*, *Aerococcus viridians*, *Nostoc* and *Chlamydomonas reinhardtii* lactate oxidases, with glycolate oxidases and hydroxy-acid oxidases of human and plants, Figure S3: Purified recombinant His-tag proteins of wild-type *At*GOX1, *At*GOX2 and *Zm*GO1 and their phospho-site mutated forms on Coomassie-stained SDS-PAGE gels, Table S1: Primers for site-directed mutagenesis and DNA sequencing.

Author Contributions: Conceptualization, M.H. and M.J.; Methodology, M.H. and M.J.; Formal Analysis, M.H. and M.J.; Investigation, M.J., Y.L. and S.M.; Resources, M.H.; Data Curation, M.H. and M.J.; Writing & Editing, M.H. and M.J.; Supervision, M.H. and M.J.; Project Administration, M.H.; Funding Acquisition, M.H. and M.J. All authors have read and agreed to the published version of the manuscript.

Funding: This work was supported by a public grant overseen by the French National Research Agency (ANR) as part of the «Investissement d'Avenir» program, through the "Lidex-3P" project and a French State grant (ANR-10-LABX-0040-SPS) funded by the IDEX Paris-Saclay, ANR-11-IDEX-0003-02. This work was also supported by the ANR-14-CE19-0015 grant REGUL3P awarded to M.H., Y.L. was supported by a PhD fellowship from the China Scholarship Council.

Acknowledgments: We would like to thank Younès Dellero (INRA Rennes, France) who helped to initiate recombinant GOX studies in the team.

Conflicts of Interest: The authors declare no conflict of interest.

References

1. Bauwe, H.; Hagemann, M.; Fernie, A.R. Photorespiration: Players, partners and origin. *Trends Plant Sci.* **2010**, *15*, 330–336. [CrossRef] [PubMed]
2. South, P.F.; Cavanagh, A.P.; Liu, H.W.; Ort, D.R. Synthetic glycolate metabolism pathways stimulate crop growth and productivity in the field. *Science* **2019**, *363*, eaat9077. [CrossRef] [PubMed]
3. Shen, B.R.; Wang, L.M.; Lin, X.L.; Yao, Z.; Xu, H.W.; Zhu, C.H.; Teng, H.Y.; Cui, L.L.; Liu, E.E.; Zhang, J.J.; et al. Engineering a New Chloroplastic photorespiratory bypass to increase photosynthetic efficiency and productivity in rice. *Mol. Plant* **2019**, *12*, 199–214. [CrossRef]
4. Flügel, F.; Timm, S.; Arrivault, S.; Florian, A.; Stitt, M.; Fernie, A.R.; Bauwe, H. The photorespiratory metabolite 2-Phosphoglycolate regulates photosynthesis and starch accumulation in Arabidopsis. *Plant Cell* **2017**, *29*, 2537–2551. [CrossRef] [PubMed]
5. Dellero, Y.; Jossier, M.; Schmitz, J.; Maurino, V.G.; Hodges, M. Photorespiratory glycolate-glyoxylate metabolism. *J. Exp. Bot.* **2016**, *67*, 3041–3052. [CrossRef]
6. Esser, C.; Kuhn, A.; Groth, G.; Lercher, M.J.; Maurino, V.G. Plant and animal glycolate oxidases have a common eukaryotic ancestor and convergently duplicated to evolve long-chain 2-hydroxy acid oxidases. *Mol. Biol. Evol.* **2014**, *31*, 1089–1101. [CrossRef]
7. Dellero, Y.; Jossier, M.; Glab, N.; Oury, C.; Tcherkez, G.; Hodges, M. Decreased glycolate oxidase activity leads to altered carbon allocation and leaf senescence after a transfer from high CO_2 to ambient air in Arabidopsis thaliana. *J. Exp. Bot.* **2016**, *67*, 3149–3163. [CrossRef]
8. Engqvist, M.K.M.; Schmitz, J.; Gertzmann, A.; Florian, A.; Jaspert, N.; Arif, M.; Balazadeh, S.; Mueller-Roeber, B.; Fernie, A.R.; Maurino, V.G. GLYCOLATE OXIDASE3, a glycolate oxidase homolog of yeast L-lactate cytochrome c oxidoreductase, supports L-lactate oxidation in roots of arabidopsis. *Plant Physiol.* **2015**, *169*, 1042–1061. [CrossRef]
9. Rojas, C.M.; Senthil-Kumar, M.; Wang, K.; Ryu, C.-M.; Kaundal, A.; Mysore, K.S. Glycolate oxidase modulates reactive oxygen species-mediated signal transduction during nonhost resistance in Nicotiana benthamiana and Arabidopsis. *Plant Cell* **2012**, *24*, 336–352. [CrossRef]
10. Saji, S.; Bathula, S.; Kubo, A.; Tamaoki, M.; Aono, M.; Sano, T.; Tobe, K.; Timm, S.; Bauwe, H.; Nakajima, N.; et al. Ozone-sensitive arabidopsis mutants with deficiencies in photorespiratory enzymes. *Plant Cell Physiol.* **2017**, *58*, 914–924. [CrossRef]
11. Kerchev, P.; Waszczak, C.; Lewandowska, A.; Willems, P.; Shapiguzov, A.; Li, Z.; Alseekh, S.; Mühlenbock, P.; Hoeberichts, F.A.; Huang, J.; et al. Lack of GLYCOLATE OXIDASE1, but not GLYCOLATE OXIDASE2, attenuates the photorespiratory phenotype of CATALASE2-deficient Arabidopsis. *Plant Physiol.* **2016**, *171*, 1704–1719. [CrossRef]
12. Yamaguchi, K.; Nishimura, M. Reduction to below threshold levels of glycolate oxidase activities in transgenic tobacco enhances photoinhibition during irradiation. *Plant Cell Physiol.* **2000**, *41*, 1397–1406. [CrossRef]
13. Xu, H.; Zhang, J.; Zeng, J.; Jiang, L.; Liu, E.; Peng, C.; He, Z.; Peng, X. Inducible antisense suppression of glycolate oxidase reveals its strong regulation over photosynthesis in rice. *J. Exp. Bot.* **2009**, *60*, 1799–1809. [CrossRef]
14. Lu, Y.; Li, Y.; Yang, Q.; Zhang, Z.; Chen, Y.; Zhang, S.; Peng, X.X. Suppression of glycolate oxidase causes glyoxylate accumulation that inhibits photosynthesis through deactivating Rubisco in rice. *Physiol. Plant.* **2014**, *150*, 463–476. [CrossRef]
15. Zelitch, I.; Schultes, N.P.; Peterson, R.B.; Brown, P.; Brutnell, T.P. High glycolate oxidase activity is required for survival of maize in normal air. *Plant Physiol.* **2009**, *149*, 195–204. [CrossRef]
16. Hodges, M.; Dellero, Y.; Keech, O.; Betti, M.; Raghavendra, A.S.; Sage, R.; Zhu, X.-G.; Allen, D.K.; Weber, A.P.M. Perspectives for a better understanding of the metabolic integration of photorespiration within a complex plant primary metabolism network. *J. Exp. Bot.* **2016**, *67*, 3015–3026. [CrossRef]
17. Timm, S.; Florian, A.; Wittmiß, M.; Jahnke, K.; Hagemann, M.; Fernie, A.R.; Bauwe, H.; Rostock, D.; Germany, S.T. Serine acts as a metabolic signal for the transcriptional control of photorespiration-related genes. *Plant Physiol.* **2013**, *162*, 379–389. [CrossRef]
18. Sandalio, L.M.; Gotor, C.; Romero, L.C.; Romero-Puertas, M.C. Multilevel regulation of peroxisomal proteome by post-translational modifications. *Int. J. Mol. Sci.* **2019**, *20*, 4881. [CrossRef]

19. Ortega-Galisteo, A.P.; Rodríguez-Serrano, M.; Pazmiño, D.M.; Gupta, D.K.; Sandalio, L.M.; Romero-Puertas, M.C. S-Nitrosylated proteins in pea (*Pisum sativum* L.) leaf peroxisomes: Changes under abiotic stress. *J. Exp. Bot.* **2012**, *63*, 2089–2103. [CrossRef]

20. Palmieri, M.C.; Lindermayr, C.; Bauwe, H.; Steinhauser, C.; Durner, J. Regulation of plant glycine decarboxylase by S-nitrosylation and glutathionylation. *Plant Physiol.* **2010**, *152*, 1514–1528. [CrossRef]

21. Bartsch, O.; Mikkat, S.; Hagemann, M.; Bauwe, H. An autoinhibitory domain confers redox regulation to maize glycerate kinase. *Plant Physiol.* **2010**, *153*, 832–840. [CrossRef]

22. Da Fonseca-Pereira, F.; Geigenberger, P.; Thormählen, I.; Souza, P.V.L.; Nesi, A.N.; Timm, S.; Fernie, A.R.; Araújo, W.L.; Daloso, D.M. Thioredoxin h2 contributes to the redox regulation of mitochondrial photorespiratory metabolism. *Plant Cell Environ.* **2019**. [CrossRef]

23. Reinholdt, O.; Schwab, S.; Zhang, Y.; Reichheld, J.; Alisdair, R. Redox-regulation of photorespiration through mitochondrial thioredoxin o1. *Plant Physiol.* **2019**, *181*, 442–457. [CrossRef]

24. Hodges, M.; Jossier, M.; Boex-Fontvieille, E.; Tcherkez, G. Protein phosphorylation and photorespiration. *Plant Biol.* **2013**, *15*, 694–706. [CrossRef]

25. Wu, X.N.; Rodriguez, C.S.; Pertl-Obermeyer, H.; Obermeyer, G.; Schulze, W.X. Sucrose-induced receptor kinase SIRK1 regulates a plasma membrane aquaporin in Arabidopsis. *Mol. Cell. Proteom.* **2013**, *12*, 2856–2873. [CrossRef]

26. Umezawa, T.; Sugiyama, N.; Takahashi, F.; Anderson, J.C.; Ishihama, Y.; Peck, S.C.; Shinozaki, K. Genetics and phosphoproteomics reveal a protein phosphorylation network in the abscisic acid signaling pathway in Arabidopsis thaliana. *Sci. Signal.* **2013**, *6*, rs8. [CrossRef]

27. Choudhary, M.K.; Nomura, Y.; Wang, L.; Nakagami, H.; Somers, D.E. Quantitative Circadian Phosphoproteomic analysis of Arabidopsis reveals extensive clock control of key components in physiological, metabolic, and signaling pathways. *Mol. Cell. Proteom.* **2015**, *14*, 2243–2260. [CrossRef]

28. Abadie, C.; Mainguet, S.; Davanture, M.; Hodges, M.; Zivy, M.; Tcherkez, G. Concerted changes in the phosphoproteome and metabolome under different CO_2/O_2 gaseous conditions in Arabidopsis rosettes. *Plant Cell Physiol.* **2016**, *57*, 1544–1556.

29. Aryal, U.K.; Krochko, J.E.; Ross, A.R.S. Identification of phosphoproteins in arabidopsis thaliana leaves using polyethylene glycol fractionation, immobilized metal-ion affinity chromatography, two-dimensional gel electrophoresis and mass spectrometry. *J. Proteome Res.* **2012**, *11*, 425–437. [CrossRef]

30. Liu, Y.; Wu, W.; Chen, Z. Structures of glycolate oxidase from Nicotiana benthamiana reveal a conserved pH sensor affecting the binding of FMN. *Biochem. Biophys. Res. Commun.* **2018**, *503*, 3050–3056. [CrossRef]

31. Lindqvist, Y.; Brändén, C.-I. The structure of glycolate oxidase from spinach. *Proc. Natl. Acad. Sci. USA* **1985**, *82*, 6855–6859. [CrossRef] [PubMed]

32. Lindqvist, Y.; Brändén, C.I. The active site of spinach glycolate oxidase. *J. Biol. Chem.* **1989**, *264*, 3624–3628. [PubMed]

33. Macheroux, P.; Kieweg, V.; Massey, V.; Söderlind, E.; Stenberg, K.; Lindqvist, Y. Role of tyrosine 129 in the active site of spinach glycolate oxidase. *Eur. J. Biochem.* **1993**, *213*, 1047–1054. [CrossRef] [PubMed]

34. Dellero, Y.; Mauve, C.; Boex-Fontvieille, E.; Flesch, V.; Jossier, M.; Tcherkez, G.; Hodges, M. Experimental evidence for a hydride transfer mechanism in plant glycolate oxidase catalysis. *J. Biol. Chem.* **2015**, *290*, 1689–1698. [CrossRef] [PubMed]

35. Xu, Y.-P.; Yang, J.; Cai, X.-Z. Glycolate oxidase gene family in Nicotiana benthamiana: Genome-wide identification and functional analyses in disease resistance. *Sci. Rep.* **2018**, *8*, 8615. [CrossRef]

36. Chern, M.; Bai, W.; Chen, X.; Canlas, P.E.; Ronald, P.C. Reduced expression of glycolate oxidase leads to enhanced disease resistance in rice. *PeerJ* **2013**, *1*, e28. [CrossRef]

37. Fujita, Y.; Nakashima, K.; Yoshida, T.; Katagiri, T.; Kidokoro, S.; Kanamori, N.; Umezawa, T.; Fujita, M.; Maruyama, K.; Ishiyama, K.; et al. Three SnRK2 protein kinases are the main positive regulators of abscisic acid signaling in response to water stress in arabidopsis. *Plant Cell Physiol.* **2009**, *50*, 2123–2132. [CrossRef]

38. Liu, Y.; Mauve, C.; Lamothe-Sibold, M.; Guerard, F.; Glab, N.; Hodges, M.; Jossier, M. Photorespiratory serine hydroxymethyltransferase 1 activity impacts abiotic stress tolerance and stomatal closure. *Plant Cell Environ.* **2019**, *42*, 2567–2583. [CrossRef]

39. Lingner, T.; Kataya, A.R.; Antonicelli, G.E.; Benichou, A.; Nilssen, K.; Chen, X.-Y.; Siemsen, T.; Morgenstern, B.; Meinicke, P.; Reumann, S. Identification of novel plant peroxisomal targeting signals by a combination of machine learning methods and in vivo subcellular targeting analyses. *Plant Cell* **2011**, *23*, 1556–1572. [CrossRef]

40. Pan, R.; Hu, J. Proteomics of peroxisomes. *Subcell. Biochem.* **2018**, *89*, 3–45.

41. Laemmli, U.K. Cleavage of structural proteins during the assembly of the head of bacteriophage T4. *Nature* **1970**, *227*, 680–685. [CrossRef] [PubMed]

Article

Faster Removal of 2-Phosphoglycolate through Photorespiration Improves Abiotic Stress Tolerance of Arabidopsis

Stefan Timm [1,*], Franziska Woitschach [1,2], Carolin Heise [1], Martin Hagemann [1] and Hermann Bauwe [1]

[1]   Plant Physiology Department, University of Rostock, Albert-Einstein-Straße 3, D-18051 Rostock, Germany; Franziska.Woitschach@med.uni-rostock.de (F.W.); Heise_Carolin@web.de (C.H.); martin.hagemann@uni-rostock.de (M.H.); hermann.bauwe@uni-rostock.de (H.B.)

[2]   Division of Tropical Medicine and Infectious Diseases, Center of Internal Medicine II, University Medical Center Rostock, Ernst-Heydemann-Str.6, D-18057 Rostock, Germany

*   Correspondence: stefan.timm@uni-rostock.de; Tel.: +49-(0)381-4986115; Fax: +49-(0)381-4986112

Received: 19 November 2019; Accepted: 29 November 2019; Published: 2 December 2019

Abstract: Photorespiration metabolizes 2-phosphoglyolate (2-PG) to avoid inhibition of carbon assimilation and allocation. In addition to 2-PG removal, photorespiration has been shown to play a role in stress protection. Here, we studied the impact of faster 2-PG degradation through overexpression of 2-PG phosphatase (PGLP) on the abiotic stress-response of *Arabidopsis thaliana* (Arabidopsis). Two transgenic lines and the wild type were subjected to short-time high light and elevated temperature stress during gas exchange measurements. Furthermore, the same lines were exposed to long-term water shortage and elevated temperature stresses. Faster 2-PG degradation allowed maintenance of photosynthesis at combined light and temperatures stress and under water-limiting conditions. The *PGLP*-overexpressing lines also showed higher photosynthesis compared to the wild type if grown in high temperatures, which also led to increased starch accumulation and shifts in soluble sugar contents. However, only minor effects were detected on amino and organic acid levels. The wild type responded to elevated temperatures with elevated mRNA and protein levels of photorespiratory enzymes, while the transgenic lines displayed only minor changes. Collectively, these results strengthen our previous hypothesis that a faster photorespiratory metabolism improves tolerance against unfavorable environmental conditions, such as high light intensity and temperature as well as drought. In case of PGLP, the likely mechanism is alleviation of inhibitory feedback of 2-PG onto the Calvin–Benson cycle, facilitating carbon assimilation and accumulation of transitory starch.

Keywords: Arabidopsis; abiotic stress response; photosynthesis; phosphoglycolate phosphatase; photorespiration; 2-phosphoglycolate

---

1. Introduction

The characterization of a large set of photorespiratory mutants from a broad collection of phototrophs revealed photorespiration as an essential partner for oxygenic photosynthesis [1–6]. The photorespiratory pathway represents the only way to metabolize the Rubisco oxygenation reaction product 2-phosphoglycolate (2-PG; [7]) into the Calvin–Benson (CB) cycle intermediate 3-phosphoglycerate (3-PGA). Impairment of photorespiration causes 2-PG accumulation that leads to sequestration of C and P_i, severely impeding the biosynthesis of phosphorylated intermediates and triose phosphate export from the chloroplast. 2-PG has also direct inhibitory effects on CB cycle and glycolytic enzymes [8–11]. Hence, efficient 2-PG degradation through photorespiration is particularly important for photosynthesis in the presence of O_2.

Despite its essential nature, the decarboxylation of glycine in the photorespiratory pathway leads to considerable losses of freshly assimilated carbon. The magnitude of these losses depends mainly on the CO_2 and O_2 partial pressures in the chloroplast, which can dramatically change under unfavorable environmental conditions, such as high light intensity, drought and high temperatures [12–14]. It is obvious that the photorespiratory flux needs to operate at higher speed in response to such conditions in order to cope with the higher 2-PG amounts. Given the CO_2 loss during 2-PG recycling, plant research, however, aims to circumvent photorespiration in order to reduce carbon and energy losses [15–17]. In contrast, several reports demonstrated that an elevated flux through photorespiration could also increase photosynthesis under laboratory and field conditions [18–21] due to improved conversion of critical metabolites.

Apart from the central role of photorespiration supporting photosynthetic CO_2 fixation, it has been suggested that it plays also a role in the stress response of plants. For example, malfunctioning of photorespiration leads to an enhanced susceptibility of the plants against pathogen attack [22,23], and lowered tolerance towards abiotic stresses [24–27]. This is mainly because photorespiration serves as alternative energy sink under unfavorable environmental conditions regenerating ADP and NADP and, thus, decreases acceptor limitation of the light process and related ROS formation [26,28,29]. Hence, high photorespiratory flux could help to prevent the chloroplastidal electron transport chain from overreduction and, finally, photoinhibition [24]. Interestingly, NADH-dependent hydroxypyruvate reductase 1 (HPR1) protein expression significantly increases in response to water-limiting conditions [25]. This result implies that the induction of critical steps of photorespiration is a natural mechanism to enhance the metabolic flux through the pathway to dissipate excess energy under stress conditions. The efficiency of such defense mechanism might be intensified via parallel increases of the cyclic electron flow and alternative oxidase pathways [26,28,29]. Despite excess energy dissipation, it was shown intact photorespiration is required for proper stomatal regulation and as such more directly involved in stress adaptation [30–32].

Among the photorespiratory enzymes, PGLP plays a crucial role since the efficient removal of 2-PG is critical to avoid negative impacts on chloroplast function as has been demonstrated by the high oxygen sensitivity of *PGLP* knock out and knock down mutants [10]. Therefore, we aimed to gain insights if faster 2-PG removal due to *PGLP* overexpression has an impact towards the acclimation to abiotic stresses. The results obtained are discussed with respect to 2-PG toxicity on carbon utilization and allocation and the potential of faster 2-PG degradation for acclimation to abiotic environmental stresses.

2. Results

To test whether faster 2-PG removal can improve abiotic stress tolerance, we reinvestigated the previously generated *PGLP* overexpressor lines O9 and O1. Both contain decreased absolute 2-PG amounts and display lower O_2 inhibition of photosynthesis and altered stomatal features [10]. First, wild-type and *PGLP* overexpressor plants were grown under standard conditions (Figure 1A) and used for combined gas exchange and chlorophyll a fluorescence measurements to characterize their response to short-time light and temperature stresses (Figure 1B). Second, all genotypes were pre-grown under standard conditions and then exposed to two different long-term abiotic stresses, namely, water deficiency and higher temperatures (Figure 1C,D).

2.1. PGLP Overexpressors Sustain CO₂ Assimilation and Electron Transport at Combined High Light and Temperature

In a first series of experiments, we characterized the photosynthetic performance of *PGLP* overexpressors using high light intensities and elevated temperature during photosynthetic measurements. To this end, the wild type and *PGLP* overexpression lines O9 and O1 were grown under standard conditions to growth stage 5.1 [33] and then subjected to gas exchange measurements (Figure 1A,B). First, we monitored net photosynthetic CO_2 uptake rates (*A*) in response to increasing light intensities (PAR: 0, 20, 50, 100, 200, 400, 600, 800, 1200 and 1600 μmol m^{-2} s^{-1}) at the standard growth temperature of 20 °C. In parallel, stomatal conductance (g_s), transpiration (E) and relative electron transport rates (rETR) were recorded. As shown in Figure 2A, we did not detect significant differences in any of these parameters between the transgenic lines and the wild type, which is in good agreement with previous findings [10]. However, an increase of the incubation temperature from 20 °C to 30 °C caused significant and systemic short-term changes. In comparison to the wild type, we observed significantly increased A at higher light intensities, ranging from 1200 to 1600 in O9 and from 600 to 1600 μmol m^{-2} s^{-1} in our best performing line O1 (Figure 2B). Similarly, both lines were able to maintain higher g_s at the specified light intensities and displayed somewhat elevated E at 1200 and 1600 μmol^{-2} s^{-1} light (Figure 2B). Moreover, an increase in rETR was measured, but it was significant only in line O1 (400 to 1600 μmol m^{-2} s^{-1}). Collectively, our results suggest that high PGLP activity is beneficial for photosynthetic CO$_2$ fixation and chloroplastidal electron transport at high light intensities in conjunction with elevated temperatures.

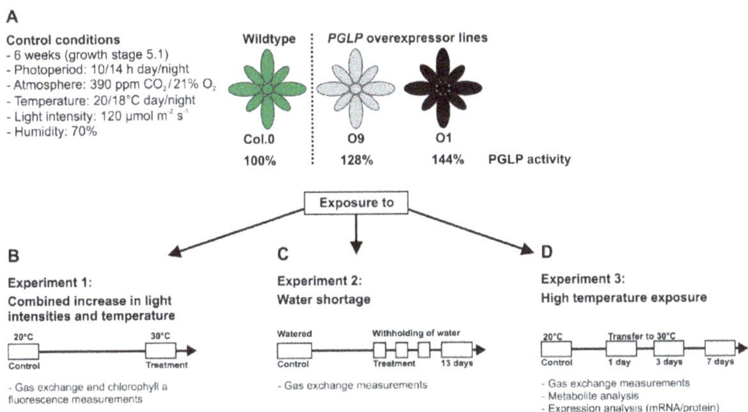

Figure 1. Scheme of the experimental strategy. (**A**) Following 6 weeks of growth under standard conditions, plants were used for (**B**) gas exchange and chlorophyll a fluorescence measurements. Light response curves were recorded at 20 °C (standard) and 30 °C (elevated temperature). (**C**) A subset of plants was exposed to water shortage and used for gas exchange measurements (A/C_i curves) after 13 days or (**D**) to elevated temperature for gas exchange measurements (A/C_i curves) after 1 and 7 days and metabolite and expression analysis.

2.2. Faster 2-PG Degradation is Beneficial for Photosynthetic CO$_2$ Assimilation under Water-Limiting Conditions

Next, we tested if the photosynthetic and stomatal features of *PGLP* overexpressors showed differences to wild-type plants under water-limiting conditions. To this end, all plants were conjointly grown under environmental controlled conditions in one pot (28 cm diameter, one plant per genotype) for 6 weeks with regular watering. Values from gas-exchange measurements at this time point served as control. Thereafter, watering was stopped, and gas-exchange measurements were repeated after 13 days under water-limiting conditions. This time point was chosen since previous studies showed upregulation of the photorespiratory pathway on the protein level 13 days after onset of water shortage [25]. As shown in Figure 3A, and in agreement with previous results [10], *PGLP* overexpressors displayed no significant change in A, the CO$_2$ compensation points (Γ), g_s and the ratio between the internal versus the external CO$_2$ concentration (C_i/C_a) under control conditions. Only minor changes were observed in the transpiration rate (E) and the water use efficiency (A_N/E) in line O9. After 13 days of withholding water, wild-type plants compromised photosynthetic performance, whereas both overexpressor lines displayed significantly higher A, g_s and E levels, while Γ was decreased (significant in line O1). The C_i/C_a and A_N/E parameters remained almost unchanged (Figure 3B).

In summary, overexpression of *PGLP* permits a longer maintenance of the photosynthetic capacity under water-limiting conditions.

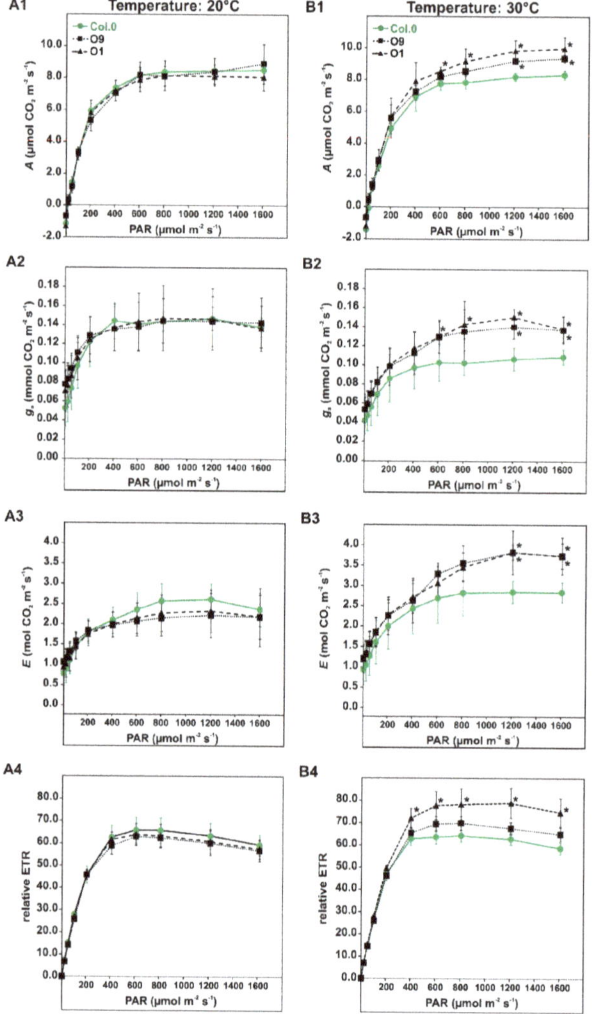

Figure 2. Light and temperature dependent gas exchange parameters of wild-type Arabidopsis and *PGLP* overexpressors. Plants were grown under standard conditions to growth stage 5.1 [33] and subsequently used for gas exchange measurements at (**A**) 20 °C and (**B**) 30 °C block temperature. Shown are (**A1,B1**) net CO_2 uptake rates (*A*), (**A2,B2**) stomatal conductance (g_s), (**A3,B3**) transpiration (*E*) and (**A4,B4**) relative electron transport rates (rETR). Values presented are mean values ± SD from at least four biological replicates (Col.0—green, solid line with circles, O9—black, dotted lines with squares and O1—black, dashed line with triangles). Asterisks indicate values statistically different from the wild type as determined by Student's *t*-test ($p < 0.05$).

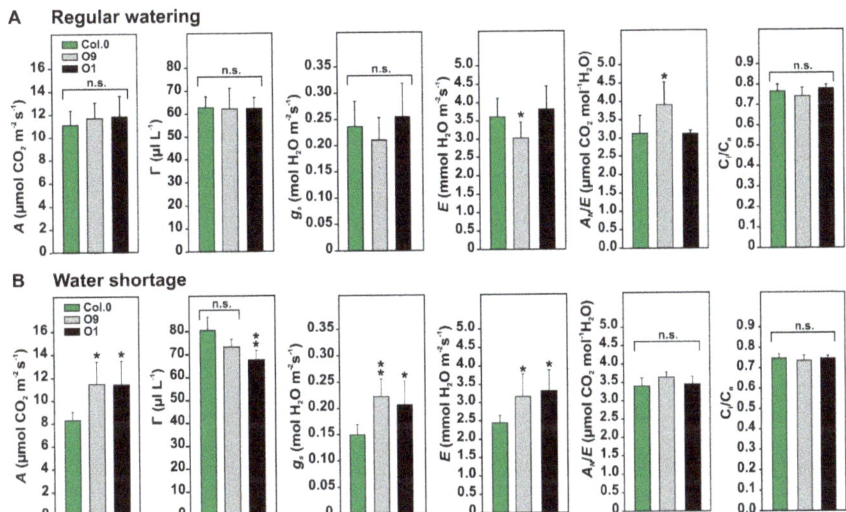

Figure 3. Photosynthetic parameters of wild-type Arabidopsis and *PGLP* overexpressors under water-limiting conditions. Wild-type and *PGLP* overexpressor (O9 and O1) plants were grown in one pot (28 cm diameter, 1 individual per genotype) for 6 weeks with regular watering (twice a week) under standard conditions. Following control measurements (**A**), watering was stopped, and plants measured after 13 days of withholding water (**B**). Shown are mean values ± SD from at least 5 individuals per genotype, grown as 5 technical replicates. Asterisks indicate values statistically different from the wild type as determined by Student's *t*-test ($p < 0.05$; n. s.—not significant). Abbreviations: A—net CO_2 uptake rates, Γ—CO_2 compensation points, g_s—stomatal conductance, E—transpiration, A_N/E—water use efficiency and C_i/C_a—ratio of internal versus external CO_2 concentration.

2.3. High PGLP Activity is Beneficial for Photosynthesis at Increased Growth Temperatures

Next, we aimed to know whether improved maintenance of photosynthesis was restricted to water-limiting conditions (Figure 3B) or if it also occurs in response to other abiotic stresses in *PGLP* overexpressor lines. Therefore, we tested increased growth temperature, which is anticipated to stimulate 2-PG production. To this end, the transgenic lines and the wild type were grown for 6 weeks under standard conditions and then exposed to elevated temperatures (30 °C; Figure 1D). As proxies for their capability to acclimate to higher temperatures, we characterized photosynthetic parameters such as A and Γ under control conditions and after one and 7 days at 30 °C. As before, we did not measure significant differences between the genotypes under control conditions. As expected, transfer of the plants to 30 °C decreased A and increased Γ. A decreased to the same extends in the transgenic lines and the wild type (Figure 4A). However, after 7 days exposure to 30 °C, Γ is significantly decreased in both transgenic lines, reaching almost control levels, while it remained at a similarly high level in wild type as found after only one day at 30 °C (Figure 4B, significant in O1). This change in line O1 was accompanied by significantly increased g_s (Col.0—0.1032 ± 0.0202 versus O1—0.1985 ± 0.0726*) and E (Col.0—1.67 ± 0.28 versus O1—2.96 ± 0.86*), which remained similar between O9 and the wild type (g_s—0.1266 ± 0.0250; E—1.92 ± 0.33). Collectively, O1, the line with lowest 2-PG contents [10], has a significant advantage over the wild type after exposure to elevated growth temperatures for one week. This beneficial effect is likely not due to enhanced CO_2 fixation, but rather due to improved carbon utilization as suggested by the lower Γ.

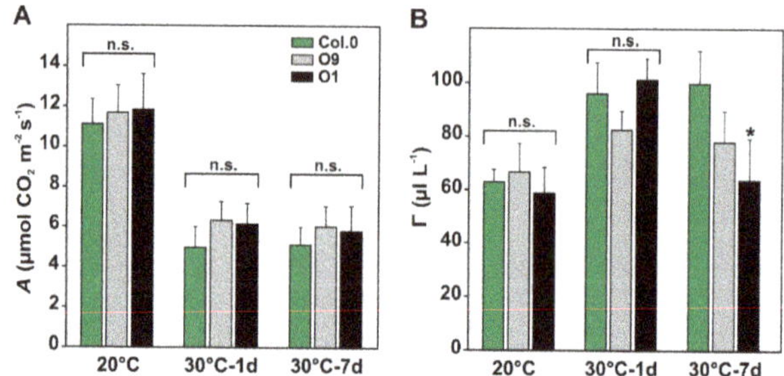

Figure 4. Photosynthetic parameters of wild-type Arabidopsis and *PGLP* overexpressors exposed to 30 °C. Wild-type and *PGLP* overexpressor (O9 and O1) plants were grown under standard conditions (20 °C) for 6 weeks following exposure to elevated temperature (30 °C). Gas exchange measurements were carried out to determine (**A**) CO_2 assimilation rates (*A*) and (**B**) net CO_2 compensation points (Γ) under control conditions (20 °C) and after 1 and 7 days after the transfer to 30 °C. Shown are mean values ± SD from at least 4 biological replicates per genotype. Asterisks indicate values statistically different from the wild type as determined by Student's *t*-test ($p < 0.05$; n. s.—not significant).

2.4. Improved 2-PG Degradation Translates to Higher Transitory Starch Stocks under Temperature Stress

Given that 2-PG levels are inversely correlated with starch accumulation [10], we quantified the amounts of starch and soluble sugars at the end of the day (EoD) in leaf-material from the temperature shift experiment. In agreement with our previous report [10], the overexpressors O9 and O1 contained significantly elevated starch contents (~27%) at EoD under control conditions compared to the wild type (Figure 5A). After the transfer to 30 °C, wild-type starch levels significantly drop compared to the control, which is in agreement with other reports [34]. Notably, both transgenic lines were able to maintain higher starch accumulation after the temperature increase, which was clearly pronounced 7 days after the shift to 30 °C. At this time point, O9 starch increased to about 46% and O1 to about 123% (Figure 5A). About the contents of soluble sugars, we did not find significant change in sucrose (Figure 5B). Glucose was similar in all plants at 20 °C. However, it significantly decreases in both lines after one to three days at 30 °C and increased again after 7 days at 30 °C (Figure 5C). Changes in fructose were only seen after one day at 30 °C, when both overexpressor lines displayed a significant drop compared to wild type (Figure 5D). Collectively, our results suggest that faster 2-PG removal and sustained photosynthesis are beneficial for carbon allocation towards transitory starch under control as well as stress conditions, without impacting steady-state sucrose amounts.

2.5. Minor Changes in the Amino and Organic Acid Contents during Temperature Stress

Considering the observed changes among carbohydrates, we were interested to which extent *PGLP* overexpression could also affect amino and organic acid levels in response to increased growth temperature. To this end, we used the same leaf-material harvested at EoD (9 h illumination, 20 °C and after 1, 3 and 7 days in 30 °C) for metabolite analysis by liquid chromatography coupled to tandem mass spectrometry (LC-MS/MS). Among the amino acid profiles (Figure 6A), we found three general accumulation patterns. Ten amino acids (group A: asparagine, glutamate, histidine, isoleucine, leucine, phenylalanine, serine, tryptophan, tyrosine and valine) increased in their abundance after the transfer to elevated temperature, with highest amounts after three days at 30 °C, but distinctly drop at day 7 (Figure 6A) to almost control levels. Six amino acids (group B: cysteine, glutamine, glycine, lysine, proline and threonine) were characterized by a gradual decrease over the course of the entire experiment. A third pattern, comprising three amino acids (group C: alanine, aspartate and methionine), showed an initial drop at day one at 30 °C and then a subsequent increase until the initial levels at 20 °C was reached (Figure 6A). In general, the observed pattern was similar in wild-type and overexpressor plants. However, glycine, an amino acid directly related to photorespiration, was found in lower amounts under control conditions and after three and 7 days at 30 °C in both overexpressor lines (Figure 6A). In addition, many amino acids showed changed levels at day 7 at 30 °C. For example, almost all of group A amino acids were decreased in the overexpressors after 7 days of temperature stress (Figure 6A), whereas group B and C representatives displayed slight increases after 7 days. Proline, as an exception, was higher in both lines one day after the shift to elevated temperature. Concerning the organic acids, we found that citrate and malate followed the same accumulation kinetics as group B amino acids, at least in the wild type (Figure 6B). Both tend to decrease over time but were significantly increased in the transgenic lines one (except O9 malate) and three days after the shift to 30 °C (Figure 6B). Compared to that, GABA and succinate fluctuate like group A amino acids, without major change in the different genotypes.

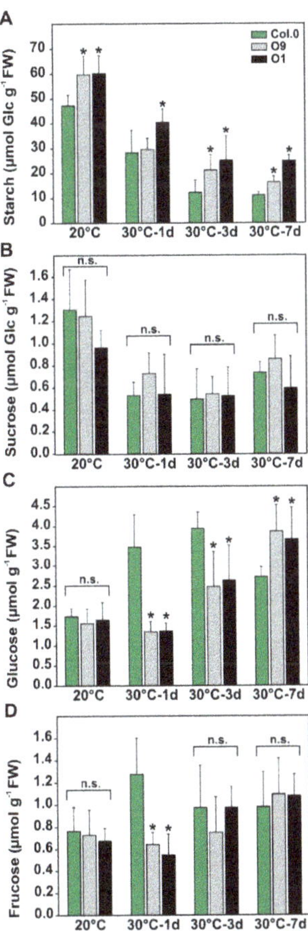

Figure 5. Carbohydrate contents of wild-type Arabidopsis and *PGLP* overexpressors exposed to 30 °C. Wild-type and *PGLP* overexpressor (O9 and O1) plants were grown in standard conditions (20 °C) for 6 weeks following exposure to elevated temperature (30 °C). Leaf material was harvested at the end of the day (9 h illumination) to determine absolute (**A**) starch, (**B**) sucrose, (**C**) glucose and (**D**) fructose contents under control conditions and after 1, 3 and 7 days after the transfer to 30 °C using gas chromatography. Shown are mean values ± SD from at least 4 biological replicates per genotype. Asterisks indicate values statistically different from the wild type as determined by Student's *t*-test ($p < 0.05$; n. s.—not significant).

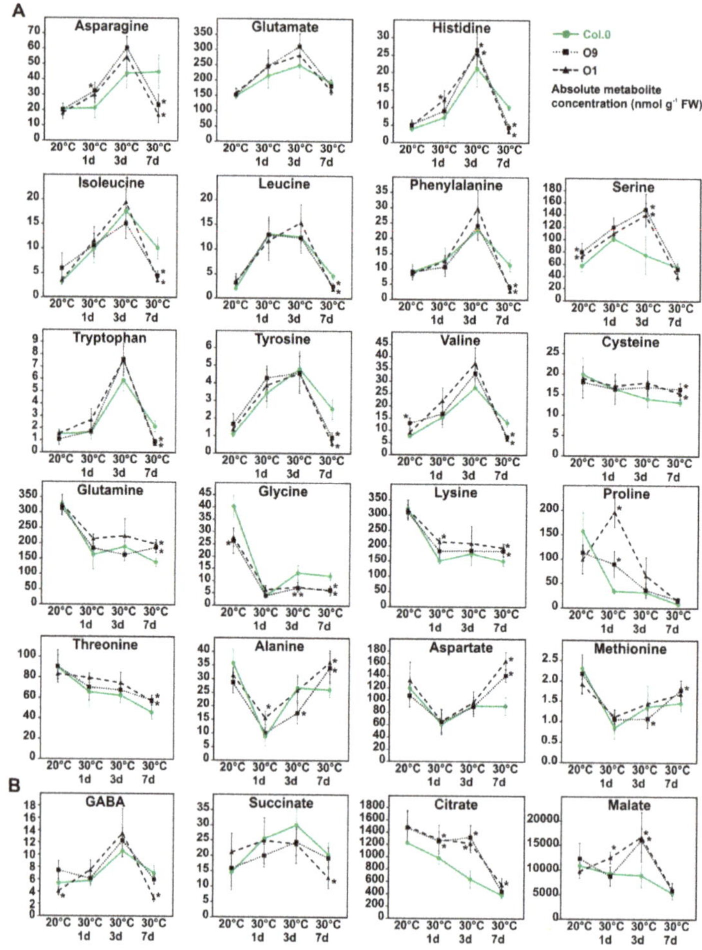

Figure 6. Amino acids and organic acid of wild-type Arabidopsis and *PGLP* overexpressors exposed to 30 °C. Wild-type and *PGLP* overexpressor (O9 and O1) plants were grown in under standard conditions (20 °C) for 6 weeks following exposure to elevated temperature (30 °C). Leaf material was harvested at the end of the day (9 h illumination) to determine absolute (**A**) amino acid and (**B**) organic acids contents under control conditions and after 1, 3 and 7 days after the transfer to 30 °C via liquid chromatography coupled to tandem mass spectrometry (LC-MS/MS). Shown are mean values ± SD of absolute metabolite concentrations (nmol g^{-1} FW) from at least 4 biological replicates per genotype (Col.0—green, solid line with circles, O9—black, dotted lines with squares and O1—black, dashed line with triangles). Asterisks indicate values statistically different from the respective wild-type time point as determined by Student's *t*-test ($p < 0.05$).

2.6. Expression of Photorespiratory Proteins Increases after High Temperature Exposure in Wild Type but not in PGLP Overexpressor Lines

Previously, it was reported that the expression and amounts of some photorespiratory genes and proteins, respectively, changed in response to abiotic stresses [25,35]. Therefore, we analyzed the expression of selected photorespiratory genes and proteins during the high temperature treatment using quantitative real-time polymerase chain reaction (qRT-PCR) and immunoblotting, respectively. As shown in Figure 7, mRNA levels of several photorespiratory genes (*PGLP1*, glutamate:glyoxylate aminotransferase 1 (*GGT1*), glycine decarboxylase P and T protein (*GDC-P* and *GDC-T*), serine hydroxymethyltransferase 1 (*SHM1*) and peroxisomal hydroxypyruvate reductase 1 (*HPR1*)) were significantly elevated in wild-type leaves after one and three days at 30 °C, while after 7 days their expression levels returned almost to the initial levels at 20 °C (Figure 7). To prove, whether these transcriptional alterations translated also to changes in the protein abundances, we exemplarily analyzed PGLP1, GDC-P, SHM1 and HPR1 protein amounts at the same time points. As observed before for mRNA expression, all four proteins increased in wild type after exposure to 30° (Figure 7). In contrast, both *PGLP* overexpression lines did not display the increased expression of photorespiratory genes and proteins. As expected, lines O9 and O1 showed significantly increased *PGLP* expression on the mRNA and protein level already under control conditions and only minor alterations in course of the temperature treatment (Figure 7A). However, all the other photorespiratory genes analyzed showed constantly lower mRNA and only very minor changes in the protein levels during the entire experiment. Only the genes *GGT1*, *GDC-T*, *GDC-P* and *HPR1* showed higher mRNA amounts in line O1 after 7 days at 30 °C (Figure 7B,D,E,F). From these results we conclude that *PGLP* overexpression somehow prepares the plants to cope with the temperature stress without a coordinative upregulation of other genes and enzymes involved in photorespiration.

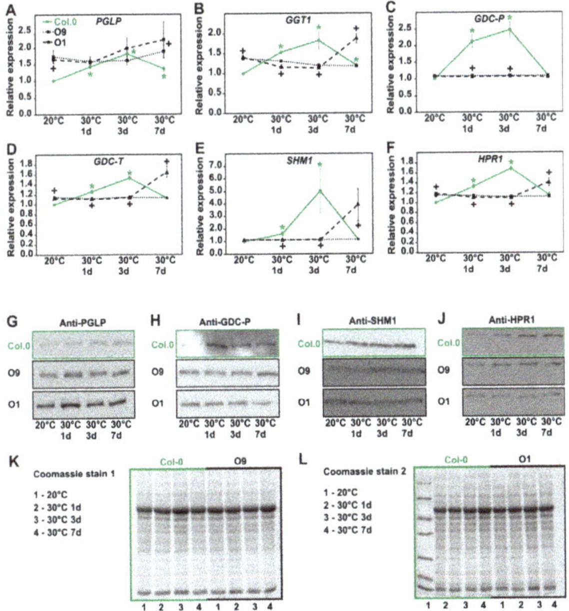

Figure 7. Expression of selected photorespiratory gene on the mRNA and protein level in wild-type Arabidopsis and *PGLP* overexpressors exposed to 30 °C. Wild-type and *PGLP* overexpressor (O9 and O1) plants were grown in under standard conditions (20 °C) for 6 weeks following exposure to elevated temperature (30 °C). Leaf material was harvested at the end of the day (9 h illumination) to analyse mRNA and protein expression of selected photorespiratory genes under control conditions (20 °C) and after 1, 3 and 7 days after the transfer to 30 °C. Shown are: mRNA expression of (**A**)—*PGLP*, (**B**)—*GGT1*, (**C**)—*GDC-P*, (**D**)—*GDC-T*, (**E**)—*SHM1*, and (**F**)—*HPR1* and protein amounts of (**G**)—PGLP, (**H**)—GDC-P, (**I**)—SHM1 and (**J**)—HPR1. Coomassie stains (10 µg protein per lane) are shown in (**K**) and (**L**) as loading controls. Values are means ± SD from at least 3 biological replicates per genotype (Col.0—green, solid line with circles, O9—black, dotted lines with squares and O1—black, dashed line with triangles). Asterisks indicate values statistically different from the wild type control at 20 °C and plusses of the transgenic lines to the respective wild-type time point as determined by Student's *t*-test ($p < 0.05$).

3. Discussion

The main function of the photorespiratory pathway is the efficient removal of critical intermediates, especially 2-PG, which severely inhibits carbon fixation and allocation [6,10]. Furthermore, it has been suggested that photorespiration plays a role in the abiotic stress response of plants [25–27]. Therefore, we hypothesized that an increased photorespiratory flux might facilitate plant abiotic stress tolerance. To test this hypothesis, we used transgenic lines overexpressing photorespiratory *PGLP*, previously shown to fix more carbon at high photorespiratory pressures and to display improved starch metabolism and stomatal movements [10].

Consistent with our hypothesis, we found improved photosynthetic performance after short-term and long-term abiotic stresses. In a first series of experiments, *PGLP* overexpression lines displayed higher CO_2 assimilation accompanied with improvements in other gas exchange parameters after increasing the temperature to 30 °C during short-term high light treatment, anticipated to promote 2-PG production (Figure 2B). The maintenance of higher photosynthesis in *PGLP* overexpressors under short-term temperature stress is likely through diminished inhibition of the CB cycle by 2-PG, but we cannot rule out that increased photorespiration eventually helps to prevent the chloroplastidal electron transport chain from overreduction as suggested before [26,28,29]. Similar results were obtained if photosynthesis of wild-type and *PGLP* overexpressor plants was characterized under water-limiting conditions. Water shortage promotes stomatal closure [36] and eventually increases 2-PG levels due to a higher RuBP fraction being oxidized. Notably, both transgenic lines showed significant improvements in photosynthetic parameters after 13 days of water shortage (Figure 3). Interestingly, we did not only measure higher CO_2 assimilation and lower CO_2 compensation points, but also higher g_s, indicating altered stomatal movements (Figure 3B). Given that photorespiration is also involved in proper guard cell metabolism [30,31], an optimized flux through the pathway in mesophyll cells could be beneficial for these specialized cells, too. Accordingly, it is likely to assume that the altered leaf-carbohydrate metabolism, in particular starch biosynthesis, can facilitate allocation of carbon from the mesophyll to the guard cells in order to enhance their energy supply. It should be noted that *PGLP* overexpression is driven through the *ST-LSI* promoter, therefore, changes in *PGLP* expression are not restricted to the mesophyll cells. Hence, changes in g_s could also be directly caused by altered PGLP activity in guard cells.

Finally, long-term exposure at elevated temperatures was analyzed in more detail. This was done for three reasons: (i) elevated temperatures favor 2-PG production, (ii) higher PGLP activities were found to be beneficial for photosynthesis on a short-term (Figure 2), and (iii) future climate change scenarios predict an increase in temperature on a global scale [13,14]. Consequently, higher *PGLP* activity could eventually be a positive trait for plant engineering. In agreement with the short-term exposure to high light and elevated temperature (Figure 2) and growth under water-limiting conditions (Figure B), the overexpressor line O1 maintained higher photosynthesis after long-term exposure to elevated temperature (Figure 4). This change was accompanied by higher g_s and increased transpiration, again suggesting changes in stomatal movements. Given the strong impact 2-PG has on CO_2 fixation and carbohydrate utilization and allocation [6,10], we quantified amounts of starch and the soluble sugars sucrose, glucose and fructose (Figure 5). In line with previous findings [10], overexpressors of *PGLP* store somewhat more starch under standard conditions, without a significant impact on sucrose synthesis. Interestingly, these trends were kept after the shift to elevated temperature until day 7 at 30 °C. These data indicate faster 2-PG removal facilitates carbon allocation to starch biosynthesis also under the stress conditions. However, we cannot neglect the possibility that starch is degraded somewhat less fast in the transgenic lines. The assumption of faster starch synthesis is supported by the decreased amounts of soluble sugars in both lines during the first days at 30 °C. However, after 7 days at 30 °C, this trend became reversed, which might be a compensatory reaction of carbon metabolism under long-term temperature stress. The effects of *PGLP* overexpression seem to be mainly restricted to alterations in photosynthesis and carbohydrate metabolism, since amino and organic acid contents showed only minor changes.

Acclimation to elevated temperatures obviously activates the photorespiratory activity in wild-type plants, because a coordinated increase of mRNAs and proteins for many photorespiratory enzymes was observed (Figure 7), which is consistent with previous reports [25]. Given the lack of response on the expression of the photorespiratory enzymes in the transgenic lines, one might speculate upregulation of *PGLP*, and in turn lowering the steady-state content of 2-PG, already acts as signal to indicate sufficient acclimation of the photorespiratory flux. Additionally, this result once more demonstrates the importance of fast 2-PG removal via photorespiration, which might be sufficient to cope with the temperature stress. Furthermore, 2-PG could also play a regulatory role in the stress acclimation as shown for serine before [37]. Direct involvement of 2-PG as inducer for the transcription of genes involved in CO_2 uptake was observed in cyanobacteria [38], but has not yet reported for plants. However, the *PGLP* knock out mutant showed strong and specific alterations in gene expression after a shift from high to low CO_2 [31]. Therefore, it cannot be excluded completely that 2-PG mediated transcriptional reprogramming mechanisms exist in plants.

4. Material and Methods

4.1. Plant Material and Standard Growth Conditions

The generation of stable T4-generations of *Arabidopsis thaliana* (Arabidopsis) lines (ecotype Columbia 0, Col-0) overexpressing the photorespiratory phosphoglycolate phosphatase 1 (*PGLP1*, At5g36700, EC 3.1.3.18) was described previously [10]. Lines with an approximately 28% (O9) and 44% (O1) increase in PGLP activity compared to the wild type were used during this study. Prior plant cultivation, seeds of all genotypes were surface sterilized with chloric acid, sown on a soil (Type Mini Tray; Einheitserdewerk, Uetersen, Germany) and vermiculite mixture (4:1) and incubated at 4 °C for at least two days to break dormancy. Subsequently, plants were grown under environmental controlled conditions in growth cabinets (SANYO, Osaka, Japan; CLF Plant Climatics, Wertingen, Germany) with the following conditions as a standard: photoperiod - 10/14 h day/night-cycle, temperature - 20/20 °C day/night-cycle, photon flux density of ~120 μmol m^{-2} s^{-1}, 70% relative humidity, 0.039% CO_2 in air (Figure 1A). During growth, plants were regularly watered with 0.2% Wuxal liquid fertilizer (Aglukon, Düsseldorf, Germany).

4.2. Stress Conditions

Wild-type and *PGLP* overexpression plants were grown under standard conditions (Figure 1A), following exposure to two different stress conditions. First, simulating water-limiting conditions, all genotypes were grown conjointly in one pot (28 cm in diameter, 1 plant per genotype, 5 technical replicates) for 6 weeks with regular water supply to allow for a high level of comparability. After control experiments were carried out, watering was stopped, and experiments performed at intervals specified in the manuscript text. Second, simulating temperature stress, all genotypes were grown under standard conditions for 6 weeks and control experiments carried out. Subsequently, all plants were exposed to elevated temperatures (30 °C) with otherwise equal conditions. Photosynthetic measurements were performed, and leaf-material harvested after 1, 3 and 7 days in 30 °C.

4.3. qRT-PCR Analysis and Immunological Studies

To follow the expression of selected photorespiratory genes and proteins we harvested leaf-material at the end of the day (9 h illumination) during temperature transition under control conditions (20 °C) and after 1, 3 and 7 days in 30 °C. For gene expression analysis total leaf RNA was extracted from ~100 mg tissue (pooled from three biological individuals) and ~2.5 μg used to synthesize cDNA (Nucleospin RNA plant kit, Macherey-Nagel; RevertAid cDNA synthesis kit, MBI Fermentas). Prior to qRT-PCR analysis, cDNA amounts were calibrated by RT-PCR according to signals from 432-bp fragments of the constitutively expressed 40S ribosomal protein *S16* gene, with oligonucleotides P444 [5′-GGC GAC ACA ACC AGC TAC TGA-3′] and P445 [5′-CGG TAA CTC TTC TGG TAA CGA-3′]. Detection and normalization of gene expression were performed as described previously (Timm et al., 2013), and mRNA amounts of *PGLP1* (P393 [5′-CAG AAT GGC GGT TGT AAG AC-3′] and P394 [5′-GGC TCC CTA ATT TGC TAT GC-3′]; 328 bp), *GGT1* (P405 [5′-CGT TGC TCA GGC TCG TTC TC-3′] and P406 [5′-CCA CCT CGC TGT CCA CAT TC-3′]; 336 bp), *GDC-P1* (P366 [5′-AGC AAA TCC GTA GCC ATC AC-3′] and P413 [5′-TAT GTC CAA TGC GTC GCT TC-3′]; 327 bp), *GDC-T* (P367 [5′-GCA ATC AAT AAC CCG TCG TC-3′] and P368 [5′-TCA ATG GCA CCT CCT TTC TC-3′]; 363 bp), *SHM1* (P395 [5′-GCC CAG TGA AGC TGT TGA TG-3′] and P396 [5′-AGT TGG CAG GAG ATC CAG AC-3′]; 365 bp) and *HPR1* (P397 [5′-GGC TGA ACT AGC TGC TTC TC-3′] and P398 [5′-GCA CCG GGT GAA GAC TTA TC-3′]; 360 bp) quantified accordingly using the oligonucleotide combinations given in brackets after each gene. Abundances of selected photorespiratory proteins were analyzed by immunoblotting. Briefly, total leaf proteins were extracted from ~100 mg leaf tissue (pooled from three biological individuals) and 10 µg separated by SDS-PAGE followed by immunoblotting according to standard protocols. Alterations in protein expression were visualized using specific antibodies against PGLP1, GDC-P, SHM1 and HPR1 [38].

4.4. Determination of Starch and Metabolite Analysis

Starch contents were measured enzymatically as described previously [39] from ~50 mg of leaf tissue harvested at EoD (9 h illumination) from at least four biological replicates per genotype. The soluble fraction of the extraction procedure was further subjected to gas chromatography (GC) analysis to quantify sucrose, glucose and fructose as described previously [40]. Amino acids and organic acids were essentially quantified on a high-performance liquid chromatograph mass spectrometer LCMS-8050 system (Shimadzu, Japan) as described recently [41] from 50 mg leaf-tissue harvested at the end of the day (9 h of illumination). The compounds were identified and quantified using the multiple reaction monitoring (MRM) values given in the LC-MS/MS method package and the LabSolutions software package (Shimadzu, Japan). Authentic standard substances (Merck, Germany) at varying concentrations were used for calibration and peak areas normalized to signals of the internal standard (2-(N-morpholino)-ethanesulfonic acid - MES).

4.5. Gas Exchange and Chlorophyll a Fluorescence Measurements

All gas exchange parameters were determined in a 6-h time period between 2 h after onset and 2 h prior offset of illumination on a Li-Cor-6400 gas exchange system (LI-COR, Lincoln, NE, USA) using fully expanded leaved from plants at growth stage 5.1 [33]. Prior the actual measurement, leaves were pre-adapted to the measuring chamber for at least 10 min. To determine net CO_2 compensation points (Γ), A/C_i curves (400, 300, 200, 100, 50, 20, 0, 400 ppm CO_2) were recorded with the following conditions: photon flux density = 1000 μmol m^{-2} s^{-1}, chamber temperature = 20 °C, flow rate = 300 μmol s^{-1} and relative humidity = 60% to 70%. Fluorescence light response curves (PAR: 1600, 1200, 800, 600, 400, 200, 100, 50, 20 and 0 μmol m^{-2} s^{-1}) were measured at two different block temperatures (20 °C and 30 °C) with the following conditions: CO_2 concentration = 400 ppm; flow rate = 300 μmol s^{-1} and relative humidity = 60 to 70%. Relative rates of electron transport around PSII at a given light intensity (PAR) were assessed by the formula ETR = $Y_{PSII} \times PAR \times 0.84 \times 0.5$. The factors are based on the assumptions that 84% of the incident quanta are absorbed by the leaf (factor 0.84) and that the transport of one electron by the two photosystems requires the absorption of two quanta (factor 0.5). Y_{PSII} (effective quantum yield of photosystem II) was calculated as described previously [42] using the following formula:

4.6. Statistical Analysis

If values were described to be significantly different from the control within the text, the differences have been determined due to the performance of the two tailed Student's *t*-test algorithm incorporated into Microsoft Excel 10.0 (Microsoft, Seattle, WA, USA).

4.7. Accession Numbers

The Arabidopsis Genome Initiative database contains sequence data from this article under the following accession numbers: *PGLP1* (At5g36700), *GDC-P1* (At4g33010), *GDC-T* (At1g11680), *SHM1* (At4g37930), *HPR1* (At1g68010), *GGT1* (AT1G23310), and 40S ribosomal protein *S16* (At2g09990).

5. Conclusions

In summary, we provided evidence that upregulation of PGLP activity leading to lowered 2-PG contents, is beneficial for maintaining photosynthesis under abiotic stresses. This is likely due to the removal of 2-PG-mediated negative metabolic feedback on central enzymes of the CB cycle and carbon export from the chloroplast. We suggest that an optimized photorespiratory flux can thus stabilize the production and allocation of organic carbon under unfavorable environmental conditions. The strategy presented here might have further implications for plant engineering approaches to generate highly productive and more stress-resistant plants for future climate scenarios.

Author Contributions: H.B. and S.T. conceived and supervised the project. C.H., S.T., and F.W. performed the research and analyzed data. M.H. provided experimental equipment and tools. S.T. wrote the article with additions and revisions from all authors. All authors have read and approved the final version of the manuscript.

Funding: This work was financially supported by the Deutsche Forschungsgemeinschaft (Research Unit FOR 1186 Promics, BA 1177/12 to H.B.) and the University of Rostock (to M.H. and S.T.). The LC-MS equipment at the University of Rostock was financed through a grant of the Hochschulbauförderungsgesetz (HBFG) program (INST 264/125-1 FUGG to H.B.).

Acknowledgments: We gratefully acknowledge Friedrich Kirsch (University of Rostock) for help with the GC analysis and Saskia Schwab (University of Rostock) for valuable technical assistance.

Conflicts of Interest: The authors declare no conflicts of interest.

References

1. Eisenhut, M.; Ruth, W.; Haimovich, M.; Bauwe, H.; Kaplan, A.; Hagemann, M. The photorespiratory glycolate metabolism is essential for cyanobacteria and might have been conveyed endosymbiontically to plants. *Proc. Natl. Acad. Sci. USA* **2008**, *105*, 17199–17204. [CrossRef] [PubMed]

2. Zelitch, I.; Schultes, N.P.; Peterson, R.B.; Brown, P.; Brutnell, T.P. High glycolate oxidase activity is required for survival of maize in normal air. *Plant Physiol.* **2009**, *149*, 195–204. [CrossRef] [PubMed]

3. Bauwe, H.; Hagemann, M.; Kern, R.; Timm, S. Photorespiration has a dual origin and manifold links to central metabolism. *Curr. Opin. Plant Biol.* **2012**, *15*, 269–275. [CrossRef] [PubMed]

4. Timm, S.; Mielewczik, M.; Florian, A.; Frankenbach, S.; Dreissen, A.; Hocken, N.; Fernie, A.R.; Walter, A.; Bauwe, H. High-to-low CO_2 acclimation reveals plasticity of the photorespiratory pathway and indicates regulatory links to cellular metabolism of *Arabidopsis*. *PLoS ONE* **2012**, *7*, e42809. [CrossRef]

5. Rademacher, N.; Kern, R.; Fujiwara, T.; Mettler-Altmann, T.; Miyagishima, S.Y.; Hagemann, M.; Eisenhut, M.; Weber, A.P. Photorespiratory glycolate oxidase is essential for the survival of the red alga *Cyanidioschyzon merolae* under ambient CO_2 conditions. *J. Exp. Bot.* **2016**, *67*, 3165–3175. [CrossRef]

6. Levey, M.; Timm, S.; Mettler-Altmann, T.; Borghi, G.L.; Koczor, M.; Arrivault, S.; Weber, A.P.M.; Bauwe, H.; Gowik, U.; Westhoff, P. Efficient 2-phosphoglycolate degradation is required to maintain carbon assimilation and allocation in the C_4 plant *Flaveria bidentis*. *J. Exp. Bot.* **2019**, *70*, 575–587. [CrossRef]

7. Bowes, G.; Ogren, W.L.; Hageman, R.H. Phosphoglycolate production catazyded by ribulose diphosphate carboxylase. *Biochem. Biophys. Res. Commun.* **1971**, *45*, 716–722. [CrossRef]

8. Anderson, L.E. Chloroplast and cytoplasmic enzymes. 2. Pea leaf triose phosphate isomerases. *Biochim. Biophys. Acta* **1971**, *235*, 237–244. [CrossRef]

9. Kelly, G.J.; Latzko, E. Inhibition of spinach-leaf phosphofructokinase by 2-phosphoglycollate. *FEBS Lett.* **1976**, *68*, 55–58. [CrossRef]

10. Flügel, F.; Timm, S.; Arrivault, S.; Florian, A.; Stitt, M.; Fernie, A.R.; Bauwe, H. The photorespiratory metabolite 2-phosphoglycolate regulates photosynthesis and starch accumulation in *Arabidopsis*. *Plant Cell* **2017**, *29*, 2537–2551. [CrossRef]

11. Li, J.; Weraduwage, S.M.; Peiser, A.L.; Tietz, S.; Weise, S.E.; Strand, D.D.; Froehlich, J.E.; Kramer, D.M.; Hu, J.; Sharkey, T.D. A Cytosolic Bypass and G6P Shunt in Plants Lacking Peroxisomal Hydroxypyruvate Reductase. *Plant Physiol.* **2019**, *180*, 783–792. [CrossRef] [PubMed]

12. Sharkey, T.D. Estimating the rate of photorespiration in leaves. *Physiol. Plant* **1988**, *73*, 147–152. [CrossRef]

13. Walker, B.J.; VanLoocke, A.; Bernacchi, C.J.; Ort, D.R. The costs of photorespiration to food production now and in the future. *Annu. Rev. Plant Biol.* **2016**, *67*, 107–129. [CrossRef] [PubMed]

14. Slattery, A.R.; Ort, D.R. Carbon assimilation in crops at high temperature. *Plant Cell Environ.* **2019**, *42*, 2750–2758. [CrossRef]

15. Peterhänsel, C.; Maurino, V.G. Photorespiration redesigned. *Plant Physiol.* **2011**, *155*, 49–55. [CrossRef]

16. Xin, C.P.; Tholen, D.; Devloo, V.; Zhu, X.G. The benefits of photorespiratory bypasses: How can they work? *Plant Physiol.* **2015**, *167*, 574–585. [CrossRef]

17. South, P.F.; Cavanagh, A.P.; Lopez-Calcagno, P.E.; Raines, C.A.; Ort, D.R. Optimizing photorespiration for improved crop productivity. *J. Integr. Plant Biol.* **2018**, *60*, 1217–1230. [CrossRef]

18. Timm, S.; Florian, A.; Arrivault, S.; Stitt, M.; Fernie, A.R.; Bauwe, H. Glycine decarboxylase controls photosynthesis and plant growth. *FEBS Lett.* **2012**, *586*, 3692–3697. [CrossRef]

19. Timm, S.; Wittmiß, M.; Gamlien, S.; Ewald, R.; Florian, A.; Frank, M.; Wirtz, M.; Hell, R.; Fernie, A.R.; Bauwe, H. Mitochondrial dihydrolipoyl dehydrogenase activity shapes photosynthesis and photorespiration of *Arabidopsis thaliana*. *Plant Cell* **2015**, *27*, 1968–1984. [CrossRef]

20. Simkin, A.; Lopez-Calcagno, P.; Davey, P.; Headland, L.; Lawson, T.; Timm, S.; Bauwe, H.; Raines, C. Simultaneous stimulation of the SBPase, FBP aldolase and the photorespiratory GDC-H protein increases CO_2 assimilation, vegetative biomass and seed yield in *Arabidopsis*. *Plant Biotechnol. J.* **2017**, *15*, 805–816. [CrossRef]

21. López-Calcagno, P.E.; Fisk, S.; Brown, K.L.; Bull, S.E.; South, P.F.; Raines, C.A. Overexpressing the H-protein of the glycine cleavage system increases biomass yield in glasshouse and field-grown transgenic tobacco plants. *Plant Biotechnol. J.* **2019**, *17*, 141–151. [CrossRef]

22. Rojas, C.M.; Senthil-Kumar, M.; Wang, K.; Ryu, C.M.; Kaundal, A.; Mysore, K.S. Glycolate oxidasemodulates reactive oxygen species-mediated signaltransduction during nonhost resistance inNicotianabenthamianaandArabidopsis. *Plant Cell* **2012**, *24*, 336–352. [CrossRef] [PubMed]

23. Sørhagen, K.; Laxa, M.; Peterhänsel, C.; Reumann, S. The emerging role of photorespiration and non-photorespiratory peroxisomal metabolism in pathogen defense. *Plant Biol.* **2013**, *15*, 723–736. [CrossRef] [PubMed]

24. Kozaki, A.; Takeba, G. Photorespiration protects C3 plants from photooxidation. *Nature* **1996**, *384*, 557–560. [CrossRef]

25. Wingler, A.; Lea, P.J.; Quick, W.P.; Leegood, R.C. Photorespiration: Metabolic pathways and their role in stress protection. *Philos. Trans. R. Soc. Lond. B Biol. Sci.* **2000**, *355*, 1517–1529. [CrossRef]

26. Voss, I.; Sunil, B.; Scheibe, R.; Raghavendra, A.S. Emerging concept for the role of photorespiration as an important part of abiotic stress. *Plant Biol.* **2013**, *15*, 713–722. [CrossRef]

27. Saji, S.; Bathula, S.; Kubo, A.; Tamaoki, M.; Aono, M.; Sano, T.; Tobe, K.; Timm, S.; Bauwe, H.; Nakajima, N.; et al. Ozone-Sensitive Arabidopsis Mutants with Deficiencies in Photorespiratory Enzymes. *Plant Cell Physiol.* **2017**, *58*, 914–924. [CrossRef]

28. Huang, W.; Yang, Y.J.; Wang, J.H.; Hu, H. Photorespiration is the major alternative electron sink under high light in alpine evergreen sclerophyllous Rhododendron species. *Plant Sci.* **2019**. [CrossRef]

29. Sunil, B.; Saini, D.; Bapatla, R.B.; Aswani, V.; Raghavendra, A.S. Photorespiration is complemented by cyclic electron flow and alternative oxidase pathway to optimize photosynthesis and protect against abiotic stress. *Photosynth. Res.* **2019**, *139*, 67–79. [CrossRef]

30. Igamberdiev, A.U.; Mikkelsen, T.N.; Ambus, P.; Bauwe, H.; Lea, P.J.; Gardeström, P. Photorespiration Contributes to Stomatal Regulation and Carbon Isotope Fractionation: A Study with Barley, Potato and *Arabidopsis* Plants Deficient in Glycine Decarboxylase. *Photosynth. Res.* **2004**, *81*, 139–152. [CrossRef]

31. Eisenhut, M.; Bräutigam, A.; Timm, S.; Florian, A.; Tohge, T.; Fernie, A.R.; Bauwe, H.; Weber, A. Photorespiration is crucial to the dynamic response of photosynthetic metabolism to altered CO_2 availability. *Mol. Plant* **2017**, *10*, 437–461. [CrossRef] [PubMed]

32. Liu, Y.; Mauve, C.; Lamothe-Sibold, M.; Guérard, F.; Glab, N.; Hodges, M.; Jossier, M. Photorespiratory serine hydroxymethyltransferase 1 activity impacts abiotic stress tolerance and stomatal closure. *New Phytol.* **2019**, *42*, 2567–2583. [CrossRef] [PubMed]

33. Boyes, D.C.; Zayed, A.M.; Ascenzi, R.; McCaskill, A.J.; Hoffman, N.E.; Davis, K.R.; Gorlach, J. Growth stage-based phenotypic analysis of *Arabidopsis*: A model for high throughput functional genomics in plants. *Plant Cell* **2001**, *13*, 1499–1510. [CrossRef] [PubMed]

34. Florian, A.; Nikoloski, Z.; Sulpice, R.; Timm, S.; Araújo, W.L.; Tohge, T.; Bauwe, H.; Fernie, A. Analysis of short-term metabolic alterations in *Arabidopsis* following changes in the prevailing environmental conditions. *Mol. Plant* **2014**, *7*, 893–911. [CrossRef]

35. Li, J.; Hu, J. Using Co-Expression Analysis and Stress-Based Screens to Uncover Arabidopsis Peroxisomal Proteins Involved in Drought Response. *PLoS ONE* **2015**, *10*, e0137762. [CrossRef]

36. Buckley, T.N. How stoma respond to water status? *New Phytol.* **2019**, *224*, 21–36. [CrossRef]

37. Timm, S.; Florian, A.; Wittmiß, M.; Jahnke, K.; Hagemann, M.; Fernie, A.R.; Bauwe, H. Serine acts as a metabolic signal for the transcriptional control of photorespiration-related genes in Arabidopsis. *Plant Physiol.* **2013**, *162*, 379–389. [CrossRef]

38. Jiang, Y.L.; Wang, X.P.; Sun, H.; Han, S.J.; Li, W.F.; Cui, N.; Lin, G.M.; Zhang, J.Y.; Cheng, W.; Cao, D.D.; et al. Coordinating carbon and nitrogen metabolic signaling through the cyanobacterial global repressor NdhR. *Proc. Natl. Acad. Sci. USA* **2018**, *115*, 403–408. [CrossRef]

39. Cross, J.M.; von Korff, M.; Altmann, T.; Bartzetko, L.; Sulpice, R.; Gibon, Y.; Palacios, N.; Stitt, M. Variation of enzyme activities and metabolite levels in 24 Arabidopsis accessions growing in carbon-limited conditions. *Plant Physiol.* **2006**, *142*, 1574–1588. [CrossRef]

40. Sievers, N.; Muders, K.; Henneberg, M.; Klähn, S.; Effmert, M.; Junghans, H.; Hagemann, M. Establishing glucosylglycerol synthesis in potato (*Solanum tuberosum* L. cv. albatros) by expression of the *ggpPS* gene from *Azotobacter vinelandii*. *J. Plant Sci. Mol. Breed.* **2013**, *2*, 1. [CrossRef]

41. Reinholdt, O.; Schwab, S.; Zhang, Y.; Reichheld, J.P.; Fernie, A.R.; Hagemann, M.; Timm, S. Redox-regulation of photorespiration through mitochondrial thioredoxin o1. *Plant Physiol.* **2019**, *181*, 442–457. [CrossRef] [PubMed]

42. Genty, B.; Briantais, J.M.; Baker, N.R. The relationship between the quantum yield of photosynthetic transport and quenching of chlorophyll fluorescence. *Biochim. Biophys. Acta* **1989**, *990*, 87–92. [CrossRef]

Review

Plant Mitochondrial Carriers: Molecular Gatekeepers That Help to Regulate Plant Central Carbon Metabolism

M. Rey Toleco [1,2], Thomas Naake [1], Youjun Zhang [1,3], Joshua L. Heazlewood [2] and Alisdair R. Fernie [1,3,*]

[1] Max-Planck-Institut für Molekulare Pflanzenphysiologie, Am Mühlenberg 1, 14476 Potsdam-Golm, Germany; mtoleco@student.unimelb.edu.au (M.R.T.); Naake@mpimp-golm.mpg.de (T.N.); yozhang@mpimp-golm.mpg.de (Y.Z.)
[2] School of BioSciences, the University of Melbourne, Victoria 3010, Australia; jheazlewood@unimelb.edu.au
[3] Center of Plant Systems Biology and Biotechnology, 4000 Plovdiv, Bulgaria
*   Correspondence: fernie@mpimp-golm.mpg.de

Received: 19 December 2019; Accepted: 15 January 2020; Published: 17 January 2020

Abstract: The evolution of membrane-bound organelles among eukaryotes led to a highly compartmentalized metabolism. As a compartment of the central carbon metabolism, mitochondria must be connected to the cytosol by molecular gates that facilitate a myriad of cellular processes. Members of the mitochondrial carrier family function to mediate the transport of metabolites across the impermeable inner mitochondrial membrane and, thus, are potentially crucial for metabolic control and regulation. Here, we focus on members of this family that might impact intracellular central plant carbon metabolism. We summarize and review what is currently known about these transporters from in vitro transport assays and *in planta* physiological functions, whenever available. From the biochemical and molecular data, we hypothesize how these relevant transporters might play a role in the shuttling of organic acids in the various flux modes of the TCA cycle. Furthermore, we also review relevant mitochondrial carriers that may be vital in mitochondrial oxidative phosphorylation. Lastly, we survey novel experimental approaches that could possibly extend and/or complement the widely accepted proteoliposome reconstitution approach.

Keywords: MCF; TCA cycle; oxidative phosphorylation; mitochondrial carriers; transporters

1. Introduction

An ancestral endosymbiotic α-proteobacteria fused with an ancestral eukaryotic cell to give rise to the mitochondria approximately 1.5 billion years ago [1]. Since then, metabolic processes involving multiple compartments of the cell are facilitated by specific transmembrane transporter proteins [2]. The intracellular transport of mitochondrial metabolites plays an important role in cellular respiration via the processes of the tricarboxylic acid (TCA) cycle, oxidative phosphorylation, amino acids biosynthesis [3], fatty acids biosynthesis [4], photorespiration [5], and C4 photosynthesis [6]. Metabolic intermediates of these pathways pass through the double membrane of mitochondria that delineates four different sub-compartments: the outer mitochondrial membrane (OMM), the intermembrane space (IMS), the inner mitochondrial membrane (IMM), and the mitochondrial matrix (MM) [7]. The OMM is highly permissive to the passage of ions and small uncharged molecules (<5 kDa) through pore-forming membrane proteins (porins), such as the voltage-dependent anion channels [8]. Larger molecules, especially proteins, must be imported by specialized translocases. By contrast, the IMM is a more stringent molecular barrier allowing only specific metabolites to cross from or into the MM [9]. The highly impermeable IMM is required to establish an electrochemical gradient via the activities of the oxidative phosphorylation membrane protein complexes needed for ATP biosynthesis [10]. Thus, an array of nuclear-encoded mitochondrial carrier (MC) proteins are responsible for the transport of a wide range of substrates shuttled across the IMM. MCs belong to a superfamily of transporters called the mitochondrial carrier family (MCF). Various aspects of plant MCFs have been extensively reviewed by Palmieri et al. (2011) [11], Monné et al. (2019) [3], Haferkamp and Schmitz-Esser (2012) [12], and Lee and Millar (2016) [13].

Most MCF members are relatively small, ranging from 30 to 35 kDa, around 300 amino acids in length, and have a conserved six transmembrane α-helices region [14]. Most of the primary structure of MCs is comprised of three homologous regions, each approximately 100 amino acids in length [15], and both the N and C termini face the IMS [16]. Each repeat region is comprised of two transmembrane segments flanking a short helical region that is oriented parallel to the lipid bilayer [17]. Furthermore, each repeat region is comprised of two hydrophobic transmembrane α-helices connected by a long hydrophilic matrix loop [3,18] and bears the mitochondrial carrier domain superfamily motif (IPR023395). In the odd-numbered α-helices is the conserved PX[DE]XX[RK] motif, the charged residues in this motif can form inter-domain salt bridges (also called matrix salt-bridge network) [19–21]. Residues of the salt-bridge form a hydrogen bond with a proximal glutamine residue stabilizing the network (also called Q brace) [19–21]. Moreover, in the even-numbered α-helices is another conserved [YF][DE]xx[KR] motif that can form another salt bridge called the cytoplasmic salt-bridge network [19–21]. This network is stabilized by the Y brace, hydrogen bonds formed by the tyrosine residue in the motif [19–21].

Decades of biochemical and structural data have shed light on the transport mechanism of MCs [21,22]. This transport cycle involves what is now referred to as the alternating access mechanism [20] but was previously described by a number of groups as "gated pore model" [23] or "ping-pong mechanism" [24]. The mechanism describes the process wherein a substrate binds to the transporter in its c-state (cytoplasmic-side open and matrix-side closed state), undergoes a conformational change to a transition state and to an m-state (matrix-side open and cytoplasmic-side closed state), followed by the release of the substrate into the MM [17,20,23,25–27]. The counter-substrate binds to the transporter in the m-state, undergoes a conformational change eventually leading to the c-state, and releases the counter-substrate into the cytosol [17,20,23,25–27]. Upon release of the counter-substrate, the transporter is ready to begin a new transport cycle. The c-state confirmation was confirmed by the crystal structure of the ADP/ATP carrier in complex with the inhibitor carboxyatractyloside [17]. Recently, an ADP/ATP carrier in its m-state was crystallized in complex with another inhibitor, bongkrekic acid [20]. The transport cycle mechanism of MCFs has recently been reviewed [21,28].

While ADP/ATP carriers have been studied extensively, much is yet to be learned from other members of the family, particularly among plant MCF members. In *Arabidopsis thaliana*, 58 MCF proteins have been identified, but in vitro studies indicate broad substrate specificities and their physiological function *in planta* is largely unknown [11–13]. This lack of specificity among plant MCFs is surprising given the metabolic control expected to exist at the IMM. Among the substrates that have been demonstrated to be transported by Arabidopsis MCs across the IMM are: nucleotides and dinucleotides (ATP, ADP, AMP, NAD$^+$, FAD/folate); di-/tricarboxylates (malate, succinate, 2-oxoglutarate (2-OG), oxaloacetate (OAA), fumarate, citrate, isocitrate); amino acids (glutamate, aspartate, *S*-adenosylmethionine); cofactors (coenzyme A, thiamine diphosphate); and ions (phosphates, protons, Fe$^{2+/3+}$) [3,12]. Most MC proteins are localized to the IMM, while a few are localized elsewhere. For example, AT5G17400 (ER-ANT) is targeted to the endoplasmic reticulum [29] and some members like AT4G32400 (AtBrittle1) are dual targeted to the mitochondria and plastids [30]. Of the 58 MCF members in *A. thaliana*, only 28 have thus far been confirmed to localize to the mitochondria by organellar proteomics and localization by fluorescent protein tagging, while a total of 12 MCF members have been reported to localize elsewhere [13]. Here, we focus on a subset of MCs that are potentially involved in cellular respiration either in the transport of TCA cycle intermediates or in transport processes relevant to mitochondrial oxidative phosphorylation.

2. Mitochondrial Organic Acid Transporters Are Important to Central Carbon Metabolism

Metabolite partitioning between the cytosol and mitochondria, mediated by mitochondrial carriers, is a distinguishing feature of eukaryotic metabolism that necessitates flexibility in the IMM transport direction and kinetics. These transport processes are primarily driven by an electrochemical gradient or proton motive force (PMF), which is comprised of a proton gradient (ΔpH) and membrane potential ($\Delta\Psi$) generated across the IMM by proton pumps of the electron transport chain [13]. The location of mitochondrial carriers in three-dimensional space in the context of physical compartmentalization of plant metabolism makes them strategic points for metabolic regulation and control, e.g., in the TCA cycle. In vitro biochemical data regarding mitochondrial organic acid transporters showed that these mitochondrial carriers are most likely responsible for shuttling of TCA cycle intermediates, i.e., between the cytosol and the mitochondria. This strongly suggests that mitochondrial organic acid transporters play a role in central carbon metabolism. However, *in planta* interrogations of the physiological roles of these mitochondrial organic acid transporters are yet to be achieved. To date, there are no reported changes to plant metabolism when mitochondrial carrier loss-of-function plants have been characterized. Recent work on the Arabidopsis TCA cycle interactome shows that a putative phosphate transporter interacted with TCA cycle enzymes [31,32]. However, to date, the physiological significance of these protein–protein interactions remains unknown.

In plants, there are three mitochondrial carriers most likely to be relevant to TCA cycle operation under different flux modes. These are: (1) dicarboxylate carriers (DICs); (2) dicarboxylate/tricarboxylate carriers (DTC); and (3) succinate/fumarate carrier (SFC). To recapitulate the evolutionary relationships among these three transporters in the context of the entire mitochondrial carrier superfamily, we sampled sequences that showed similarity to an amino acid profile of MCF sequences. To this end, we aligned previously known MCF protein sequences and some close paralogs and using MUSCLE [33] built a protein profile using hmmbuild [34] after selecting conserved regions of the alignment via GBLOCKS. The HMM protein profile was queried against the complete proteome files of 69 species to detect protein sequences with similarity to the MCF profile. The resulting matches were aligned (hmmalign) against the protein profile. Homosites with more than 20% missing values, as well as the misaligned N- and C- terminus regions, were removed from the alignment. The phylogenetic relationships were inferred based on Maximum Likelihood using RAxML [35], and branch supports were calculated using BOOSTER [36]. Our analysis showed that DICs, DTC, and SFC are not monophyletic (Figure 1). Mitochondrial organic acid transporter formed two distinct clades. In the first clade, DICs and DTC grouped with 2-OG carriers (OGCs). The SFC formed the second organic acid clade with other non-plant organic acid transporters including oxodicarboxylate carriers (ODCs), citrate carriers (CiCs), and yeast suppressor of HM (histone-like proteins in yeast mitochondria) mutant 2 (YHM2). While these non-plant organic acid transporters likely play a vital role in these species, they will not be discussed in this review. Biochemical data would insinuate that DTC and CiC must be closely related as they both transport citrate; phylogenetic analysis revealed SFC and not DTC, is more similar to CiC (Figure 1). Based on available biochemical data, it appears that transport functions of CiC and DTC have evolved independently but perhaps convergently.

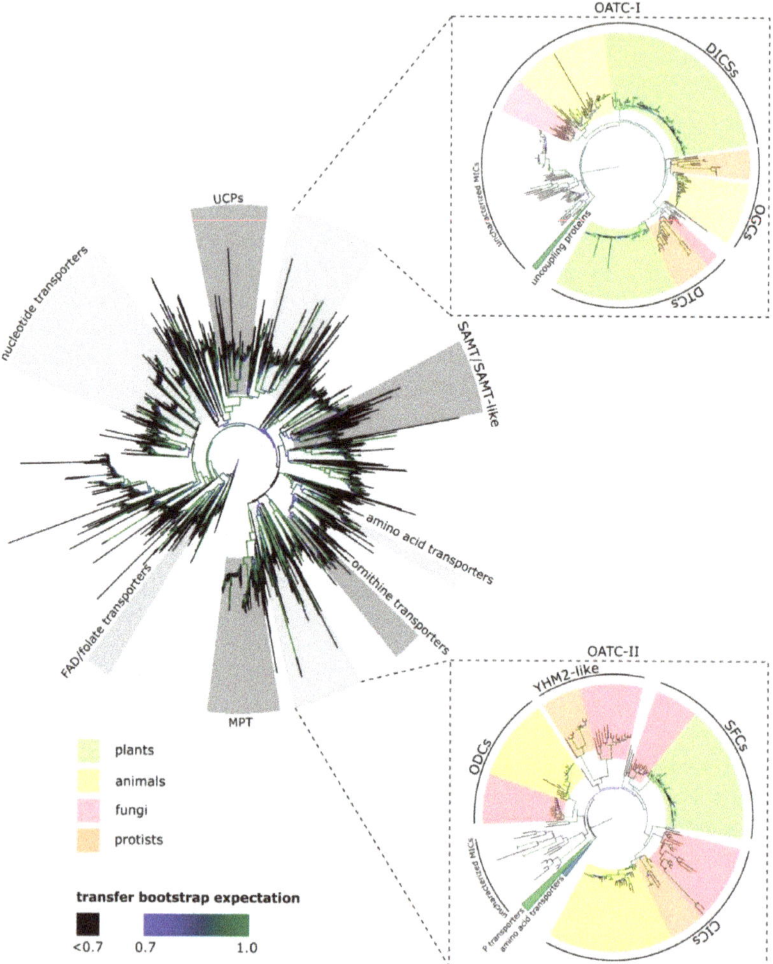

Figure 1. An unrooted phylogenetic tree of mitochondrial carrier families (MCFs) across Domain Eukaryota. The tree was inferred based on Maximum Likelihood using the RAxML software and visualized using iTOL (https://itol.embl.de/), see text for details. UCP: uncoupling proteins, SAMT: S-adenosyl methionine transporter, OATC-I/II: organic acid transporters clade I/II, DICs: dicarboxylate transporters, OGCs: 2-oxoglutarate carriers, DTCs: dicarboxylate/tricarboxylate carriers, ODC: oxodicarboxylate carriers, YHM2-like: yeast HM mutant 2-like transporters, SFCs: succinate/fumarate carriers, CiCs: citrate transporters.

It has been established that the plant TCA cycle can also operate distinctly from the textbook cyclic mode [13,37] (Figure 2). The well-established cyclic mode of TCA flux most often associated with non-photosynthetic organisms is most likely to operate in leaves in the dark when there is a high demand for ATP through cellular respiration (Figure 2A) [13]. Import of pyruvate may be exclusively attributed to the mitochondrial pyruvate carriers (MPCs) [38]. However, malate/2-OG exchange could be undertaken by either DICs and/or DTC. On the other hand, fumarate efflux is probably catalyzed by SFC using 2-OG as counter-substrate as there is no net flux of succinate reported. The observation that these different non-cyclic modes are dictated by cellular metabolic demands suggests some level of control. However, whether regulation of the activity and/or expression of these transporters exist is still an open question. Based on the available biochemical data on these relevant transporters (see below), we can begin to put forward some theories regarding their potential roles in central carbon metabolism. Several metabolites associated with the TCA cycle have been proposed to exchange across the IMM and thereby link the operation of several enzymes in mitochondrial to those in other cellular compartments. For example, citrate is suggested to be exported to the cytosol and then converted to 2-OG for redistribution to either chloroplasts or mitochondria [39]. Similarly, the malate-OAA shuttle can possibly mediate photorespiration through the mitochondrial malate dehydrogenase (mMDH) catalyzed reversible reaction after OAA import into the mitochondria [39,40].

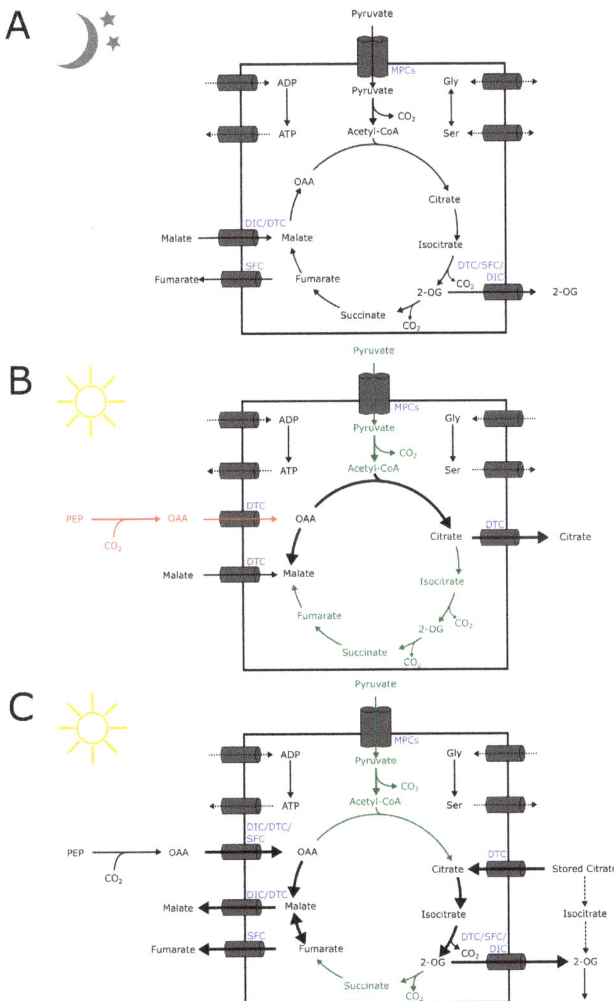

Figure 2. The plant central metabolism related to the mitochondrial Organic Acid Transporters (**A**) In the dark, the TCA cycle most likely operate in the familiar cyclic mode. (**B**) Non-cyclic TCA flux mode has been modeled based on enzymatic kinetic analysis during the day [37,41]; flux in red is based on non-modeling literature. (**C**) A non-cyclic flux mode of the TCA cycle based on isotope labeling studies [42], an alternative metabolic route of stored citrate is shown as broken black arrows. Fluxes shown in green are either inactive or significantly reduced.

During the day, ATP demand from cellular respiration is low and enzyme kinetic analysis suggests that this results in the operation of a non-cyclic flux mode [37] (Figure 2). In this model, pyruvate entry into the TCA is reduced since the pyruvate dehydrogenase complex is inactivated in the light by phosphorylation. The model, furthermore, predicts that there is a net influx of malate accompanied by a net efflux of citrate, a metabolite exchange that is in congruence with the in vitro transport activity of DTC. Net import of OAA by the mitochondria during the day has been proposed on the basis of experiments described in the literature [13]. Such a net influx of OAA could also be mediated by DTC in exchange with citrate. We propose that DTC is sufficient to support this flux, partly explaining the very high DTC protein abundance in an average Arabidopsis mitochondrion [43] (Figure 2B). A non-cyclic flux mode of the TCA cycle has been proposed based on evidence from isotope labeling studies in illuminated leaves of *Xanthium strumarium* [42]. Here, there is a net influx of OAA as well as citrate. However, this influx has also proposed to be accompanied by net efflux of malate, fumarate, and 2-OG. In this model, the metabolite fluxes can only be supported by the concerted activities of DICs, DTC, and SFC. Citrate influx is mediated by DTC in exchange with OAA, malate, or 2-OG, while fumarate is transported by SFC using OAA or 2-OG as counter-substrate. DICs may also play a role in the transport of malate, 2-OG, and OAA. Alternatively, stored citrate can be converted to 2-OG via isocitrate in the cytosol [37]. In this case, the remaining metabolic fluxes can be attributed to the activities of DICs and SFC. The overlap of substrates allowed for transport by these transporters makes it challenging to pin down the responsible transporter for a specific metabolite exchange. While absolute quantification of the subcellular levels of organic acids remains a major analytical challenge, there is an acute need to define the *in planta* transport substrates and the directionality of the transport carried out by these MCs in order to complete our understanding of mitochondrial metabolism as they potentially dictate the metabolic fates of the TCA intermediates.

2.1. Dicarboxylate Carriers (DICs)

DICs are members of the MCF reported to facilitate the transport of dicarboxylates such as malate and related compounds as well as phosphate, sulfate, and thiosulfate across the IMM [44]. Phylogenetic analysis showed that distinct DIC kingdom-level subclades can be distinguished clearly, separating those of fungal, animal, or plant origins (Figure 1). All higher land plants in our analysis possess at least one copy of DIC. However, there is no DIC homolog in algae and it seems to have first evolved in bryophytes (*Selaginella moellendorffii* and *Physcomitrella patens*), which have three and five DIC homologs, respectively. This is consistent with the analysis of MCFs in *Ostreococcus lucimarinus* where a DIC homolog could not be identified [11]. DIC gene duplication seemed to be more common in plants than in animals. Consistent with our analysis, *A. thaliana* was reported to have three DIC homologs—AtDIC1 (AT2G22500), AtDIC2 (AT4G24570), and AtDIC3 (AT5G09470) [44]. *Populus trichocarpa* has eight DICs that all belong to the AtDIC2 subgroup. Cassava (*Manihot esculenta*) possesses five DICs, clustering with the AtDIC2 subgroup, and two additional DICs in the AtDIC3 subgroup.

Arabidopsis DICs have been characterized in vitro and were shown to have varying transport kinetics but similar substrate selectivity, transporting mainly malate, OAA, succinate, maleate, malonate, phosphate, sulfate, and thiosulfate [44]. In the case of malate homo exchange, the V_{max} for AtDIC3 (2.21 ± 0.31 mmol min^{-1} g protein^{-1}) was at least double of AtDIC2 (1.01 ± 0.11 mmol min^{-1} g protein^{-1}) and at least seven times compared to that of AtDIC1 (0.29 ± 0.06 mmol min^{-1} g protein^{-1}). While AtDIC3 seems to be the most efficient transporter for malate, its transcript could not be detected in any of the tissues tested [44]. Recent Arabidopsis mitochondrial proteomic surveys support the observation that AtDIC3 does not appear to be highly expressed [43]. By contrast, AtDIC1 (59 protein copies per mitochondria) was found to be slightly more abundant than AtDIC2 (21 protein copies per mitochondria) [43]. However, transcript levels of *AtDIC1* and 2 appear to vary across tissues [44]. The mRNA level of the former was found to be higher in roots and in flowers, while the latter was more abundant in leaves, stems, and seedlings. It should be noted that the precise roles of the AtDIC homologs remain to be investigated *in planta*. Thus, it is still unknown whether these homologs are functionally redundant or if indeed they can transport such a wide array of substrates. Furthermore, K_m values were reported to be in the millimolar range, e.g., 0.4 ± 0.09 mM for AtDIC1 during malate/malate exchange [44] as opposed to the ADP/ATP carrier (AtAAC1) that was shown to be in the millimolar range [45]. The orders of magnitude of the K_m and V_{max} values are similar to ones measured for DIC homologs in *Drosophila* [46] rat and *C. elegans* [47]. We hypothesize that this might be a form of metabolic control, i.e., metabolites are exported/imported only when a certain threshold concentration is achieved. For example, malate synthesized in the mitochondria is not exported as soon as it is formed by the action of DICs, rather, only when a critical concentration is reached would DICs transport the metabolites in excess. The transport process is also rapid as reflected by the V_{max} values consistent with our hypothesis of *relief* from metabolite accumulation.

2.2. Dicarboxylate/Tricarboxylate Carrier (DTC)

DTCs facilitate the transport of dicarboxylates such as malate and 2-OG and tricarboxylates such as citrate [48]. Our phylogenetic analysis revealed that DTC homologs are present in all included plant species and in some fungal species (Figure 1). DTCs and OGCs share a common ancestor. The relative position of the two subgroups indicates that the ancestral gene is closely related to DICs. We hypothesize that the common ancestral gene was duplicated and underwent neofunctionalization upon speciation. The animal copy became OGC while in plants it became DTC. Alternatively, neofunctionalization happened in the last common ancestor and the plant and animal lineages received DTC and OGC, respectively, upon speciation. Hints to support the second hypothesis are provided by the proteomes of some protists. There are three species that contain both OGC-like and DTC-like proteins: *Aureococcus anophagefferens* (Class Pelagophyceae), *Emiliania huxleyi* (Class Prymnesiophyceae), and *Tetrahymena thermophila* (Class Oligohymenophorea). However, other protists possess only OGC-like transporters as in *Leishmania major* and *Trypanosoma brucei* and yet other protists, such as *Plasmodium falciparum*, only have DTC-like transporters.

DTCs are among the most abundant MC proteins in the Arabidopsis IMM comprising 0.8% of the total IMM area, i.e., 6836 protein copies per mitochondria [43]. Unlike the other three highly abundant MC proteins: ADP/ATP carriers (AtAAC1-3; 6.2% of IMM area; 53,065 protein copies/mitochondria); mitochondrial phosphate carriers (AtMPT2-3; 2.5% of IMM area; 21,325 protein copies/mitochondria); and, uncoupling proteins (AtUCP1-3, 1.0% of IMM area; 8595 protein copies/mitochondria), there is only one DTC homolog in Arabidopsis. DTCs have been studied in a few plants including Arabidopsis and *Nicotiana tabacum* [48], *Vitis vinifera* (grapes) [49], *Helianthus tuberosus* (Jerusalem artichoke) [50], and *Citrus junos* (yuzo) [51]. The purification and characterization of a citrate transporter in maize have been described [52]. The reported activity may represent the MzDTC homolog since the transport substrates of the maize citrate transporter closely resemble that of AtDTC. In the plant kingdom, the numbers of DTC homologs vary without any clear pattern. A single homolog was found in *Chlamydomonas reinhardtii*, while there are two and three homologs in the mosses *S. moellendorffii* and *P. patens*, respectively. In the *Brassica* genus, the number of DTC homologs varies from one in *A. thaliana*, Arabidopsis *lyrata*, and *Capsella rubella*, two in *Brassica oleracea*, and three in *Brassica rapa*. *N. tabacum* has four homologs (NtDTC1-4) consistent with that reported in literature [48]. Regalado et al. (2013) [49] reported that there were three DTC homologs in *V. vinifera* (VvDTC1-3) and that VvDTC2 and VvDTC3 reached high transcript levels in the berry mesocarp at the onset of ripening. However, the phylogenetic analysis here showed that only two of these homologs clustered with plant DTCs. Our analysis revealed that VvDTC1 (vitvi_GSVIVT01025463001) is similar to AtDTC. An investigation of VvDTC2 and VvDTC3 revealed that these two proteins have the same locus tag indicating they are likely the same gene, and this would correspond to the second VvDTC observed in our analysis. Finally, a single DTC homolog was cloned from *C. junos* [51] and, similarly, we were able to detect only one DTC homolog in the close relative *Citrus sinensis*.

AtDTC and NtDTCs have been assigned a transport function that involves an obligatory electroneutral exchange of dicarboxylates such as malate and 2-OG and tricarboxylates such as citrate [48]. It was demonstrated that DTCs were able to catalyze homoexchange transport, i.e., dicarboxylate/dicarboxylate and tricarboxylate/tricarboxylate on top of the dicarboxylate/tricarboxylate transport modality. It remains to be seen, however, which of these modalities are relevant *in planta*. It is clear from the in vitro transport data [48] that DTCs are promiscuous in terms of transport substrate. DTCs can transport almost all the intermediates of the TCA cycle except fumarate and succinyl-CoA for which there is no available data. For AtDTC, the homoexchange kinetic constants measured for different substrates in two different pH values showed that regardless of the substrate, the K_m and V_{max} changed as a function of pH. The K_m values generally increased at pH 7, indicating that substrate affinities were decreased; V_{max} values were also decreased at pH 7. These changes in the transport kinetics as a function of pH is critical because it has been shown that in Arabidopsis, the pH of the MM is slightly basic (pH 8.1) and that cytosolic pH is close to neutral, pH 7.3 [53]. It is, however, important to note that these in vitro data were not obtained under conditions representative of physiological conditions. To our knowledge, there are no reports where the external and the internal pH mimic physiological values in DTC transport measurements.

2.3. Succinate/Fumarate Carriers (SFC)

SFCs facilitate the exchange of succinate and fumarate [54]. In plants, SFC is present in all lineages, in the green alga, *Chlamydomonas*, but not in the red alga *Cyanidioschyzon merolae*. Our analysis also indicates that animals do not have the SFC homolog.

The gene encoding SFC was first described in the yeast mutant *arc1* while screening for a mutant unable to utilize ethanol as the sole carbon source [55]. The yeast SFC was shown to transport fumarate, succinate, methylfumarate, 2-OG, and OAA against [^{14}C]oxoglutarate [54]. SFC was further shown to prefer succinate and fumarate as substrates since the presence of either substrate almost completely inhibits fumarate/[^{14}C]oxoglutarate exchange [54]. Catoni et al. (2003) [56] complemented the *sfc* yeast mutant (*arc1*) with an Arabidopsis SFC homolog (AT5G01340) that was 35% similar to the ScSFC gene. This resulted in the re-establishment of the yeast to grow in minimal media with ethanol as the sole carbon source. However, the transport behavior of AtSFC is yet to be described.

In yeast, SFC may play a role in shuttling cytosolic succinate from the glyoxylate cycle [54,57,58]. The transported mitochondrial fumarate can be acted upon by cytosolic fumarase to yield malate that could ultimately be used for gluconeogenesis [54,57]. Thus, SFC may potentially link TCA, glyoxylate cycle, and gluconeogenesis [56]. In plants, the glyoxylate cycle is needed in lipid mobilization especially during the early stages of germination. The β-glucuronidase (GUS) reporter system has been used to show that the SFC promoter was active in the cotyledons, hypocotyls, and root tips consistent with the hypothesized role in early germination [56]. The *AtSFC* promoter was also active in pollen and during in vitro germination of pollen tubes, which is also consistent with upregulation of glyoxylate cycle genes during pollen development. Catoni et al. (2003) [56] also reported that the *AtSFC* promoter is active in specific regions of mature leaves, i.e., patches of veins and trichomes. Moreover, a recent report on the Arabidopsis mitochondrial proteome showed that an average mitochondrion has only 73 SFC protein copies, which are around the same number as DIC1 and DIC2 combined [43]. Based on the TCA flux mode during the day inferred from enzyme kinetic analysis (Figure 2B), SFC is unlikely to play a role in TCA metabolite shuttling. However, TCA flux model based on isotope labeling experiments (Figure 2C) and during the night (Figure 2A) indicated that there is a net efflux of fumarate from the mitochondria. We can hypothesize that perhaps SFC plays a role in this transport process since DICs and DTC are not able to efficiently transport fumarate [44,48]. It has not escaped our attention that there has been no report or model prediction showing net influx of succinate to the mitochondria, which would have been expected as succinate is the preferred substrate of non-plant SFCs. It is thus likely that SFC is using another metabolite as a counter-substrate to facilitate fumarate transport, possibly OAA and 2-oxogulatrate, in congruence with yeast SFC transport assays [54].

3. Mitochondrial Carriers Relevant to Mitochondrial Oxidative Phosphorylation

Oxidative phosphorylation (electron transport-linked phosphorylation) is the route by which ATP is formed as a result of the transfer of electrons from NADH or $FADH_2$ to O_2 by a series of electron carriers in the IMM [59]. The oxidative phosphorylation system is composed of the mitochondrial electron transport chain (ETC) and the ATP synthase complex (Complex V) from bacteria to higher eukaryotes [59]. To drive favorable translocation of charged species across the IMM, the $\Delta\Psi m$ and ΔpH generated through the action of the mitochondrial respiratory electron transfer chain could influence the transporter activity. The electrochemical gradient is ultimately used by the ATP synthase to produce ATP. Therefore, the state of oxidative phosphorylation would be expected to affect the kinetic behavior, specificity, transport orientation, and cooperativity of IMM transporters. There are three transport processes directly or indirectly related to mitochondrial oxidative phosphorylation that is mediated by MCs: (1) transport of NAD^+, (2) proton translocation, and (3) transport of adenine nucleotides (Figure 3).

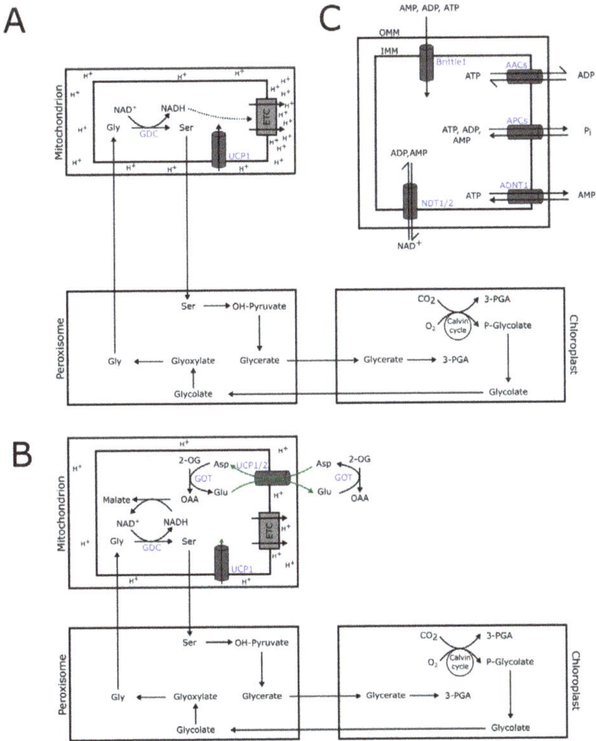

Figure 3. Mitochondrial carriers relevant to mitochondrial oxidative phosphorylation. (**A**) Sweetlove et al. [60] proposed that the function of UCP1 in photorespiration is in the dissipation of protons generated by the increased flux of NADH to the complex 1 of the electron transport chain (ETC). (**B**) Monne et al. [61] showed that UCP1 and 2 mediate aspartate/glutamate exchange. It was proposed that UCP1/2 play a role in dissipating reducing equivalents across the mitochondrial membrane as part of the malate/aspartate shuttle during photorespiration. (**C**) NDT1 and 2 facilitate the transport of NAD^+ from the cytosol in exchange with ADP or AMP from the mitochondria. ADP/ATP carriers, APCs, ADNT1, and AtBrittle1 mediate the transport of adenine nucleotides. UCP1/2: uncoupling protein 1/2, GDC: glycine decarboxylase complex, GOT: glutamic oxaloacetic transaminase, AACs: ADP/ATP carriers, APCs: ATP/Pi carriers, ADNT1: adenine nucleotide transporter 1, NDT1/2: NAD^+ transporter 1/2.

3.1. NAD⁺ Transport

As a coenzyme for redox processes, NAD$^+$ is playing important roles in the operation of a wide range of dehydrogenase activities, signaling pathways through their interaction with reactive oxygen species (ROS) and generation of NADH from oxidative phosphorylation. NAD$^+$ is essential for several metabolic pathways including glycolysis, TCA cycle, glycine decarboxylation, the Calvin–Benson cycle, and β-oxidation in peroxisomes [62]. Hence, the movement of NAD$^+$ from different subcellular compartments is mediated by different subcellular NAD$^+$ transporters. In plants, both de novo and salvage NAD$^+$ biosynthetic pathways culminate in the synthesis of nicotinate mononucleotide (NaMN) [63]. The salvage pathway starts with nicotinamide (NAM) or nicotinic acid (NA), while the de novo pathway starts in plastids using aspartate or tryptophan as precursors. Both metabolic fluxes converge in the formation of nicotinic acid mononucleotide (NAMN), which, in turn, gives rise to NAD$^+$. Since the last step of NAD$^+$ synthesis takes place in the cytosol, NAD$^+$ must be imported into the mitochondria to allow TCA cycle metabolism and oxidative phosphorylation [64]. In Arabidopsis, there are three MCF members responsible for NAD$^+$ transport, AtNDT1 (AT2G47490) and AtNDT2 (AT1G25380), targeted to the IMM, and AtPXN (AT2G39970), located in the peroxisomal membrane [64,65]. Although previous research suggested that AtNDT1 is targeted to the inner membrane of chloroplasts [65], recent subcellular localization experiments and proteomics data revealed that AtNDT1 locates exclusively to the IMM [64]. Both AtNDT1 and AtNAD2 are able to complement the phenotype of a yeast mutant lacking NAD$^+$ transport [65]. Interestingly, both AtNDT1 and AtNDT2 have similar substrate specificity; importing NAD$^+$ against ADP or AMP; they do not accept NADH, nicotinamide, nicotinic acid, NADP$^+$, or NADPH as transport substrates [65]. The AtPXN transporter has a more versatile transport behavior, able to accept NAD$^+$, NADH, and CoA in vitro [66–68]. In addition, as a pyridine nucleotide, NAD$^+$ is involved in the transport of electrons within oxidation–reduction reactions as well as being a highly important component of cellular signaling [63]. Given that the redox status regulates the plant TCA cycle [69], NAD$^+$ import not only provides co-enzymes but might also act as a signal that regulates central metabolism.

3.2. Uncoupling Proteins (UCPs)

Uncoupling proteins (UCPs) are suggested to mediate a non-phosphorylating free fatty acid-activated proton re-entry into the MM leading to a thermogenic dissipation of proton gradients, thereby uncoupling oxidative phosphorylation [61,70]. In animals, the protonophoretic function of UCP1 in brown adipose tissues leading to thermogenesis is of great importance among newborns, cold-acclimated, and hibernating mammals [71,72]. While initial reports in Arabidopsis indicated the presence of 6 UCPs [73], current studies consider the genome to encode three UCP homologs (UCP1-3: AT3G54110, AT5G58970, and AT1G14140) [11]. AtUCP1 was shown to be needed for efficient photosynthesis [60]. During photorespiration, the mitochondrial conversion of glycine to serine by glycine decarboxylase leads to an accumulation of NADH. Subsequently, the NADH is used by MDH or is funneled to the electron transport chain. Since there is a lower demand for mitochondria-derived ATP during the day, photorespiration leads to a substantial increase in the proton gradient by virtue of increased NADH oxidation in Complex I. Without ATP synthesis-coupled proton re-entry into the MM, UCPs may thus provide a mechanism to dissipate proton gradient build-up so as not to restrict electron flow to regenerate NAD$^+$ needed in photorespiration [60]. This hypothesis is supported by the observation that *ucp1* mutants showed a decreased photorespiratory flux from glycine to serine. However, there was no concomitant accumulation of glycine accumulation in *ucp1* mutants. It was suggested that there might be a slight overcompensation in upstream photorespiratory regulatory mechanisms [60]. UCPs are known to curb mitochondrial reactive oxygen species production, which was observed in *ucp1* mutants alongside a significant decrease in the activities of two mitochondrial enzymes, malic enzyme and aconitase, both are particularly sensitive to oxidative inactivation [60]. It should, however, be noted that proton transport in this study was not directly measured. For example, 6-methoxy-*N*-(3-sulfopropyl)-quinolinium (SPQ) can be used to monitor proton flux and has been used for the functional characterization of UCP homologs [74]. Recently, the function of AtUCP1 and 2 has been reported as the transporter of amino acids (aspartate, glutamate, cysteine sulfinate, and cysteate), dicarboxylates (malate, OAA, and 2-OG), phosphate, sulfate, and thiosulfate [61]. The function of AtUCP1 and AtUCP2 is also suggested to catalyze an aspartate out/glutamate in exchange across the mitochondrial membrane and, thereby, contribute to the export of reducing equivalents from the mitochondria in photorespiration [61]. AtUCP1 and AtUCP2 thus have very broad substrate specificities compared to most MCs thus far characterized [61], especially the dicarboxylates of TCA cycle. To reconcile with the earlier proposal regarding the role of AtUCP1 in photorespiration [60], it has been suggested that the role of AtUCP1 and AtUCP2 may be in the glycolate pathway for the shuttling of redox equivalents across the mitochondria as part of the malate/aspartate shuttle [61].

3.3. Adenylate Transporters

In plant mitochondria, a suite of proteins transports various forms of adenine nucleotides. These include ADP/ATP carriers (AAC1-3: AT3G08580, AT5G13490, AT4G28390), adenedine nucleotide transporter (ADNT1: AT4G01100), adenine nucleotide/phosphate carriers (APC1-3: AT5G61810, AT5G51050, AT5G07320), and AtBrittle1 (AT4G32400). Of these, AAC1-3 occupy 6.2% of the total IMM surface area, and together with APCs and ADNT1 comprise 54,114 proteins copies of the total 1,390,777 proteins in the mitochondrial proteome [43].

Our current understanding of MCFs is mostly due to work done on ADP/ATP carriers [20,75]. Nonetheless, it is still surprising that there is only one report on a plant ADP/ATP carrier homolog, which could be explained by their presumed functional redundancies. AtAAC1 and AtAAC2 were functionally expressed and inserted in the *E. coli* cellular membrane and both preferentially export ATP [45]. However, to date, there is no *in planta* work undertaken on this set of plant transporters. The expression of ADP/ATP carriers can vary in response to various stresses as analyzed from transcriptomic datasets [76]. For example, *AtAAC3* is upregulated in roots under osmotic, salt, oxidative, heat, UV-B irradiation, and mechanical stresses. On the other hand, *AtAAC1* expression remained invariant [76]. This suggests differential transcriptional regulation of the various ADP/ATP carriers, especially under stress.

Functional characterization of AtAPCs indicated that unlike their closest human orthologs, these adenylate transporters do not show a preference for Mg-ATP as a substrate [77–79]. AtAPCs were able to accept inorganic phosphate, AMP, ADP, ATP, or adenosine 5′-phosphosulfate as well as ATP in complex with metal divalent ions (Ca^{2+}, Mn^{2+}, Fe^{2+}, and Zn^{2+}) as a transport substrate, and, interestingly, AtAPCs were not completely inhibited by either carboxyatractyloside or bongkrekic acid, both of which are powerful inhibitors of ADP/ATP carriers [22,77,79]. The transport activities of AtAPCs were shown to be Ca^{2+}-dependent, probably due to the presence of a calcium-ion binding motif in the N-terminal [77–79]. Thus, the transporter activities were abolished in the presence of chelating agents such as EDTA and EGTA [77,78]. Since calcium is an important plant signaling molecule, it is possible that mitochondrial energy metabolism under stress is mediated by regulating expression of *AtAPCs* via Ca^{2+}. Transcriptomic datasets have shown that *AtAPC1* is highly induced under drought, salt, or osmotic stress [76].

AtADNT1 was described to facilitate the exchange of AMP and ATP [80]. GUS expression reporter assays showed that the *AtADNT1* promoter was highly active in seedling roots, radicles, cotyledon vascular tissues, and in leaf primordia [80]. Furthermore, lower promoter activity was observed in the leaves of adult plants but higher promoter activity in the vascular tissues of sepals. The physiological role of AtADNT1 is yet to be elucidated. However, it was observed that *adnt1* mutants had significantly reduced root growth rates accompanied by impaired root respiration [80]. It was hypothesized that AtADNT1 is relevant in the mitochondrial uptake of AMP especially during exposure to stress such as hypoxia whereupon AMP tend to accumulate in the cytosol [80]. However, *AtADNT1* transcripts were not markedly increased during hypoxic, cold, heat, drought, salt, or oxidative stresses [76]. It was, however, noted that the expression of *AtADNT1* was upregulated during senescence [76,80]

AtBrittle1 (AtBt1) localizes to the endoplasmic reticulum when transiently expressed in Arabidopsis protoplasts [81] but localized to chloroplasts when transiently expressed in tobacco leaf protoplasts [82]. Finally, when fused with GFP and stably expressed in Arabidopsis, it was found to be dual targeted to both mitochondria and chloroplasts [30]. GUS reporter expression assays showed that *AtBt1* was highly expressed in root tips and germinating pollen [82]. When expressed in intact *E. coli* cells, AtBt1 facilitated a unidirectional uptake of AMP, ADP, and ATP [82]. It was hypothesized that AtBt1 is crucial for shuttling newly synthesized adenylates from the plastids [82]. Homozygous *atbt1* lines are highly compromised, and if they survive they do not produce fertile seeds [82]. This suggests that the physiological role of AtBt1 is not compensated by the other adenylate transporters in its absence. Bahaji et al. (2011) [83] showed that when AtBt1 was targeted specifically to the mitochondria, the construct was able to rescue the growth and sterility phenotypes of *atbt1* mutants, indicating that the mitochondrial role of AtBt1 is important for normal growth and development through a mechanism yet to be understood.

4. Novel Approaches to Investigate MCFs

The most widely used system for studying the transport properties of MCFs is through expression–purification–reconstitution assays [18]. In this approach, the candidate gene is cloned and homologously [84] or heterologously expressed in either *E. coli* [85,86], *Lactococcus lactis* [87], or yeast [85,87]. The overexpressed protein is isolated and reconstituted into liposomes and the transport activity is measured by direct uptake or export of radioactively labeled substrates. Although the assay seems straightforward, its execution is technically challenging. Much of the current knowledge regarding MCF members was gained through the expression–purification–reconstitution assay system. One limitation of the expression–purification–reconstitution assay is the use of radiolabeled substrates. Aside from the usual prohibitive costs, risks, and regulatory hurdles associated with the use of radioactive materials, the range of substrates that can be directly tested is severely hampered by the availability of radioactive substrates. Most MCF research groups circumvent this obstacle by co-incubating the radioactive substrate with the unlabeled candidate substrate and measure the inhibitory effect compared to the assay measurements without the competing substrate. While not a direct evidence of transport, this approach show competition for the binding site. For example, the substrates for 3′-phosphoadenosine 5′-phosphosulfate Transporter 1 and 2 (PASPT1 [85] and PAPST2 [86]) were assessed in this manner. Still, the better way of doing the transport assay is to directly monitor the uptake of the labeled substrate. To overcome issues associated with the use of radioactive substrates, Rautengarten and colleagues [88] have developed a transport assay system that combines liposome reconstitution and mass spectrometry to directly analyze transport properties of nucleotide sugar transporters in Arabidopsis. Theoretically, this system could be used to study the transport behavior of MCF proteins. Monne and colleagues [79] have already demonstrated the use of inductively coupled plasma mass spectrometry to measure divalent ions to characterize APCs from Arabidopsis and humans.

Protein-effector thermostability shift assays were recently introduced to screen for potential substrates. This is based on the principle that physiologically meaningful interactions result in more stable configurations. Mechanistic insights into the interaction of MC proteins with their substrates, inhibitors, and lipids can be evaluated by monitoring their thermostability [89]. While not direct evidence of transport, the results give a strong indication of the binding of a potential substrate, inhibitor, or other effectors. The assay uses a thiol-specific fluorochrome *N*-[4-(7-diethylamino-4-methyl-3-coumarinyl) phenylmaleimide (CPM), which forms fluorescent adducts with cysteine residues in the protein. As the protein is thermally denatured, the buried cysteine residues become available to CPM, which can be taken as a readout for the denaturation process [90]. Thus, shifts in the thermostability of the protein in the presence of an effector molecule can be measured fluorometrically. The approach has been extended to widen the chemical search space for screening potential MCF substrates that would otherwise be largely delimited by using radioactive substrates [91].

While not members of the MCF, recent work on yeast mitochondrial pyruvate carriers (MPCs) showed that by mimicking the physiological pH gradient between the mitochondria and the cytosol has resulted in quantifiable pyruvate transport [92]. In the absence of the pH gradient, no transport can be detected [92]. This could also be true among MCF members, and, thus, it may be worthwhile to run the MCF transport assays under physiological pH conditions. Perhaps the transporter activity of the MCFs proteins can also be modulated by forming complexes with interacting partner/s. The glutathione transport function of OGC could not be replicated using the typical expression–purification–reconstitution assay approach. An anti-apoptotic protein, Bcl-2, was found to be an interacting protein partner of rat OGC and when co-expressed with OGC in CHO cells, the total mitochondrial glutathione content was significantly increased 24 h post-transfection [93].

5. Perspectives

The transport activities of many plant MCs have been characterized in vitro [11,12]; however, most are yet to be interrogated *in planta*. Specifically, transporters such as DICs, SFC, and DTC, which have been shown to transport TCA cycle intermediates in vitro, have not had their physiological functions clearly elucidated *in planta*. A common feature is their broad substrate specificities in vitro. Whether a more stringent gating mechanism exists *in planta* is one of the more interesting questions. Thus far, there is no experimental evidence regarding how these transporters might impact cellular respiration, carbon metabolism, or cellular redox poise. For example, 2-OG shuttling is critical as it integrates carbon and nitrogen metabolism [39,40], while data indicates that some MCs interact with TCA cycle enzymes [31,32]. To date, the physiological significance of these potential protein–protein interactions is yet to be elucidated. Perhaps, these MCs participate in the formation of TCA metabolon serving as membrane anchors. Data from yeast MPCs suggested that formation of complexes modulate transporter activity. Whether canonical MCF members' activity can also be modulated by the formation of protein complexes and by extension, impose a stricter substrate specificity is an attractive area of research. Recently, it was shown that the transport activity of the human CiC seemed to be modulated via acetylation of a lysine residue in response to glucose supply [94]. Moreover, the transport activity of the rat carnitine/acylcarnitine carrier was shown to be modulated by glutathionylation [95]. It will be interesting to ascertain how other post-translational modifications might modulate MC transport activity and whether this can also be observed *in planta*. More recently, it was shown the circadian protein CLOCK (Circadian Locomotor Output Cycles Kaput) was found to bind the human and mouse DIC, suggesting that a possible connection between circadian rhythm and mitochondrial metabolism may be mediated by dicarboxylate carriers [96]. To date, there have been no such reports on the connection between plant circadian rhythm and plant MCs and how this might be relevant in the diurnal regulation of plant metabolism. While these questions are no doubt interesting, the major challenge in our understanding of the function of these carriers is presented by their apparent functional redundancy. This feature renders it difficult to tease apart their physiological function—the broader adoption of CRISPR/Cas9 based approaches in plants [97] and natural variance screening methods [98] may address this challenge. That said, we still lack biochemical data for many of the members of the plant MCF. It is likely that only as a result of data coming from multiple approaches, we will be able to fully comprehend their physiological importance in the regulation of plant central carbon metabolism.

Author Contributions: M.R.T., Y.Z. and A.R.F. conceived the topic and wrote the manuscript. T.N. and M.R.T. did the phylogenetic analysis. T.N., J.L.H., and A.R.F. edited and provided valuable comments to improve the manuscript. All authors have read and agreed to the published version of the manuscript.

Funding: M.R.T. is supported by the Melbourne-Potsdam PhD Program (MelPoPP). T.N. is supported by the International Max Planck Research School 'Primary Metabolism and Plant Growth' (IMPRS-PMPG) PhD program. A.R.F. and Y.Z. would like to thank the European Union's Horizon 2020 research and innovation program, project PlantaSYST (SGA-CSA No. 739582 under FPA No. 664620) for supporting their research. J.L.H. is supported by an Australian Research Council Discovery Project [DP180102630].

Conflicts of Interest: The authors declare no conflict of interest.

References

1. Gray, M.W.; Burger, G.; Lang, B.F. The origin and early evolution of mitochondria. *Genome Biol.* **2001**, *2*, reviews1018. [CrossRef] [PubMed]
2. Sweetlove, L.J.; Fernie, A.R. The spatial organization of metabolism within the plant cell. *Annu. Rev. Plant Biol.* **2013**, *64*, 723–746. [CrossRef] [PubMed]
3. Monné, M.; Vozza, A.; Lasorsa, F.M.; Porcelli, V.; Palmieri, F. Mitochondrial carriers for aspartate, glutamate and other amino acids: A review. *Int. J. Mol. Sci.* **2019**, *20*, 4456. [CrossRef] [PubMed]
4. Gueguen, V.; Macherel, D.; Jaquinod, M.; Douce, R.; Bourguignon, J. Fatty acid and lipoic acid biosynthesis in higher plant mitochondria. *J. Biol. Chem.* **2000**, *275*, 5016–5025. [CrossRef] [PubMed]
5. Douce, R.; Neuburger, M. Biochemical dissection of photorespiration. *Curr. Opin. Plant Biol.* **1999**, *2*, 214–222. [CrossRef]
6. Edwards, G.E.; Franceschi, V.R.; Ku, M.S.B.; Voznesenskaya, E.V.; Pyankov, V.I.; Andreo, C.S. Compartmentation of photosynthesis in cells and tissues of C4 plants. *J. Exp. Bot.* **2001**, *52*, 577–590. [CrossRef]
7. Demine, S.; Reddy, N.; Renard, P.; Raes, M.; Arnould, T. Unraveling biochemical pathways affected by mitochondrial dysfunctions using metabolomic approaches. *Metabolites* **2014**, *4*, 831–878. [CrossRef]
8. Bayrhuber, M.; Meins, T.; Habeck, M.; Becker, S.; Giller, K.; Villinger, S.; Vonrhein, C.; Griesinger, C.; Zweckstetter, M.; Zeth, K. Structure of the human voltage-dependent anion channel. *Proc. Natl. Acad. Sci. USA* **2008**, *105*, 15370–15375. [CrossRef]
9. Nury, H.; Dahout-Gonzalez, C.; Trézéguet, V.; Lauquin, G.J.M.; Brandolin, G.; Pebay-Peyroula, E. Relations between structure and function of the mitochondrial ADP/ATP carrier. *Annu. Rev. Biochem.* **2006**, *75*, 713–741. [CrossRef]
10. Kühlbrandt, W. Structure and function of mitochondrial membrane protein complexes. *BMC Biol.* **2015**, *13*, 89. [CrossRef]
11. Palmieri, F.; Pierri, C.L.; De Grassi, A.; Nunes-Nesi, A.; Fernie, A.R. Evolution, structure and function of mitochondrial carriers: A review with new insights. *Plant J.* **2011**, *66*, 161–181. [CrossRef] [PubMed]
12. Haferkamp, I.; Schmitz-Esser, S. The plant mitochondrial carrier family: Functional and evolutionary aspects. *Front. Plant Sci.* **2012**, *3*, 2. [CrossRef] [PubMed]
13. Lee, C.P.; Millar, A.H. The plant mitochondrial transportome: Balancing metabolic demands with energetic constraints. *Trends Plant Sci.* **2016**, *21*, 662–676. [CrossRef] [PubMed]
14. Palmieri, F. The mitochondrial transporter family SLC25: Identification, properties and physiopathology. *Mol. Asp. Med.* **2013**, *34*, 465–484. [CrossRef]
15. Saraste, M.; Walker, J.E. Internal sequence repeats and the path of polypeptide in mitochondrial ADP/ATP translocase. *FEBS Lett.* **1982**, *144*, 250–254. [CrossRef]
16. Capobianco, L.; Bisaccia, F.; Michel, A.; Sluse, F.E.; Palmieri, F. The N- and C-termini of the tricarboxylate carrier are exposed to the cytoplasmic side of the inner mitochondrial membrane. *FEBS Lett.* **1995**, *357*, 297–300. [CrossRef]
17. Pebay-Peyroula, E.; Dahout-Gonzalez, C.; Kahn, R.; Trézéguet, V.; Lauquin, G.J.M.; Brandolin, G. Structure of mitochondrial ADP/ATP carrier in complex with carboxyatractyloside. *Nature* **2003**, *426*, 39–44. [CrossRef]
18. Palmieri, F.; Monné, M. Discoveries, metabolic roles and diseases of mitochondrial carriers: A review. *BBA-Mol. Cell Res.* **2016**, *1863*, 2362–2378. [CrossRef]
19. Robinson, A.J.; Overy, C.; Kunji, E.R.S. The mechanism of transport by mitochondrial carriers based on analysis of symmetry. *Proc. Natl. Acad. Sci. USA* **2008**, *105*, 17766–17771. [CrossRef]
20. Ruprecht, J.J.; King, M.S.; Zögg, T.; Aleksandrova, A.A.; Pardon, E.; Crichton, P.G.; Steyaert, J.; Kunji, E.R.S. The molecular mechanism of transport by the mitochondrial ADP/ATP carrier. *Cell* **2019**, *176*, 435–447.e415. [CrossRef]
21. Ruprecht, J.J.; Kunji, E.R.S. The SLC25 mitochondrial carrier family: Structure and mechanism. *Trends Biochem. Sci.* **2019**. [CrossRef] [PubMed]
22. Klingenberg, M. The ADP and ATP transport in mitochondria and its carrier. *BBA-Biomembranes* **2008**, *1778*, 1978–2021. [CrossRef] [PubMed]
23. Klingenberg, M. The ADP,ATP shuttle of the mitochondrion. *Trends Biochem. Sci.* **1979**, *4*, 249–252. [CrossRef]

24. Indiveri, C.; Tonazzi, A.; Palmieri, F. The reconstituted carnitine carrier from rat liver mitochondria: Evidence for a transport mechanism different from that of the other mitochondrial translocators. *BBA-Biomembranes* **1994**, *1189*, 65–73. [CrossRef]

25. Ruprecht, J.J.; Hellawell, A.M.; Harding, M.; Crichton, P.G.; McCoy, A.J.; Kunji, E.R. Structures of yeast mitochondrial ADP/ATP carriers support a domain-based alternating-access transport mechanism. *Proc. Natl. Acad. Sci. USA* **2014**, *111*, E426–E434. [CrossRef]

26. Pietropaolo, A.; Pierri, C.L.; Palmieri, F.; Klingenberg, M. The switching mechanism of the mitochondrial ADP/ATP carrier explored by free-energy landscapes. *BBA-Bioenergetics* **2016**, *1857*, 772–781. [CrossRef]

27. Springett, R.; King, M.S.; Crichton, P.G.; Kunji, E.R.S. Modelling the free energy profile of the mitochondrial ADP/ATP carrier. *BBA-Bioenergetics* **2017**, *1858*, 906–914. [CrossRef]

28. Ruprecht, J.J.; Kunji, E.R. Structural changes in the transport cycle of the mitochondrial ADP/ATP carrier. *Curr. Opin. Struc. Biol.* **2019**, *57*, 135–144. [CrossRef]

29. Leroch, M.; Neuhaus, H.E.; Kirchberger, S.; Zimmermann, S.; Melzer, M.; Gerhold, J.; Tjaden, J. Identification of a novel adenine nucleotide transporter in the endoplasmic reticulum of Arabidopsis. *Plant Cell* **2008**, *20*, 438–451. [CrossRef]

30. Bahaji, A.; Ovecka, M.; Bárány, I.; Risueño, M.C.; Muñoz, F.J.; Baroja-Fernández, E.; Montero, M.; Li, J.; Hidalgo, M.; Sesma, M.T.; et al. Dual targeting to mitochondria and plastids of AtBT1 and ZmBT1, two members of the mitochondrial carrier family. *Plant Cell Physiol.* **2011**, *52*, 597–609. [CrossRef]

31. Zhang, Y.; Swart, C.; Alseekh, S.; Scossa, F.; Jiang, L.; Obata, T.; Graf, A.; Fernie, A.R. The extra-pathway interactome of the TCA cycle: Expected and unexpected metabolic interactions. *Plant Physiol.* **2018**, *177*, 966–979. [CrossRef] [PubMed]

32. Zhang, Y.; Beard, K.F.M.; Swart, C.; Bergmann, S.; Krahnert, I.; Nikoloski, Z.; Graf, A.; George Ratcliffe, R.; Sweetlove, L.J.; Fernie, A.R.; et al. Protein-protein interactions and metabolite channelling in the plant tricarboxylic acid cycle. *Nat. Commun.* **2017**, *8*, 15212. [CrossRef] [PubMed]

33. Edgar, R.C. MUSCLE: A multiple sequence alignment method with reduced time and space complexity. *BMC Bioinform.* **2004**, *5*, 113. [CrossRef] [PubMed]

34. Finn, R.D.; Clements, J.; Eddy, S.R. HMMER web server: Interactive sequence similarity searching. *Nucleic Acids Res.* **2011**, *39*, W29–W37. [CrossRef] [PubMed]

35. Stamatakis, A. RAxML version 8: A tool for phylogenetic analysis and post-analysis of large phylogenies. *Bioinformatics* **2014**, *30*, 1312–1313. [CrossRef] [PubMed]

36. Lemoine, F.; Domelevo Entfellner, J.B.; Wilkinson, E.; Correia, D.; Dávila Felipe, M.; De Oliveira, T.; Gascuel, O. Renewing Felsenstein's phylogenetic bootstrap in the era of big data. *Nature* **2018**, *556*, 452–456. [CrossRef]

37. Sweetlove, L.J.; Beard, K.F.M.; Nunes-Nesi, A.; Fernie, A.R.; Ratcliffe, R.G. Not just a circle: Flux modes in the plant TCA cycle. *Trends Plant Sci.* **2010**, *15*, 462–470. [CrossRef]

38. Bricker, D.K.; Taylor, E.B.; Schell, J.C.; Orsak, T.; Boutron, A.; Chen, Y.-C.; Cox, J.E.; Cardon, C.M.; Vraken, J.G.V.; Dephoure, N.; et al. A mitochondrial pyruvate carrier required for pyruvate uptake in yeast, *Drosophila* and humans. *Science* **2012**, *337*, 96–100. [CrossRef]

39. Zhang, Y.; Fernie, A.R. On the role of the tricarboxylic acid cycle in plant productivity. *J. Integr. Plant Biol.* **2018**, *60*, 1199–1216. [CrossRef]

40. Fernie, A.R.; Zhang, Y.; Sweetlove, L.J. Passing the baton: Substrate channelling in respiratory metabolism. *Research* **2018**, *2018*, 1–16. [CrossRef]

41. Steuer, R.; Nesi, A.N.; Fernie, A.R.; Gross, T.; Blasius, B.; Selbig, J. From structure to dynamics of metabolic pathways: Application to the plant mitochondrial TCA cycle. *Bioinformatics* **2007**, *23*, 1378–1385. [CrossRef] [PubMed]

42. Tcherkez, G. In vivo respiratory metabolism of illuminated leaves. *Plant Physiol.* **2005**, *138*, 1596–1606. [CrossRef] [PubMed]

43. Fuchs, P.; Rugen, N.; Carrie, C.; Elsässer, M.; Finkemeier, I.; Giese, J.; Hildebrandt, T.M.; Kühn, K.; Maurino, V.G.; Ruberti, C.; et al. Single organelle function and organization as estimated from Arabidopsis mitochondrial proteomics. *Plant J.* **2019**. [CrossRef] [PubMed]

44. Palmieri, L.; Picault, N.; Arrigoni, R.; Besin, E.; Palmieri, F.; Hodges, M. Molecular identification of three *arabidopsis thaliana* mitochondrial dicarboxylate carrier isoforms: Organ distribution, bacterial expression, reconstitution into liposomes and functional characterization. *Biochem. J.* **2008**, *410*, 621–629. [CrossRef] [PubMed]

45. Haferkamp, I.; Hackstein, J.H.P.; Voncken, F.G.J.; Schmit, G.; Tjaden, J. Functional integration of mitochondrial and hydrogenosomal ADP/ATP carriers in the *Escherichia coli* membrane reveals different biochemical characteristics for plants, mammals and anaerobic chytrids. *Eur. J. Biochem.* **2002**, *269*, 3172–3181. [CrossRef]

46. Iacopetta, D.; Madeo, M.; Tasco, G.; Carrisi, C.; Curcio, R.; Martello, E.; Casadio, R.; Capobianco, L.; Dolce, V. A novel subfamily of mitochondrial dicarboxylate carriers from *Drosophila melanogaster*: Biochemical and computational studies. *BBA-Bioenergetics* **2011**, *1807*, 251–261. [CrossRef]

47. Fiermonte, G.; Palmieri, L.; Dolce, V.; Lasorsa, F.M.; Palmieri, F.; Runswick, M.J.; Walker, J.E. The sequence, bacterial expression, and functional reconstitution of the rat mitochondrial dicarboxylate transporter cloned via distant homologs in yeast and *Caenorhabditis elegans*. *J. Biol. Chem.* **1998**, *273*, 24754–24759. [CrossRef]

48. Picault, N.; Palmieri, L.; Pisano, I.; Hodges, M.; Palmieri, F. Identification of a novel transporter for dicarboxylates and tricarboxylates in plant mitochondria: Bacterial expression, reconstitution, functional characterization, and tissue distribution. *J. Biol. Chem.* **2002**, *277*, 24204–24211. [CrossRef]

49. Regalado, A.; Pierri, C.L.; Bitetto, M.; Laera, V.L.; Pimentel, C.; Francisco, R.; Passarinho, J.; Chaves, M.M.; Agrimi, G. Characterization of mitochondrial dicarboxylate/tricarboxylate transporters from grape berries. *Planta* **2013**, *237*, 693–703. [CrossRef]

50. Spagnoletta, A.; Santis, A.D.; Tampieri, E.; Baraldi, E.; Bachi, A.; Genchi, G. Identification and kinetic characterization of HtDTC, the mitochondrial dicarboxylate-tricarboxylate carrier of Jerusalem artichoke tubers. *J. Bioenerg. Biomembr.* **2006**, *38*, 57–65. [CrossRef]

51. Deng, W.; Luo, K.; Li, Z.; Yang, Y. Molecular cloning and characterization of a mitochondrial dicarboxylate/tricarboxylate transporter gene in *Citrus junos* response to aluminum stress. *Mitochondrial DNA* **2008**, *19*, 376–384. [CrossRef]

52. Genchi, G.; Spagnoletta, A.; De Santis, A.; Stefanizzi, L.; Palmieri, F. Purification and characterization of the reconstitutively active citrate carrier from maize mitochondria. *Plant Physiol.* **1999**, *120*, 841–847. [CrossRef] [PubMed]

53. Shen, J.; Zeng, Y.; Zhuang, X.; Sun, L.; Yao, X.; Pimpl, P.; Jiang, L. Organelle pH in the Arabidopsis endomembrane system. *Mol. Plant* **2013**, *6*, 1419–1437. [CrossRef] [PubMed]

54. Palmieri, L.; Lasorsa, F.M.; De Palma, A.; Palmieri, F.; Runswick, M.J.; Walker, J.E. Identification of the yeast ACR1 gene product as a succinate-fumarate transporter essential for growth on ethanol or acetate. *FEBS Lett.* **1997**, *417*, 114–118. [CrossRef]

55. Fernández, M.; Fernández, E.; Rodicio, R. ACR1, a gene encoding a protein related to mitochondrial carriers, is essential for acetyl-CoA synthetase activity in *Saccharomyces cerevisiae*. *Mol. Gen. Genet.* **1994**, *242*, 727–735. [CrossRef] [PubMed]

56. Catoni, E.; Schwab, R.; Hilpert, M.; Desimone, M.; Schwacke, R.; Flügge, U.I.; Schumacher, K.; Frommer, W.B. Identification of an Arabidopsis mitochondrial succinate-fumarate translocator. *FEBS Lett.* **2003**, *534*, 87–92. [CrossRef]

57. Kunze, M.; Pracharoenwattana, I.; Smith, S.M.; Hartig, A. A central role for the peroxisomal membrane in glyoxylate cycle function. *BBA-Mol. Cell Res.* **2006**, *1763*, 1441–1452. [CrossRef]

58. Lee, Y.J.; Jang, J.W.; Kim, K.J.; Maeng, P.J. TCA cycle-independent acetate metabolism via the glyoxylate cycle in *Saccharomyces cerevisiae*. *Yeast* **2011**, *28*, 153–166. [CrossRef]

59. Meyer, E.H.; Welchen, E.; Carrie, C. Assembly of the complexes of the oxidative phosphorylation system in land plant mitochondria. *Annu. Rev. Plant Biol.* **2019**, *70*, 23–50. [CrossRef]

60. Sweetlove, L.J.; Lytovchenko, A.; Morgan, M.; Nunes-Nesi, A.; Taylor, N.L.; Baxter, C.J.; Eickmeier, I.; Fernie, A.R. Mitochondrial uncoupling protein is required for efficient photosynthesis. *Proc. Natl. Acad. Sci. USA* **2006**, *103*, 19587–19592. [CrossRef]

61. Monné, M.; Daddabbo, L.; Gagneul, D.; Obata, T.; Hielscher, B.; Palmieri, L.; Miniero, D.V.; Fernie, A.R.; Weber, A.P.M.; Palmieri, F. Uncoupling proteins 1 and 2 (UCP1 and UCP2) from *Arabidopsis thaliana* are mitochondrial transporters of aspartate, glutamate, and dicarboxylates. *J. Biol. Chem.* **2018**, *293*, 4213–4227. [CrossRef]

62. Geigenberger, P.; Fernie, A.R. Metabolic control of redox and redox control of metabolism in plants. *Antioxid. Redox Signal.* **2014**, *21*, 1389–1421. [CrossRef]

63. Gakière, B.; Hao, J.; de Bont, L.; Pétriacq, P.; Nunes-Nesi, A.; Fernie, A.R. NAD$^+$ biosynthesis and signaling in plants. *Crit. Rev. Plant Sci.* **2018**, *37*, 259–307. [CrossRef]

64. De Souza Chaves, I.; Araújo, E.F.; Florian, A.; Medeiros, D.B.; da Fonseca-Pereira, P.; Charton, L.; Heyneke, E.; Apfata, J.A.C.; Pires, M.V.; Mettler-Altmann, T. Mitochondrial NAD$^+$ transporter (NDT1) plays important roles in cellular NAD$^+$ homeostasis in *Arabidopsis thaliana*. *Plant J.* **2019**, *100*, 487–504. [CrossRef]

65. Palmieri, F.; Rieder, B.; Ventrella, A.; Blanco, E.; Do, P.T.; Nunes-Nesi, A.; Trauth, A.U.; Fiermonte, G.; Tjaden, J.; Agrimi, G.; et al. Molecular identification and functional characterization of *Arabidopsis thaliana* mitochondrial and chloroplastic NAD$^+$ carrier proteins. *J. Biol. Chem.* **2009**, *284*, 31249–31259. [CrossRef] [PubMed]

66. Van Roermund, C.W.; Schroers, M.G.; Wiese, J.; Facchinelli, F.; Kurz, S.; Wilkinson, S.; Charton, L.; Wanders, R.J.; Waterham, H.R.; Weber, A.P. The peroxisomal NAD carrier from Arabidopsis imports NAD in exchange with AMP. *Plant Physiol.* **2016**, *171*, 2127–2139. [CrossRef] [PubMed]

67. Agrimi, G.; Russo, A.; Pierri, C.L.; Palmieri, F. The peroxisomal NAD$^+$ carrier of *Arabidopsis thaliana* transports coenzyme A and its derivatives. *J. Bioenerg. Biomembr.* **2012**, *44*, 333–340. [CrossRef] [PubMed]

68. Bernhardt, K.; Wilkinson, S.; Weber, A.P.; Linka, N. A peroxisomal carrier delivers NAD$^+$ and contributes to optimal fatty acid degradation during storage oil mobilization. *Plant J.* **2012**, *69*, 1–13. [CrossRef] [PubMed]

69. Daloso, D.M.; Müller, K.; Obata, T.; Florian, A.; Tohge, T.; Bottcher, A.; Riondet, C.; Bariat, L.; Carrari, F.; Nunes-Nesi, A.; et al. Thioredoxin, a master regulator of the tricarboxylic acid cycle in plant mitochondria. *Proc. Natl. Acad. Sci. USA* **2015**, *112*, E1392–E1400. [CrossRef]

70. Ježek, P.; Holendová, B.; Garlid, K.D.; Jabůrek, M. Mitochondrial uncoupling proteins: Subtle regulators of cellular redox signaling. *Antioxid. Redox Signal.* **2018**, *29*, 667–714. [CrossRef]

71. Ježek, P.; Jabůrek, M.; Porter, R.K. Uncoupling mechanism and redox regulation of mitochondrial uncoupling protein 1 (UCP1). *BBA-Bioenergetics* **2019**, *1860*, 259–269. [CrossRef] [PubMed]

72. Echtay, K.S. Mitochondrial uncoupling proteins-What is their physiological role? *Free Radic. Biol. Med.* **2007**, *43*, 1351–1371. [CrossRef] [PubMed]

73. Vercesi, A.E.; Borecky, J.; Maia Ide, G.; Arruda, P.; Cuccovia, I.M.; Chaimovich, H. Plant uncoupling mitochondrial proteins. *Annu. Rev. Plant Biol.* **2006**, *57*, 383–404. [CrossRef] [PubMed]

74. Jabůrek, M.; Vařecha, M.; Gimeno, R.E.; Dembski, M.; Ježek, P.; Zhang, M.; Burn, P.; Tartaglia, L.A.; Garlid, K.D. Transport function and regulation of mitochondrial uncoupling proteins 2 and 3. *J. Biol. Chem.* **1999**, *274*, 26003–26007. [CrossRef]

75. Kunji, E.R.S.; Aleksandrova, A.; King, M.S.; Majd, H.; Ashton, V.L.; Cerson, E.; Springett, R.; Kibalchenko, M.; Tavoulari, S.; Crichton, P.G.; et al. The transport mechanism of the mitochondrial ADP/ATP carrier. *BBA-Mol. Cell Res.* **2016**, *1863*, 2379–2393. [CrossRef]

76. Da Fonseca-Pereira, P.; Neri-Silva, R.; Cavalcanti, J.H.F.; Brito, D.S.; Weber, A.P.M.; Araújo, W.L.; Nunes-Nesi, A. Data-mining bioinformatics: Connecting adenylate transport and metabolic responses to stress. *Trends Plant Sci.* **2018**, *23*, 961–974. [CrossRef]

77. Monné, M.; Miniero, D.V.; Obata, T.; Daddabbo, L.; Palmieri, L.; Vozza, A.; Nicolardi, M.C.; Fernie, A.R.; Palmieri, F. Functional characterization and organ distribution of three mitochondrial ATP–Mg/Pi carriers in *Arabidopsis thaliana*. *BBA-Bioenergetics* **2015**, *1847*, 1220–1230. [CrossRef]

78. Lorenz, A.; Lorenz, M.; Vothknecht, U.C.; Niopek-Witz, S.; Neuhaus, H.E.; Haferkamp, I. In vitro analyses of mitochondrial ATP/phosphate carriers from *Arabidopsis thaliana* revealed unexpected Ca^{2+}-effects. *BMC Plant Biol.* **2015**, *15*, 238. [CrossRef]

79. Monne, M.; Daddabbo, L.; Giannossa, L.C.; Nicolardi, M.C.; Palmieri, L.; Miniero, D.V.; Mangone, A.; Palmieri, F. Mitochondrial ATP-Mg/phosphate carriers transport divalent inorganic cations in complex with ATP. *J. Bioenerg. Biomembr.* **2017**, *49*, 369–380. [CrossRef]

80. Iacobazzi, V.; Infantino, V.; Palmieri, F. Epigenetic mechanisms and Sp1 regulate mitochondrial citrate carrier gene expression. *Biochem. Biophys. Res. Commun.* **2008**, *376*, 15–20. [CrossRef]

81. Inan, G.; Goto, F.; Jin, J.B.; Rosado, A.; Koiwa, H.; Shi, H.; Hasegawa, P.M.; Bressan, R.A.; Maggio, A.; Li, X. Isolation and characterization of *shs1*, a sugar-hypersensitive and ABA-insensitive mutant with multiple stress responses. *Plant Mol. Biol.* **2007**, *65*, 295–309. [CrossRef] [PubMed]

82. Kirchberger, S.; Tjaden, J.; Ekkchard Neuhaus, H. Characterization of the Arabidopsis Brittle1 transport protein and impact of reduced activity on plant metabolism. *Plant J.* **2008**, *56*, 51–63. [CrossRef] [PubMed]

83. Bahaji, A.; Muñoz, F.J.; Ovecka, M.; Baroja-Fernández, E.; Montero, M.; Li, J.; Hidalgo, M.; Almagro, G.; Sesma, M.T.; Ezquer, I.; et al. Specific delivery of AtBT1 to mitochondria complements the aberrant growth and sterility phenotype of homozygous AtBT1 Arabidopsis mutants. *Plant J.* **2011**, *68*, 1115–1121. [CrossRef] [PubMed]

84. Kadooka, C.; Izumitsu, K.; Onoue, M.; Okutsu, K.; Yoshizaki, Y.; Takamine, K.; Goto, M.; Tamaki, H.; Futagami, T. Mitochondrial citrate transporters CtpA and YhmA are required for extracellular citric acid accumulation and contribute to cytosolic acetyl coenzyme A generation in *Aspergillus luchuensis* mut. *kawachii*. *Appl. Environ. Microbiol.* **2019**, *85*, e03136-18. [CrossRef]

85. Gigolashvili, T.; Geier, M.; Ashykhmina, N.; Frerigmann, H.; Wulfert, S.; Krueger, S.; Mugfor, S.G.; Kopriv, S.; Haferkamp, I.; Flügge, U.I. The Arabidopsis thylakoid ADP/ATP carrier TAAC Has an additional role in supplying plastidic phosphoadenosine 5′-phosphosulfate to the cytosol. *Plant Cell* **2012**, *24*, 4187–4204. [CrossRef]

86. Ashykhmina, N.; Lorenz, M.; Frerigmann, H.; Koprivova, A.; Hofsetz, E.; Stührwohldt, N.; Flügge, U.I.; Haferkamp, I.; Kopriva, S.; Gigolashvilia, T. PAPST2 plays critical roles in removing the stress signaling molecule 3′-phosphoadenosine 5′-phosphate from the cytosol and its subsequent degradation in plastids and mitochondria. *Plant Cell* **2019**, *31*, 231–249. [CrossRef]

87. Boulet, A.; Vest, K.E.; Maynard, M.K.; Gammon, M.G.; Russell, A.C.; Mathews, A.T.; Cole, S.E.; Zhu, X.; Phillips, C.B.; Kwong, J.Q.; et al. The mammalian phosphate carrier SLC25A3 is a mitochondrial copper transporter required for cytochrome c oxidase biogenesis. *J. Biol. Chem.* **2018**, *293*, 1887–1896. [CrossRef]

88. Rautengarten, C.; Ebert, B.; Moreno, I.; Temple, H.; Herter, T.; Link, B.; Donas-Cofre, D.; Moreno, A.; Saez-Aguayo, S.; Blanco, F.; et al. The Golgi localized bifunctional UDP-rhamnose/UDP-galactose transporter family of Arabidopsis. *Proc. Natl. Acad. Sci. USA* **2014**, *111*, 11563–11568. [CrossRef]

89. Crichton, P.G.; Lee, Y.; Ruprecht, J.J.; Cerson, E.; Thangaratnarajah, C.; King, M.S.; Kunji, E.R.S. Trends in thermostability provide information on the nature of substrate, inhibitor, and lipid interactions with mitochondrial carriers. *J. Biol. Chem.* **2015**, *290*, 8206–8217. [CrossRef]

90. Alexandrov, A.I.; Mileni, M.; Chien, E.Y.T.; Hanson, M.A.; Stevens, R.C. Microscale fluorescent thermal stability assay for membrane proteins. *Structure* **2008**, *16*, 351–359. [CrossRef]

91. Majd, H.; King, M.S.; Palmer, S.M.; Smith, A.C.; Elbourne, L.D.H.; Paulsen, I.T.; Sharples, D.; Henderson, P.J.F.; Kunji, E.R.S. Screening of candidate substrates and coupling ions of transporters by thermostability shift assays. *eLife* **2018**, *7*, e38821. [CrossRef]

92. Tavoulari, S.; Thangaratnarajah, C.; Mavridou, V.; Harbour, M.E.; Martinou, J.C.; Kunji, E.R. The yeast mitochondrial pyruvate carrier is a hetero-dimer in its functional state. *EMBO J.* **2019**, *38*, 1–13. [CrossRef] [PubMed]

93. Wilkins, H.M.; Marquardt, K.; Lash, L.H.; Linseman, D.A. Bcl-2 is a novel interacting partner for the 2-oxoglutarate carrier and a key regulator of mitochondrial glutathione. *Free Radic. Biol. Med.* **2012**, *52*, 410–419. [CrossRef] [PubMed]

94. Palmieri, E.M.; Spera, I.; Menga, A.; Infantino, V.; Porcelli, V.; Iacobazzi, V.; Pierri, C.L.; Hooper, D.C.; Palmieri, F.; Castegna, A. Acetylation of human mitochondrial citrate carrier modulates mitochondrial citrate/malate exchange activity to sustain NADPH production during macrophage activation. *BBA-Bioenergetics* **2015**, *1847*, 729–738. [CrossRef]

95. Giangregorio, N.; Palmieri, F.; Indiveri, C. Glutathione controls the redox state of the mitochondrial carnitine/acylcarnitine carrier Cys residues by glutathionylation. *BBA-Gen. Subj.* **2013**, *1830*, 5299–5304. [CrossRef]

96. Cai, T.; Hua, B.; Luo, D.; Xu, L.; Cheng, Q.; Yuan, G.; Yan, Z.; Sun, N.; Hua, L.; Lu, C. The circadian protein CLOCK regulates cell metabolism via the mitochondrial carrier SLC25A10. *BBA-Mol. Cell Res.* **2019**, *1866*, 1310–1321. [CrossRef]

97. Chen, K.; Wang, Y.; Zhang, R.; Zhang, H.; Gao, C. CRISPR/Cas genome editing and precision plant breeding in agriculture. *Annu. Rev. Plant Biol.* **2019**, *70*, 667–697. [CrossRef]

98. Sew, Y.S.; Stroher, E.; Holzmann, C.; Huang, S.; Taylor, N.L.; Jordana, X.; Millar, A.H. Multiplex micro-respiratory measurements of *Arabidopsis* tissues. *New Phytol.* **2013**, *200*, 922–932. [CrossRef]

Review

Ascorbate and Thiamin: Metabolic Modulators in Plant Acclimation Responses

Laise Rosado-Souza, Alisdair R. Fernie * and Fayezeh Aarabi *

Max-Planck-Institut für Molekulare Pflanzenphysiologie, Am Mühlenberg 1, 14476 Potsdam-Golm, Germany;
Rosado@mpimp-golm.mpg.de
* Correspondence: Fernie@mpimp-golm.mpg.de (A.R.F.); Arabi@mpimp-golm.mpg.de (F.A.)

Received: 20 December 2019; Accepted: 10 January 2020; Published: 13 January 2020

Abstract: Cell compartmentalization allows incompatible chemical reactions and localised responses to occur simultaneously, however, it also requires a complex system of communication between compartments in order to maintain the functionality of vital processes. It is clear that multiple such signals must exist, yet little is known about the identity of the key players orchestrating these interactions or about the role in the coordination of other processes. Mitochondria and chloroplasts have a considerable number of metabolites in common and are interdependent at multiple levels. Therefore, metabolites represent strong candidates as communicators between these organelles. In this context, vitamins and similar small molecules emerge as possible linkers to mediate metabolic crosstalk between compartments. This review focuses on two vitamins as potential metabolic signals within the plant cell, vitamin C (L-ascorbate) and vitamin B_1 (thiamin). These two vitamins demonstrate the importance of metabolites in shaping cellular processes working as metabolic signals during acclimation processes. Inferences based on the combined studies of environment, genotype, and metabolite, in order to unravel signaling functions, are also highlighted.

Keywords: metabolite signaling/acclimation; TCA cycle; Calvin–Benson cycle; photoperiodic changes; photosynthesis; redox-regulation; environmental adaptation

---

1. Introduction

The different cellular compartments such as chloroplasts, mitochondria, and the nucleus require a tightly orchestrated coordination of metabolic activity within each compartment by anterograde and retrograde signaling pathways [1,2]. Anterograde signaling is essentially a top-down regulatory pathway originated in the nucleus and sent to the organelles. Retrograde signaling, on the other hand, is the ability of organelles to coordinate, via signaling molecules, the expression of nuclear genes [3,4].

During the past years, significant advances in uncovering specific signals from plastids and their mechanisms of action have been made [4,5]. Several classes of factors including organelle gene expression, redox status, accumulation of pigment precursors like tetrapyrroles, reactive oxygen species (ROS), and metabolites have been proposed to act as plastidial signals [1]. They are sensed by plastidial factors including Executer1, Executer2, Genomes Uncoupled 1 (GUN1), and a thylakoid protein kinase (STN7), to initiate signaling cascades [1,4]. Some nuclear factors such as the transcription factor abscisic acid insensitive 4 (ABI4) have additionally been identified to be involved in retrograde signaling [6].

Much less is known about plant mitochondrial retrograde regulation/signaling [7,8]. Current research focuses on the response to a dysfunctional mitochondrial electron transport chain (ETR) and activation of genes encoding enzymes associated with the recovery of mitochondrial function, such as alternative oxidase (AOX) and alternative NAD(P)H dehydrogenases. It is also known that genes encoding proteins associated with the maintenance of the redox homeostasis, such as glutathione reductase, catalases, ascorbate peroxidases, and superoxide dismutases, are affected [9]. Recently an additional key player in the coordination between chloroplast and mitochondrial signaling pathways has been identified by Shapiguzov and coworkers (2019); their results suggest that the nuclear protein radical-induced cell death1 (RCD1) combines the signaling from both organelles in order to govern transcriptional and metabolic process within each organelle [10,11]. RCD1 mediates this regulation by suppressing the abscisic-acid-responsive NAC (ANAC) transcription factors ANAC013 and ANAC017, known as regulators of the mitochondrial dysfunction stimulon (MDS) genes, and also by receiving the ROS signals from the chloroplast underging protein modifications [10,11].

Many studies have emphasized the high degree of interrelationship between photosynthesis and respiration, the major energy production pathways that are confined to the chloroplast and mitochondria, respectively [12–14]. Metabolite signals are now frequently proposed as potential signals for inter-organellar communication and possible modulators to support photosynthesis during acclimation to fluctuating environments [3,15,16].

The focus of this review is on two vitamins, vitamin C (L-ascorbate) and vitamin B_1 (thiamin), as potential metabolic signals within the plant cell, and to summarize recent advances on their roles in plant acclimation responses.

Vitamins in general are essential for plant metabolism, because many of them display important redox chemistry and antioxidant potential or are used as cofactors in several enzymatic reactions. Cartenoids (Pro-vitamin A), ascorbate, vitamin E (both tocopherols and tocotrienols), and vitamin B compounds (such as thiamin) are known to have predominant antioxidant roles in plants under oxidative stresses. Plastids are organelles highly exposed to oxidative stress because of oxygenic photosynthesis, and thus are protected by antioxidant vitamins, as reviewed in Asensi-Fabado and Munne-Bosch, 2010 [17].

Ascorbate is known as the most abundant and ubiquitous cellular antioxidant and is present in most cellular compartments [18]. The antioxidant function of ascorbate is mainly attributed to its action as a substrate for the ascorbate–glutathione cycle in scavenging hydrogen peroxide [19]. Ascorbate is also used as a cofactor for the violaxanthin de-epoxidase (VDE) enzyme, a critical component of the non-photochemical quenching (NPQ) [18]. Having the profound antioxidant functions to scavenge ROS renders ascorbate an important metabolite in the plant acclimation responses to changing environments [20,21]. For instance, ascorbate has been demonstrated to accumulate in Arabidopsis leaves during the acclimation process following the transition from low to high light conditions [21], as well as in the leaves of highland species and pea acclimated to high light and low temperature [22]. Ascorbate is connected to the mitochondria and the respiration processes, because the last enzyme of the pathway is located in the inner membrane of the mitochondria; however, it is found to be almost ubiquitously scattered in all cellular compartments, including chloroplast [23]. Further, ascorbate is known as a key component of the redox hub in balancing redox homeostasis in cellular compartments [19], and owing to the fact that redox equivalents can also be transferred between cellular compartments, ascorbate is, therefore, assumed as part of the inter-organellar communication.

Thiamin (or thiamine), also known as vitamin B_1, is one of the water-soluble B-complex vitamins. The term refers to the three vitamers forms, free thiamin; thiamin monophosphate (TMP); and thiamin pyrophosphate (TPP, or thiamin diphosphate, TDP), which is the active form. TPP works as an essential coenzyme for enzymes involved in photosynthesis in chloroplasts, in ATP synthesis in the participation in oxidative decarboxylation of pyruvate, and in the tricarboxylic acid cycle in mitochondrial central metabolism, as well as in the pentose phosphate pathway and alcoholic fermentation in cytoplasm [24–27]. Thiamin has also been shown to be involved in the acclimation responses to abiotic stresses and photoperiod [28–32]. It plays important roles, working directly as an antioxidant, scavenging ROS, and protection molecule, and indirectly by contributing to the cell energy poll, conferring the cell the necessary metabolic flexibility to acclimate to new conditions [17,32,33].

In the next sections, a detailed description of each pathway and their roles in plant acclimation responses to environmental cues, in particular high light and photoperiod acclimation, are discussed.

2. Ascorbate Biosynthesis and Subcellular Distribution

Ascorbate is present in Arabidopsis leaves as one of the most abundant primary metabolites [18]. In plants, ascorbate is generally synthesized through one dominant pathway, so-called the D-mannose/L-galactose (Smirnoff–Wheeler) pathway [18]. Three other pathways, so-called the *Myo*-inositol [34], L-gulose [35], and L-galacturonate [36], have been also suggested as alternative routes for ascorbate biosynthesis [37]. However, strong evidence on the existence of all these alternative pathways has not yet been reported. Moreover, it appears that even if the alternative pathways exist, their roles in ascorbate biosynthesis should be minor at least in Arabidopsis, where ascorbate loss in the *vtc2vtc5* double mutants appears not to be compensated by the other pathways [38].

Therefore, here, we only demonstrate the details of the major pathway (Figure 1). In this pathway, D-glucose-6-phosphate is converted to ascorbate in nine enzymatic reactions, as depicted in Figure 1, with the last step catalyzed by L-galactono-1,4-lactone dehydrogenase (GLDH), located in the inner membrane of the mitochondria in which L-galactono-1, 4-lactone, the direct precursor of ascorbate, is converted to ascorbate.

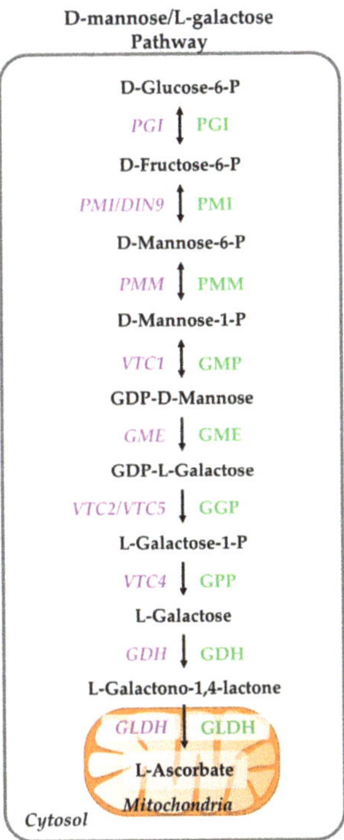

Figure 1. D-Mannose/L-Galactose pathway of ascorbate biosynthesis in plants. The genes of the pathway are highlighted in purple and written in italics. The enzymes are highlighted in green. Phosphoglucose isomerase (PGI), phosphomannose isomerase (PMI), and phosphomannomutase (PMM) are responsible for the conversion of D-glucose-6-P to D-mannose-1-P, the direct precursor of GDP-D-mannose pyrophosphorylase (GMP), the first committed enzyme of the pathway encoded by *VTC1*. GDP-mannose-3′-5′-epimerase (GME), GDP-L-galactose transferase (GGP), L-galactose-1-phosphate phosphatase (GPP), L-galactose dehydrogenase (GDH), and L-galactono-1,4-lactone dehydrogenase (GLDH) are the next enzymes of the pathway. GGP is the key enzyme of the pathway encoded by *VTC2* and *VTC5* paralogs. This enzyme undergoes feedback regulation by ascorbate pool size. GLDH is located in the intermembrane of mitochondria and is connected to the mitochondria respiratory chain.

Ascorbate specific immunogold labelling and quantitative transmission electron microscopy showed that ascorbate was found in most cellular organelles, including cytosol, nuclei, peroxisomes, vacuoles, mitochondria, and chloroplasts, but not in cell walls and intercellular spaces. Moreover, it has been shown that, despite showing a strong increase in chloroplasts (104%) under high light conditions (700 μmol m^{-2}s^{-1}), vacuoles even demonstrated a stronger ascorbate specific labeling (395%) than chloroplasts. This highlights the relevance of vacuoles in ascorbate metabolism in response to high light acclimation, which deserves further investigations [23].

Given that ascorbate distributes across all the cellular compartments, despite exclusive production in mitochondria [39], the involvement of ascorbate transporters is necessary for its function. The identification of ascorbate transporters has long been considered as a difficult task [40] but eventually, a phosphate transporter 4 family protein (*At*PHT4;4) was identified as an ascorbate transporter located at the chloroplast envelope membrane [41]. However, transporters localized to other membranes remain unknown and information concerning subcellular ascorbate concentration is rare and normally confined to single environmental conditions.

3. Role of Ascorbate in Light Acclimation

Under the light acclimation process, the chloroplast undergoes coordinated metabolic adjustments with extra-chloroplastic metabolism in order to maintain the overall fitness of plants and avoid damage [42]. Several metabolites produced in the plastids and motochondria are subsequently transmitted to the nucleus and modulate nuclear gene expression. This phenomenon is termed as retrograde signaling and known as a critical component of plant acclimation responses. Several light-shift experiments have been conducted to unravel the early and late metabolic responses to different light intensities. Changes in light intensity rapidly manipulate the electron pressure generated in the photosynthetic electron transport chain (pETC); therefore, the ROS signals generated from pETC are considered as important retrograde signals for short- and long-term acclimation [42]. Ascorbate and glutathione are known as redox signals, playing roles on longer time scales [42]. It has been revealed that total ascorbate levels increased after an hour in plants exposed to high light (800 μmol photons m^{-2} s^{-1}). This increase is even delayed by 3 h in plants transferred to high light following acclimation to low light intensity (8 μmol photons m^{-2} s^{-1}) [43].

One important function of ascorbate is as a cofactor in the xanthophyll cycle, in which the excess excitation energy is dissipated as heat from excited chlorophylls to xanthophyll carotenoids, a photoprotection mechanism termed as non-photochemical fluorescence quenching (NPQ) [44]. In this cycle, the violaxanthin de-epoxidase (VDE) enzyme, localized in the thylakoid lumen, uses ascorbate as a cofactor to reduce the epoxide group of the substrate violaxanthin and converts it to antheraxanthin and zeaxanthin [45]. In a light-shift experiment (from 160 μmol photons m^{-2} s^{-1} to 1800 μmol photons m^{-2} s^{-1}), roles of ascorbate in light acclimation were investigated using the Arabidopsis mutant deficient in VDE enzyme (*npq1*), ascorbate deficient mutant (*vtc2*), along with *vtc2npq1* double knockouts [46]. It has been revealed that the *vtc2* mutants, having 10%–30% of the wild type (WT) ascorbate levels, lost their acclimation capacity after long-term exposure to high light (up to five days at 1800 μmol photons m^{-2} s^{-1}). In contrast to the *npq1* single mutants, deficient in zeaxanthin, which were slightly more sensitive to high light than the WTs, *vtc2* and *vtc2npq1* double mutants showed an increased degree of bleached leaves, lipid peroxidation, and photoinhibition (increased degree of damage to (Photosystem II) PSII, measured by Fv/Fm). These data confirmed the importance of ascorbate in light acclimation responses and also showed that ascorbate has even more important roles than other photoprotective metabolites such as xanthophylls in acclimation to high light stress. Further, loss of PSII efficiency was not observed after short-term high light exposure (up to 2 h) in *vtc2* mutants, however, the conversion rate of violaxanthin to zeaxanthin was reduced owing to the dependency of VDE to ascorbate [47]. These data further corroborated the importance of ascorbate on long-term acclimation to high light rather than short-term.

In a subsequent study in which they investigated the thylakoid-associated proteome of Arabidopsis WT and *vtc2* after transition to high light (1000 μmol photons m^{-2} s^{-1}), differential protein accumulation could be observed in a number of stress-associated proteins between WT and *vtc2* including Fe-superoxide dismutase (Fe-SOD), Cu, Zn-SOD, HSP70s (cpHSP70-1 and 2), PsbS protein, and a chloroplast-localized glyoxalate I [48]. SODs are metalloenzymes, which have been long known as stable markers for abiotic stress tolerance against ROS [49]. Also, it has been shown that HSP70-2 in *Chlamydomonas reinhardtii* chloroplasts has photoprotective roles for PSII reaction centers during photoinhibition and PSII repair [48,50]. Apart from the xanthophyll zeaxanthin, PsbS is known as another component of NPQ [51]. PsbS-dependent quenching site has been recently deciphered to be in Light-harvesting complex II (LHCII), and in the PSII core, most likely in the core antenna complexes CP43 and/or CP47 [52]. In the study of Giacomelli and coworkers, PsbS protein was up-regulated more than twofold upon transition to high light, however, it remained unchanged in the *vtc2*, which is in line with the observation that *vtc2* mutants have reduced levels of non-photochemical quenching [47]. This study shows that ascorbate has a significant impact on chloroplast proteome linking to oxidative stress and quenching, however, it cannot be entirely ruled out that these changes are the consequences of a direct or indirect effect of ascorbate deficiencies in the *vtc2* mutants. Moreover, the ascorbate deficient mutants, *vtc1*, *vtc2*, and *vtc3*, were found to accumulate visibly and quantitatively less anthocyanin compared with the wild types during the high light treatment in several studies [48,53,54]. *vtc1* and *vtc2* mutants were also unable to induce the expression of anthocyanin biosynthesis enzymes, and the corresponding transcription factors of the pathway, PAP1, GL3, and EGL3 under high light acclimation [54], whereas the transcripts related to anthocyanin biosynthesis and regulation are known to be up-regulated rapidly by high light in Arabidopsis WT plants [54,55]. Further, given the fact that both ascorbate and anthocyanin have been shown to accumulate in a similar time-scale (days) and in similar ranges of light intensities after high light exposure, and that the *vtc* mutants had defects in the accumulation of anthocyanin, the existing interconnection between them has been proposed in the study of Page and coworkers [54]. The authors observed a tight correlation of ascorbate and anthocyanin levels across six different Arabidopsis ecotypes under normal and high light conditions, which adds further proof to the relationship between them [54]. More investigation should be done to explore the co-regulatory mechanism of ascorbate levels with anthocyanin under high light acclimation.

4. Light Regulation of Ascorbate

Despite existing cumulative evidence on the importance of ascorbate on light acclimation responses, regulatory mechanisms of the ascorbate pool size by light remained poorly understood. It appears that ascorbate pool size is highly sensitive to both the light intensity and time of the day because transcript profiles of the genes encoding the enzymes of the pathway behaved unpredictably in different light-shift experiments and vary between species. Therefore, conclusions on the correlation between the gene expression, activity of the corresponding enzymes of the pathway, and the ascorbate pool size are inconsistent between studies. That being said, the comparison of multiple light-shift experiments revealed GDP-L-galactose phosphorylase (GGP) as the key enzyme of the pathway controlling the ascorbate levels under high light [54,56–59]. The corresponding genes encoding this enzyme, the first committed step of the ascorbate biosynthetic pathway, are *VTC2* and *VTC5* paralogs, which were identified to be induced in concert upon 24 h exposure to high light, leading to a 20-fold increase in the activity of the corresponding enzyme, and an increase in ascorbate levels [56].

In a study where the authors explored the transcriptional regulation of ascorbate by RNA-seq following a step change of light intensity in Arabidopsis, *VTC2* and, to lesser extent, *VTC5* were validated as regulatory points in light accumulation of ascorbate, the expression of both genes were correlated with different light intensities, however, a minor change in GDP mannose pyrophosphorylase (GMP) could also be observed [57]. Moreover, GGP has been proposed as a key rate-limiting step for ascorbate biosynthesis not only in Arabidopsis [60], but also in other species including tobacco [61], apple [62], and kiwifruit [60,63]. Besides, the light-responsiveness of *VTC2* expression has been observed in tomato fruits following an observation on *VTC2* reduction under a continuous shading [64]. GGP has also been validated as a highly regulated enzyme in the green algea, *Chlamydomonas reinhardtii*, where it is thought to exhibit protective function against oxidative stress [65,66]. Further, a potential regulatory role for GGP has been proposed owing to the evidence on nuclear localization of the protein; however, as yet, no mechanistic evidence proposed has been supposed as a hypothesis [58].

Besides GGP, GLDH has been also suggested as an important controlling point for light regulation of ascorbate biosynthesis at the level of the enzyme activity [67,68]. Arabidopsis plants, grown under high light after supplementation with L-galactone-1,4-lactone (L-Gal; the precursor of ascorbate), accumulated up to twofold ascorbate levels and had twice as high GLDH activities of the low-light grown plants, assumed as higher respiration rates [69]. GLDH is located in the inner membrane of the mitochondria, which carries a redox-sensitive thiol residue (Cys-340), critical for the conversion of L-Gal into ascorbate [70]. This residue has been validated to be irreversibly oxidized by H_2O_2, inactivating GLDH [70], and has been suggested to be responsible for regulation of GLDH activity during the early stages of heat stress produced programmed cell death [67,71]. Moreover, Arabidopsis *GLDH* overexpressing lines accumulated higher ascorbate levels and demonstrated higher chlorophyll fluorescence parameters after exposure to high light for 14 days, which led them to have lower sensitivity to light stress [72].

It should be noted that, despite observing multiple studies on light effects on the ascorbate biosynthetic pathway, so far, few reports exist concerning the effects of light on the components of ascorbate recycling and turnover [73]. One report, however, does demonstrate that the activities of dehydroascorbate reductase and monodehydroascorbate reductase are enhanced in the Arabidopsis plants, grown under high light [69].

4.1. Transcriptional Regulation

Despite observing alterations in gene expression patterns of the ascorbate biosynthetic pathway under light stress in multiple studies, the upstream signal transduction pathway controlling this phenomenon is largely unknown. Studies were performed to decipher light-regulated *cis*-elements in rice [74] and subsequently in Arabidopsis [75]. The conserved sequences (the GT1 box and the TGACG motif) in the promoter regions of the *L-galactose-1-phosphate phosphatase* (*GPP*) and *GLDH* genes were found to be responsible for light induction of these genes in rice [74]. Further efforts have been made to find such consensus elements in Arabidopsis, however, authors identified a different, but critical region for light regulation of *VTC2*, in –40 to –70 bp of its promoter [75]. Information on whether such a casual promoter region exist upstream of other genes of the pathway is not yet available.

Ascorbic acid mannose pathway regulator 1 (AMR1) has been identified as a negative regulator of multiple genes encoding early and late enzymes of the Man/L-Gal pathway, including GMP, GME, GGP, GPP, GDH, and GLDH, with the highest effect on GME and GGP [76]. The expression of AMR1 has been validated to be decreased by light and to be accompanied by an increase in ascorbate levels [76]. In contrast to AMR1, the ethylene response factor98 (*At*ERF98) has been identified as a positive regulator of D-Man/L-Gal pathway, by directly binding to the promoter of *VTC1*; encoding GMP; and also enhancing the expression of multiple genes of the pathway including, *VTC1*, *VTC2*, *GDH*, and *GLDH* [77]. Although the essential role of ERF98 has been revealed under the salt stress [77], no investigations have been done to verify its role under the light stress.

Recently, implementing genome wide association study (GWAS) on 302 tomato accessions identified a basic helix–loop–helix (bHLH) transcription factor, *Sl*bHLH59, which positively regulates ascorbate content in tomato fruits [78]. The most similar protein to *Sl*bHLH59 in Arabidopsis appears to be unfertilized embryo sac 12 (UNE12), regulating fertilization [78]. Further investigations are, however, needed to clarify whether this protein has links to the accumulation of ascorbate in Arabidopsis. The schematic representation of the regulatory factors is depicted in Figure 2.

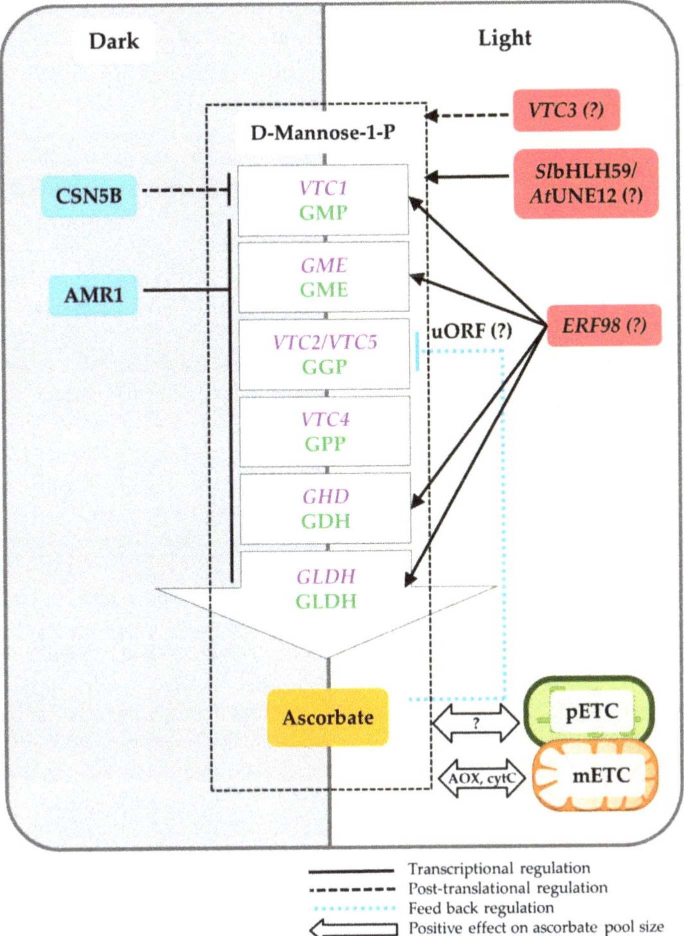

Figure 2. Overview of light regulation on ascorbate biosynthesis. Ascorbate biosynthesis is regulated transcriptionally by the ascorbic acid mannose pathway regulator 1 (AMR1), ethylene response factor 98 (ERF98), and *Solanum lycopersicum* basic helix–loop–helix (bHLH) transcription factor 59 (*sl*bHLH59; tomato-specific). AMR1 has negative effects on *GMP, GME, GGP, GPP, GDH*, and *GLDH* in Arabidopsis. *AMR1* expression decreases rapidly under light; thereafter, ascorbate levels increase. *At*ERF98 is the positive regulator of the ascorbate pathway by directly binding to the promoter of *VTC1*, and up-regulating the expression of *VTC1, VTC2, GDH*, and *GLDH*. The light-specific functionality of ERF98 has yet to be investigated. *sl*bHLH59 activates the genes of the pathway in tomato fruits and increases ascorbate levels under the light. The close homolog of this transcription factor (TF) in Arabidopsis is unfertilized embryo sac 12 (UNE12), however, its role for ascorbate biosynthesis has yet to be investigated. Ascorbate undergoes post-translational regulation via constitutive photomorphogenic9-signalosome subunit 5B (CSN5B), VTC3, and the feedback regulation of *VTC2*. CSN5B binds to VTC1 and promotes its degradation under the dark. VTC3 is a putative kinase/phosphatase for light regulation of ascorbate. Feedback regulation of ascorbate is controlled by an unusual open reading frame (uORF), located upstream of the *VTC2* gene. uORF functionality under the light needs further investigations. Ascorbate is also controlled via photosynthetic and mitochondria electron transport chains, designated as photosynthetic electron transport chain (pETC) and mitochondrial electron transport chain (mETC), respectively. This relationship is bidirectional.

Other components linked to ascorbate regulation have been proposed in multiple studies that are beyond the scope of this review because their roles in acclimation responses have not been fully characterized. Readers are referred to other comprehensive reviews covering the general regulatory components of ascorbate biosynthesis [67,79,80].

4.2. Post-Translational Regulation

Further studies support that light regulation of ascorbate occurs post-translationally via the three so far identified mechanisms: (i) through modulating the stability of GDP-man pyrophosphorylase (GMP/VTC1) in light/dark; (ii) through feed-back regulation of GGP (VTC2); and (iii) through a putative kinase::protein phosphatase, VTC3 (Figure 2). Constitutive photomorphogenic9-signalosome subunit 5B (CSN5B) protein has been identified as a component in light/dark regulation of ascorbate [81]. CSN5B interacts with GDP-man pyrophosphorylase (VTC1), and promotes its degradation under dark through the 26S proteasome pathway, resulting in lower ascorbate content. As mentioned above, GGP has been determined as the main regulatory point in the ascorbate biosynthetic pathway, which undergoes rapid feedback control in conditions where ascorbate levels increase [82]. An unusual open reading frame (uORF), starting with an uncommon start codon, ACG instead of AUG, has been found in the 5′-untranslated region (UTR) of GGP, which, under a high concentration of ascorbate, gets translated into a 60–65 residue peptide and further inhibits the translation of *VTC2* [82]. Interestingly, this uORF exists in a variety of plant species [82]. Although it has been hypothesized that this post-translational control might be a causative mechanism in light regulation of ascorbate, further experiments are required to validate this hypothesis. GGP has also been revealed as an important rate-limiting enzyme in *Chlamydomonas*, however, it lacks the feedback regulation mechanism existing in land plants [66].

A deeper investigation of the ascorbate deficient mutant *vtc3* led to the identification of its novel causative loci, encoding VTC3 protein, suggested as a putative kinase/phosphatase for light regulation of ascorbate [83]. Although the protein structure of the VTC3 has been proposed, harboring a kinase and a phosphatase domain at its N-terminal and C-terminal site, respectively, validation of its molecular mechanism has so far remained elusive. Given that the *vtc3* mutants were unable to accumulate ascorbate under continuous light, and localized to the chloroplast, a dual signaling function of the protein in light regulation of ascorbate was suggested by the authors. However, this remains to be experimentally validated [83].

5. Role of Sugars, Photosynthesis, and Respiration in Light Regulation of Ascorbate

Given that carbohydrates are substrates for ascorbate synthesis, attempts have been made to decipher the putative links between these metabolites in light acclimation responses. Schmitz et al. [21] examined the roles of sugar and starch metabolism in the acclimation process to high light by using Arabidopsis mutants deficient in either the triose phosphate/phosphate translocator (*tpp*) or ADP-glucose pyrophosphorylase (*AGPase*) or both of them (*adg1-1/tpt-2*). While soluble sugars, mainly glucose, accumulated in both wild type and mutant plants within four hours of high light exposure, the acclimation response was impaired in the mutants after only two days. The comparison of transcriptomic results with publicly available ones revealed a correlation between responses to high light and those to sugar levels following four hours of high light treatment, while the responses at 48 h were also similar to those of ROS accumulation. These results suggest that soluble sugars act as modulators in the short term, but this role is replaced by ROS in the long term. Ascorbate levels increased in all lines over time upon exposure to high light. Interestingly, the redox state of ascorbate was not affected in the triose phosphate/phosphate translocator (TPT) mutant, but reduced in all other lines, suggesting the effect of sugar localization on the redox state of ascorbate [21]. These results emphasize the signaling functions of soluble sugars in high light acclimation and, further, the involvement of ascorbate redox state in signaling pathways.

Although it has been identified that sugars affect the redox state of ascorbate in light acclimation, as described above, regulation of ascorbate accumulation under light appears to be independent of sugars, but dependent on the photosynthetic electron transport chain [59]. Leaf ascorbate levels and transcript levels of the ascorbate biosynthetic genes, *GMP*, *GPP*, *GDH*, and *VTC2*, were decreased, and plants were unable to accumulate ascorbate upon inhibition of photosynthetic electron transport by 3-(3,4-dichlorophenyl)-1,1-dimethylurea (DCMU) and atrazine (ATZ) treatment even under continuous light [59]. In the same study, the effect of sugars on foliar ascorbate levels was examined by transferring two-week-old Arabidopsis seedlings to the media in the presence or absence of sucrose, with subsequent transferal to the dark for 48 h to reduce the internal carbon sources and total ascorbate levels. In effect, darkness led to a significant decline in leaf sugar levels by 90% in both the presence and absence of sucrose, accompanied by a reduction in ascorbate levels. External supplementation of sucrose did not restore the leaf ascorbate pool sizes. In an opposite way, the levels of sugar increased in both the presence and absence of sucrose after transferring them from the dark to the light, and again, the levels of ascorbate could not be restored to normal levels in the sucrose supplemented plants [59]. This observation is in line with the research where the absence of a correlation between carbohydrates and ascorbate levels could be observed in the ripening period of tomato fruits under irradiances that stimulate ascorbate biosynthesis [84]. Despite a remarkable increase in ascorbate levels upon ripening under light, carbohydrate levels remained unchanged. Similarly, alteration in carbohydrate levels upon flower pruning did not show effects on ascorbate levels [84]. Furthermore, these authors confirmed in separate studies that high light has positive effects on ascorbate upregulation only in green tomato fruits, determining photosynthesis as an integral part of this mechanism [84]. Therefore, it appears that photosynthesis is a key component in controlling the leaf ascorbate pool size under light, however, carbon supply through photosynthesis appeared not to be the determinant of the ascorbate levels in Arabidopsis and tomato. Further studies are required to explore the signaling mechanisms in this process. Yet, it has been proposed that the effects of sugars on ascorbate is a genotype-specific phenomenon, which varies in different plant species [73,85]. In contrast to what was observed in the above-mentioned studies, it has been revealed that sucrose feeding in tomato fruits increased the expression of key ascorbate biosynthetic genes such as *VTC1*, *VTC2*, *GDH*, and *GLDH*, as well as of recycling and turnover genes *APX*, *MDHAR*, *DHAR*, and *GR*, pointing to yet unknown signaling components in modulating the ascorbate biosynthetic and recycling gene expression patterns mediated by sugars [86]. That being said, sucrose and glucose feeding had no effects on ascorbate levels in barley and pea embryonic axes [73]. As sucrose feeding in these plant species has not been examined in light shift experiments, it is, however, difficult to draw a solid conclusion on the role of sugars on light regulation of ascorbate.

Given that GLDH, the last enzyme of the ascorbate pathway, lies in the inner membrane of the motochondria [87], being designated as part of the complex I of the respiratory electron transport chain [88], the relationship between ascorbate biosynthesis and respiration is rendered inevitable [73]. Bartoli et al. [87] observed that isolated mitochondria from potato leaves were able to synthesize ascorbate from L-GaL, and subsequently, L-GaL stimulated mitochondrial electron transport rates. Besides, it has been demonstrated that cytochrome C (cytC), located between complexes III and IV, is the electron acceptor of GLDH and treatment of intact mitochondria with potassium cyanide (KCN, an inhibitor of respiration) blocked ascorbate production [69,87]. Following that, Bartoli et al. [69] examined the effects of high light and respiration on ascorbate synthesis in Arabidopsis WTs and transgenic plants overexpressing the mitochondrial alternative oxidase [69]. It has been observed that plants under high light had a higher amount of ascorbate, GLDH, cytochrome C, and cytochrome C oxidase (CCO) activities, accompanied by an improved capacity of the AOX and CCO electron transport rates. Furthermore, *AOX*-overexpressing lines exhibited higher ascorbate levels than WT, especially at high light [69]. AOX is an enzyme in the plant mitochondria that bypasses cytC by directly accepting the electrons from the ubiquinone pool, which prevents over-reduction of the respiratory electron transport chain, and reduces the risk of ROS overproduction [69,73,89]. These studies demonstrate an important mechanism in light regulation of ascorbate through the AOX pathway and further highlight important bidirectional interconnections between the mitochondrial electron transport chain and ascorbate biosynthesis, through both cytC and AOX respiratory pathways.

6. Role of Ascorbate in Photosynthesis Coordination of the Energy Systems of the Mitochondria and Chloroplast

Given that ascorbate production is tightly associated with the mitochondrial electron transport chain and considering the well-documented facts on having profound bidirectional relationships with the rates of photosynthesis via a range of mechanisms, we assume that ascorbate takes part in the coordination of the energy systems between the mitochondria and chloroplast.

This hypothesis is raised from validations on ascorbate roles in mitochondrial electron transport rate (mETC) discussed above, photosynthesis, and TCA cycle regulatory networks [39]. Moreover, genes of the Ascorbate-glutathione cycle (ASC–GSH) cycle are expressed in both chloroplast and mitochondria [90], thus both organelles must coordinately take part in ascorbate biosynthesis and recycling. It has been demonstrated in multiple studies that ascorbate can elevate the rate of photosynthesis by a variety of mechanisms, especially in response to acclimation to high light [46,91,92]. These mechanisms include involvement of ascorbate in protecting against photoinhibition in the water–water cycle by scavenging superoxide and hydrogen peroxide, dissipating excess energy, and contribution to thylakoid acidification leading to the control of PSII activity [93–95]. Ascorbate is also considered as an alternative electron donor for PSII, whereby it prevents photo-oxidation [96,97]. The ascorbate-redox state is also known to affect photosynthetic activity through guard cell signaling and stomatal movement and also through changing the expression of the nuclear and chloroplastic encoded genes [40]. Furthermore, feeding Arabidopsis *vtc1* mutants with ascorbate had a great impact on photosynthetic gene expression, leading to an increased and decreased expression of some of the chloroplast- and nuclear-encoded genes, respectively [98].

However, recently observed differences in the photosynthetic responses of *vtc2-1* and *vtc2-4* mutants under high light raised the hypothesis that higher susceptibility of the *vtc2-1* mutants to photoinhibition, previously reported in the literature [46,92], might not have been caused by the lower ascorbate levels in the mutants [99]. Unlike *vtc2-1*, which carries a point mutation, the *vtc2-4* mutant is a T-DNA insertion line with a complete loss of function [99]. Contrary to *vtc2-1*, the *vtc2-4* mutants have unchanged levels of zeaxanthin contents and, despite having similar levels of NPQ under high light (lower than the WT), *vtc2-1* had greater photochemical quenching in the dark (qP_d) values than the WT. Therefore, the authors suggested ascorbate as an essential component for growth, but not for photoprotection [99].

Intriguingly, observations on transgenic tomato plants antisensed in mitochondrial malate dehydrogenase (*mdh*), strengthening the hypothesis of ascorbate acting in the coordination of the energy systems of the mitochondria and chloroplast [13,100]. The reduction of the TCA cycle occurring in these transgenic lines via down-regulation of the expression of the mitochondrial MDH did not affect the respiration, but resulted in a fourfold increase in ascorbate following an upregulation of the activity of GLDH. Detailed studies revealed that this was because flux, through the GLDH activity, was upregulated in these lines, with the consequence that electrons were supplied to the mitochondrial electron transport chain [100]. Furthermore, incubating tomato leaf discs with ascorbate under constant illumination increased the amount of carbon assimilation and starch levels [100]. This further strengthens the link between chloroplast and mitochondria in ascorbate levels. Evaluation of tomato plants in which the GLDH was reduced in expression by RNA interference had dramatic consequences on both plant and fruit growth and development [101]. This observation is, however, complicated by a report of Tomaz et al. [102], in which they examined the *mdh* double mutants in Arabidopsis. The Arabidopsis *mdh* double mutants had increased levels of ascorbate, but higher levels of mitochondrial respiration and a lower activity of GLDH, contrary to what had been observed in tomato plants [102]. More investigations are needed to clarify the cause of this difference observed between the two species.

In conclusion, considering all the described effects of ascorbate on photosynthesis and vice versa, and its intimacy to the respiratory electron transport chain, ascorbate can be considered as an important component in plant central metabolism, modulating the energy systems between the chloroplast and mitochondria [39]. However, the molecular mechanisms and the signals responsible for these network of interactions need further investigations.

7. Thiamin Biosynthesis

In plants, thiamin is synthesized from pyrimidine and a thiazole moiety, both of which are synthesized in the chloroplast. Synthesis of pyrimidine moiety is catalized by thiamin C synthase (THIC) by converting 5-aminoimidazole ribonucleotide (AIR) and S-adenosylmethionine (SAM) as substrates to 4-amino-2-methyl-5-hydroxymethylpyrimidine phosphate (HMP-P) [103–105]. Thiazole moiety is synthesized by the action of 4-methyl-5-b-hydroxyethylthiazole phosphate (HET-P) synthase (THI1) [106], catalyzing the conversion of nicotinamide adenine dinucleotide (NAD+) and glycine as substrates to an adenylated thiazole intermediate (ADT) [107–109]. ADT is then hydrolyzed to HET-P, by an uncharacterized enzyme. Phosphorylation of HMP-P to HMP-PP, and subsequently condensation of HMP-PP and HET-P, is done by the action of a bifunctional enzyme, thiamin monophosphate pyrophosphorylase (TH1), which eventually leads to the formation of thiamin monophosphate (TMP) [30,110]. TMP is dephosphorylated to thiamin by a haloacid dehalogenase (HAD) family phosphatase (TH2) [111], and subsequently pyrophosphorylated to thiamin pyrophosphate (TPP) by TPP kinases (TDPKs), which are located in the cytosol (Figure 3) [110,112].

In the past years, different attempts have been made to biofortify staple crops with thiamin [112–114]. These efforts were possible thanks to a recent increase in the understanding of thiamin metabolism in plants. Several genes involved in thiamin biosynthesis, regulation, transport, and salvage have been identified owing to the availability of complete plant genome sequences [30,115,116]. Although this determined the beginning of thiamin engineering strategies, several aspects of thiamin metabolism are still unclear [114].

Regulation of Thiamin Biosynthesis

THIC promoter has been identified to be essential for the circadian clock-regulation of thiamin biosynthesis [25]. Thiamin levels are also regulated by a TPP responsive riboswitch located at the 3′ region of the THIC pre-mRNA, controlling the stability of THIC mRNA [25]. Riboswitches are mRNA sensors that bind small molecules and consecutively regulate gene expression. The TPP riboswitch mediates a feedback regulation of the thiamin biosynthesis pathway. When TPP levels are high, TPP binds to the riboswitch leading to intron splicing and instable THIC mRNA, which is consequently degraded [24,117].

Another layer of regulation relies on the activities of thiamin biosynthesis enzymes. THI1 is a single turnover protein; it loses functionality once the cysteine residue is used in the reaction [107,118], and thus a single step of thiamin biosynthesis requires high energy levels. In addition, excess HMP-PP could work as an inhibitor for TH1 activity, as suggested by *THIC*-overexpressing plants [119]. Bioengineering studies for the fortification of crops also showed that an increase in expression of *THIC* and *THI1* in Arabidopsis and rice plants does not necessarily result in an increase in thiamin production [112,114,120]; these enzymes rely on factors, such as reduction by thioredoxins for THIC and THI1, and supply of sulfur for THIC's iron-sulfur cluster [104,114,121], which could also be limiting.

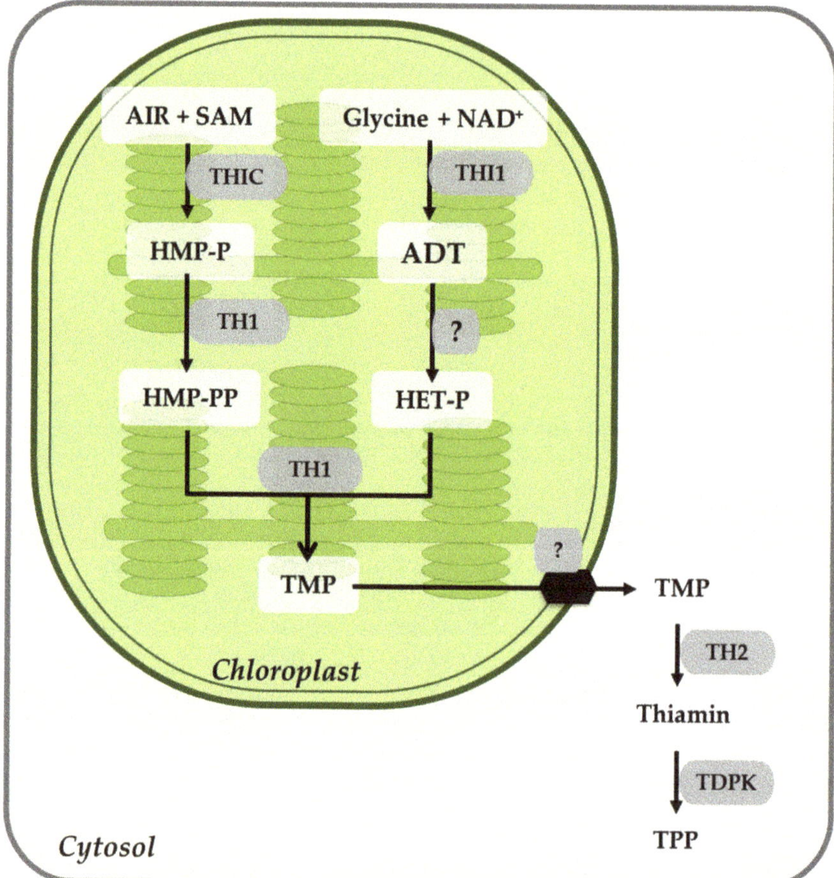

Figure 3. Thiamin biosynthesis pathway in plants. Thiamin is synthesized from pyrimidine and a thiazole moiety, both of which are synthesized in the chloroplast. The detailed description of the pathway is presented in the text. Abbreviations of the substrates: 5-aminoimidazole ribonucleotide (AIR), S-adenosylmethionine (SAM), nicotinamide adenine dinucleotide (NAD$^+$), 4-amino-2-methyl-5-hydroxymethylpyrimidine phosphate (HMP-P), adenylated thiazole intermediate (ADT), 4-methyl-5-b-hydroxyethylthiazole phosphate (HET-P), thiamin monophosphate (TMP), thiamin pyrophosphate (TPP). Abbreviations of the enzymes: thiamin C synthase (THIC), -methyl-5-b-hydroxyethylthiazole phosphate (HET-P) synthase (THI1), thiamin monophosphate pyrophosphorylase (TH1), haloacid dehalogenase (HAD) family phosphatase (TH2), TPP kinases (TDPK).

The complexity of thiamin biosynthesis regulation and the universality of thiamin-requiring enzymes across kingdoms and their association with core metabolic pathways suggest a crucial role of TPP in the regulation of cellular metabolism and the potentially prejudicial effect of inappropriate total thiamin levels. The cell localization of the main thiamine-dependent enzymes and its participation in metabolic pathways are depicted in Figure 4.

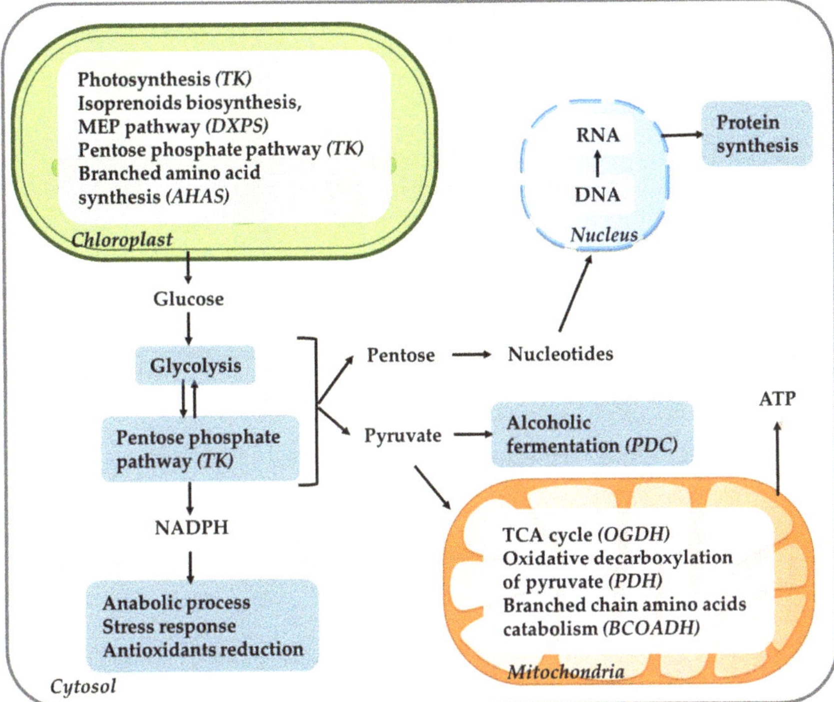

Figure 4. Cell localization of the main thiamin dependent-enzymes and its participation in metabolic pathways. Thiamin pyrophosphate (TPP), the active form of thiamin, works as an essential coenzyme for the enzymes involved in photosynthesis in chloroplasts, pentose phosphate pathway, and alcoholic fermentation in cytoplasm, as well as in ATP synthesis in the participation in oxidative decarboxylation of pyruvate and tricarboxylic acid cycle in mitochondrial central metabolism. Thiamin has also been shown to be involved in the acclimation responses to abiotic stresses; photoperiod; and working directly as an antioxidant, scavenging ROS; a protection molecule; and indirectly by contributing to the cell energy poll, conferring the cell the necessary metabolic flexibility to acclimate to new conditions. The thiamin-dependent enzymes shown are α-ketose transketolase (TK); 1-deoxy-D-xylulose-5-phosphate synthase (DXPS); acetohydroxyacid synthase (AHAS); pyruvate dehydrogenase (PDH); 2-oxoglutarate dehydrogenase (OGDH); and branched chain 2-oxoacid dehydrogenase (BCOADH). MEP, methylerythritol pathway.

8. Thiamin and Chloroplast Interactions

In the chloroplast, TPP-dependent enzymes play a role in photosynthesis (α-ketose transketolase, TK, and 1-deoxy-D-xylulose-5-phosphate synthase, DXPS), pentose phosphate pathway (TK), and branched amino acid synthesis (acetohydroxyacid synthase, AHAS). The primary pathway for carbon fixation in plants is the Calvin–Benson cycle (also known as the C_3 cycle), which is connected to several other pathways via its intermediates. TK has a central location in the Calvin–Benson cycle catalyzing the reversible transfer of a molecule with two carbons from sedoheptulose 7-phosphate to glyceraldehyde 3-phosphate (G3P), generating xylulose 5-phosphate (Xu5P) and ribose 5-phosphate, or from fructose 6-phosphate to produce Xu5P and erythrose 4-phosphate. These reactions are essential for the regeneration of ribulose 1,5-bisphosphate in the Calvin–Benson cycle maintaining active photosynthetic rates and providing precursor molecules for the shikimic acid pathway and phenylpropanoid metabolism (erythrose 4-phosphate) [122].

Additionally, TK is the main TPP-dependent enzyme in the pentose phosphate pathway. Different from the Calvin–Benson cycle, which utilizes CO_2, NADPH, and ATP to produce hexose sugars, the pentose phosphate pathway utilizes hexose substrates to produce NADPH and pentoses while releasing CO_2. Through participation in the pentose phosphate pathway, transketolase has three important functions in the metabolism of the cells: (i) provision of pentoses for the synthesis of nucleotides; (ii) provision of metabolites for glycolysis or gluconeogenesis pathways; and (iii) indirectly influencing the synthesis of NADPH, required for the anabolic processes and antioxidants reduction (glutathione, ascorbate). Consequently, an appropriate activity of transketolase is essential for the proper functioning of lipid and carbohydrate metabolism [27].

Chlorophyll and carotenoids are synthesized in the methylerythritol pathway (MEP) and the first reaction of this pathway is catalyzed by DXPS; this enzyme combines G3P from the Calvin–Benson cycle with pyruvate from the glycolytic pathway to form deoxyxylulose 5-phosphate (DXP) [26]. Additionally, DXPS is a TPP-dependent enzyme and the product of the DXPS reaction, DXP, is the first substrate in the biosynthesis of the hydroxyethylthiazole phosphate (HETP) moiety of the thiamin molecule itself, and thus TPP [123].

The acetohydroxyacid synthase (AHAS, also known as acetolactate synthase, ALS) is also a TPP-dependent enzyme. It catalyzes the first reaction in the synthesis of the branched-chain amino acids, valine, leucine, and isoleucine, which are only produced in plants [124]. This enzyme is responsible for converting two molecules of pyruvate into 2-acetolactate, the first reaction in a three-step pathway used to produce the three amino acids [124].

9. Thiamin and Mitochondria Interactions

Thiamin participates in the mitochondria central metabolism by functioning as a cofactor forpyruvate dehydrogenase (PDH), 2-oxoglutarate dehydrogenase (OGDH), and branched chain 2-oxoacid dehydrogenase (BCOADH, or branched chain ketoacid dehydrogenase, BCKDH). The PDH complex has a central role in bioenergetic processes, controlling the supply of acetyl-CoA into the TCA cycle and anabolic reactions, linking glycolysis and the TCA cycle via the oxidative decarboxylation of pyruvate [27,125]. PDH also produces acetyl-CoA from pyruvate in the chloroplast, which is used in the synthesis of fatty acids [27].

In the TCA cycle itself, OGDH catalyzes a rate-limiting step [27,125]. It converts 2-oxoglutarate, coenzyme A, and NAD^+ to succinic acid, while releasing NADH and CO_2 as part of the process. It has also been shown in several studies to be a key regulation point in plant metabolism [126–128]. Additionally, it is responsible for the distribution of succinyl-CoA and 2-oxoglutarate for substrate level phosphorylation of GDP, ADP, or for the synthesis of several amino acids and heme group [27,129].

Another TPP-dependent enzyme, BCOADH, catalyzes the breakdown of the branched chain amino acids, and studies have shown its important role in amino acid metabolism in Arabidopsis [130]. In addition, the enzyme pyruvate decarboxylase (PDC) in the cytosol also requires thiamin as a cofactor. It functions to break down pyruvate to acetaldehyde, an essential step for energy production via alcoholic fermentation under anoxia in Arabidopsis [131,132].

10. The Role of Thiamin in Mediating Plant Fitness and Acclimation

In addition to its role in central metabolism, studies in different plant species have shown that alterations in the levels of thiamin vitamers result in smaller plants, chlorosis of the leaves, growth retardation, delayed flowering, fitness cost, and an influence on yield penalty [25,26,105,111,133,134]. Some of these phenotypes, such as chlorosis and the delayed flowering, have also been shown to be dependent on the light regime [25,32].

The functions of thiamin in the regulation of the metabolic networks during photoperiod transition were deeply investigated in our group [32]. While control plants display changes in the amplitude of diurnal oscillation in the levels of metabolites, TPP riboswitch mutant plants with high levels of TPP do not show such metabolic flexibility. The results also indicate a close relationship between photorespiration and the TCA cycle as the mutant plants accumulate less photorespiratory intermediates such as glycine, serine, and glycerate [32].

Thiamin biosynthesis has also been well documented to be activated when plants are exposed to abiotic stresses [28,31,135]. Rapala-Kozik and co-authors (2008; 2012) demonstrated that thiamin biosynthesis is activated on the acclimation response of Arabidopsis to salt, osmotic, and oxidative stress [29,30]. These stresses induce the expression of genes for TPP biosynthesis and thiamin dependent-enzymes, resulting in increased levels of thiamin and TPP, those which can consequently be incorporated into the requiring enzymes associated with central metabolic pathways, as described previously. Continuous abiotic stress such as high salinity and sugar deprivation was also shown to increase thiamin biosynthesis gene expression [30,136]. Interestingly, flooding/hypoxic conditions also impact thiamin biosynthesis expression patterns in roots. Under low O_2 supply, roots switch from respiration to pyruvate fermentation, and high levels of thiamin could be required in this alternative route, as TPP acts also as a cofactor for PDC present in the fermentative metabolism [136].

Proteomics and transcriptomics studies in other species have also detected significant changes in thiamin biosynthesis and thiamin dependent enzymes during heat, drought, and cold stress conditions [137,138]. Changes in the protein levels associated with thiamin are transient, with increased abundance at early stages of stress followed by a decrease in protein levels associated with thiazole synthase in the thiamin biosynthesis pathway [137]. These conclusions are in agreement with our results describing the importance of thiamin levels for metabolic flexibility during acclimation [32]. Thiamin is also known as a potent antioxidant and a crop protection molecule in plants, playing important roles in plant acclimation [17]. It has been shown that thiamin has the antioxidant capacity by O^{2-}/OH^{-} scavenging and also recycling of vitamin C through the synthesis of NADPH [17,33]. Further, it has been revealed that paraquat-treated Arabidopsis, supplied by thiamin had reduced oxidative stress compounds, protein carbonyls, and dichlorofluorescein. It has also been shown in this study that Arabidopsis plants accumulated higher levels of TMP and TPP after exposure to high light, low temperatures, and osmotic and salt stress [31]. Yet, further investigations of thiamin in plants are needed to clarify whether it functions as an antioxidant directly or indirectly by supplying NADH and NADPH [17].

Although more scarce, studies on the function of thiamin during biotic stress have shown that treating different species with thiamin activates the systemic acquired resistance (SAR) and resistance to pathogen attack, such as fungal, bacterial, and viral infections. The thiamin treatment activates pathogen-related genes (PR) and stress signalling hormones, abscisic acid (ABA), and jasmonates, enhancing pathogen resistance in plants [135,139].

Taken together, these studies implicate that thiamin and its vitamers, despite being present at a very low concentration, play a general role in central metabolism and in acclimation responses. Indeed, their low concentrations are even representative of known signaling compounds. These results also support the idea that signaling molecules not only coordinate the expression of nuclear and organelle genes, but also maintain cellular functions at optimal levels in response to changes in environmental conditions [15,140,141].

11. Concluding Remarks

In this review, we addressed the current knowledge on the roles of ascorbate and thiamin in plant metabolism with the emphasis on plant acclimation responses, specifically to high light and photoperiod acclimation, regarding ascorbate and thiamin, respectively.

In brief, ascorbate has important roles in modulating the energy systems between the chloroplast and mitochondria during high light acclimation. Incorporation of GLDH, the ultimate enzyme of the pathway, into the mitochondrial electron transport chain is a rationale for considering the tight association of ascorbate biosynthesis and mitochondria metabolism. This relationship is found to be bidirectional as mETC has a positive regulatory role on ascorbate biosynthesis through both AOX and cytC respiratory pathways. The existence of such a relationship might guarantee the balance of the electron flow under environmental stresses [39]. Likewise, photosynthesis regulates ascorbate pool size under the light. However, carbohydrates are direct substrates for ascorbate biosynthesis; their role in light regulation of ascorbate remains ambiguous and appears to be species-specific. The relationship between ascorbate, respiration, and photosynthesis is bidirectional. In effect, ascorbate elevates the rate of photosynthesis by a variety of mechanisms in response to acclimation to high light, such as through the water–water cycle, by scavenging ROS, dissipating excess energy through the xanthophyll cycle, donating electrons to PSII, guard cell signaling, and stomatal movement. It also regulates the nuclear and chloroplastic encoded genes of the photosynthesis.

Furthermore, in this review, we addressed the complexity of thiamin biosynthesis regulation, the universality of thiamin-requiring enzymes across kingdoms, and their association with core metabolic pathways. Therefore, a general role of TPP in the regulation of cellular metabolism and the acclimation process can be considered. A schematic description of the organellar communication and the involvement of ascorbate and thiamin is depicted in Figure 5.

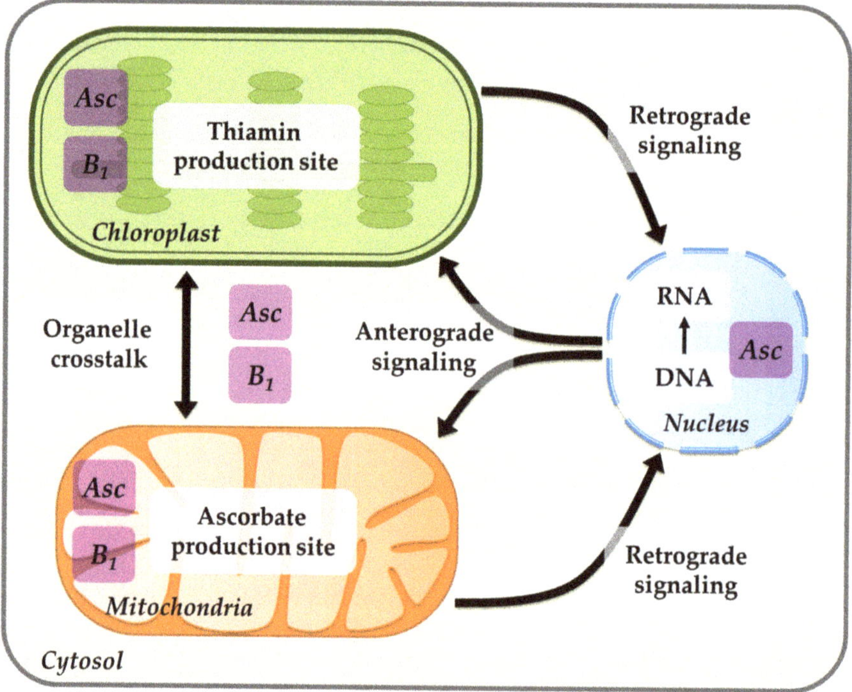

Figure 5. Schematic illustration of plant intracellular communication. Anterograde (nucleus to organelle) and retrograde (organelle to nucleus) signaling pathways, as well as the main active site of ascorbate (Asc) and thiamin (B$_1$) as signaling molecules, are shown. The ubiquitous existence of ascorbate and thiamin in cellular organelles, as well as the tight interconnection of the two vitamins between chloroplast and mitochondria, points to their important roles in the crosstalk between the two organelles.

Despite the thus far identified physiological dependencies of the two vitamins on the key players of metabolism, photosynthesis, and respiration and their roles in optimizing their activities, the underlying signaling and genetic factors in this process have remained a challenge for future research. QTL mapping and GWAS can be considered as alternatives to fill this gap.

Author Contributions: L.R.-S. and F.A. wrote the manuscript with supervision and contribution of A.R.F. All authors have read and agreed to the published version of the manuscript.

Funding: This research was funded by Collaborative Research Centers, SFB (Sonderforschungsbereich, Grant TRR 175/1) to A.R.F and F.A.

References

1. Pesaresi, P.; Schneider, A.; Kleine, T.; Leister, D. Interorganellar communication. *Curr. Opin. Plant Biol.* **2007**, *10*, 600–606. [CrossRef] [PubMed]

2. Kleine, T.; Leister, D. Retrograde signaling: Organelles go networking. *Biochim. Biophys. Acta* **2016**, *1857*, 1313–1325. [CrossRef] [PubMed]

3. Leister, D. Genomics-based dissection of the cross-talk of chloroplasts with the nucleus and mitochondria in arabidopsis. *Gene* **2005**, *354*, 110–116. [CrossRef] [PubMed]

4. Woodson, J.D.; Chory, J. Coordination of gene expression between organellar and nuclear genomes. *Nat. Rev. Genet.* **2008**, *9*, 383–395. [CrossRef] [PubMed]

5. Bobik, K.; Burch-Smith, T.M. Chloroplast signaling within, between and beyond cells. *Front. Plant Sci.* **2015**, *6*, 781. [CrossRef] [PubMed]

6. Koussevitzky, S.; Nott, A.; Mockler, T.C.; Hong, F.; Sachetto-Martins, G.; Surpin, M.; Lim, J.; Mittler, R.; Chory, J. Signals from chloroplasts converge to regulate nuclear gene expression. *Science* **2007**, *316*, 715–719. [CrossRef]

7. Rhoads, D.M.; Subbaiah, C.C. Mitochondrial retrograde regulation in plants. *Mitochondrion* **2007**, *7*, 177–194. [CrossRef]

8. Schwarzlander, M.; Finkemeier, I. Mitochondrial energy and redox signaling in plants. *Antioxid. Redox Signal.* **2013**, *18*, 2122–2144. [CrossRef]

9. Dojcinovic, D.; Krosting, J.; Harris, A.J.; Wagner, D.J.; Rhoads, D.M. Identification of a region of the arabidopsis ataox1a promoter necessary for mitochondrial retrograde regulation of expression. *Plant Mol. Biol.* **2005**, *58*, 159–175. [CrossRef]

10. Shapiguzov, A.; Nikkanen, L.; Fitzpatrick, D.; Vainonen, J.P.; Tiwari, A.; Gossens, R.; Alseekh, S.; Aarabi, F.; Blokhina, O.; Panzarová, K.; et al. Increased expression of mitochondrial dysfunction stimulon genes affects chloroplast redox status and photosynthetic electron transfer in arabidopsis. *bioRxiv* **2019**. [CrossRef]

11. Shapiguzov, A.; Vainonen, J.P.; Hunter, K.; Tossavainen, H.; Tiwari, A.; Jarvi, S.; Hellman, M.; Aarabi, F.; Alseekh, S.; Wybouw, B.; et al. Arabidopsis rcd1 coordinates chloroplast and mitochondrial functions through interaction with anac transcription factors. *eLife* **2019**, 8. [CrossRef]

12. Nunes-Nesi, A.; Araujo, W.L.; Fernie, A.R. Targeting mitochondrial metabolism and machinery as a means to enhance photosynthesis. *Plant Physiol.* **2011**, *155*, 101–107. [CrossRef] [PubMed]

13. Nunes-Nesi, A.; Carrari, F.; Gibon, Y.; Sulpice, R.; Lytovchenko, A.; Fisahn, J.; Graham, J.; Ratcliffe, R.G.; Sweetlove, L.J.; Fernie, A.R. Deficiency of mitochondrial fumarase activity in tomato plants impairs photosynthesis via an effect on stomatal function. *Plant J.* **2007**, *50*, 1093–1106. [CrossRef] [PubMed]

14. Nunes-Nesi, A.; Sulpice, R.; Gibon, Y.; Fernie, A.R. The enigmatic contribution of mitochondrial function in photosynthesis. *J. Exp. Bot.* **2008**, *59*, 1675–1684. [CrossRef]

15. Chi, W.; Feng, P.; Ma, J.; Zhang, L. Metabolites and chloroplast retrograde signaling. *Curr. Opin. Plant Biol.* **2015**, *25*, 32–38. [CrossRef]

16. Caldana, C.; Fernie, A.R.; Willmitzer, L.; Steinhauser, D. Unraveling retrograde signaling pathways: Finding candidate signaling molecules via metabolomics and systems biology driven approaches. *Front. Plant Sci.* **2012**, *3*, 267. [CrossRef]

17. Asensi-Fabado, M.A.; Munne-Bosch, S. Vitamins in plants: Occurrence, biosynthesis and antioxidant function. *Trends Plant Sci.* **2010**, *15*, 582–592. [CrossRef]

18. Smirnoff, N. Ascorbic acid metabolism and functions: A comparison of plants and mammals. *Free Radic. Biol. Med.* **2018**, *122*, 116–129. [CrossRef]

19. Foyer, C.H.; Noctor, G. Ascorbate and glutathione: The heart of the redox hub. *Plant Physiol.* **2011**, *155*, 2–18. [CrossRef]

20. Suzuki, N.; Mittler, R. Reactive oxygen species and temperature stresses: A delicate balance between signalling and destruction. *Physiol. Plant.* **2006**, *126*, 41–51. [CrossRef]

21. Schmitz, J.; Heinrichs, L.; Scossa, F.; Fernie, A.R.; Oelze, M.-L.; Dietz, K.-J.; Rothbart, M.; Grimm, B.; Flügge, U.-I.; Häusler, R.E. The essential role of sugar metabolism in the acclimation response of arabidopsis thaliana to high light intensities. *J. Exp. Bot.* **2014**, *65*, 1619–1636. [CrossRef] [PubMed]

22. Streb, P.; Aubert, S.; Gout, E.; Bligny, R. Reversibility of cold- and light-stress tolerance and accompanying changes of metabolite and antioxidant levels in the two high mountain plant species soldanella alpina and ranunculus glacialis. *J. Exp. Bot.* **2003**, *54*, 405–418. [CrossRef] [PubMed]

23. Zechmann, B. Subcellular distribution of ascorbate in plants. *Plant Signal. Behav.* **2014**, *6*, 360–363. [CrossRef] [PubMed]

24. Bocobza, S.E.; Aharoni, A. Switching the light on plant riboswitches. *Trends Plant Sci.* **2008**, *13*, 526–533. [CrossRef] [PubMed]

25. Bocobza, S.E.; Malitsky, S.; Araújo, W.L.; Nunes-Nesi, A.; Meir, S.; Shapira, M.; Fernie, A.R.; Aharoni, A. Orchestration of thiamin biosynthesis and central metabolism by combined action of the thiamin pyrophosphate riboswitch and the circadian clock in arabidopsis. *Plant Cell* **2013**, *25*, 288–307. [CrossRef] [PubMed]

26. Khozaei, M.; Fisk, S.; Lawson, T.; Gibon, Y.; Sulpice, R.; Stitt, M.; Lefebvre, S.C.; Raines, C.A. Overexpression of plastid transketolase in tobacco results in a thiamine auxotrophic phenotype. *Plant Cell* **2015**, *27*, 432–447. [CrossRef]

27. Tylicki, A.; Łotowski, Z.; Siemieniuk, M.; Ratkiewicz, A. Thiamine and selected thiamine antivitamins—Biological activity and methods of synthesis. *Biosci. Rep.* **2018**, *38*. [CrossRef]

28. Sayed, S.A.; Gadallah, M.A.A. Effects of shoot and root application of thiamin on salt-stressed sunflower plants. *Plant Growth Regul.* **2002**, *36*, 71–80. [CrossRef]

29. Rapala-Kozik, M.; Kowalska, E.; Ostrowska, K. Modulation of thiamine metabolism in zea mays seedlings under conditions of abiotic stress. *J. Exp. Bot.* **2008**, *59*, 4133–4143. [CrossRef]

30. Rapala-Kozik, M.; Wolak, N.; Kujda, M.; Banas, A.K. The upregulation of thiamine (vitamin b1) biosynthesis in *arabidopsis thaliana* seedlings under salt and osmotic stress conditions is mediated by abscisic acid at the early stages of this stress response. *BMC Plant Biol.* **2012**, *12*, 2. [CrossRef]

31. Tunc-Ozdemir, M.; Miller, G.; Song, L.; Kim, J.; Sodek, A.; Koussevitzky, S.; Misra, A.N.; Mittler, R.; Shintani, D. Thiamin confers enhanced tolerance to oxidative stress in arabidopsis. *Plant Physiol.* **2009**, *151*, 421–432. [CrossRef] [PubMed]

32. Rosado-Souza, L.; Proost, S.; Moulin, M.; Bergmann, S.; Bocobza, S.E.; Aharoni, A.; Fitzpatrick, T.B.; Mutwil, M.; Fernie, A.R.; Obata, T. Appropriate thiamin pyrophosphate levels are required for acclimation to changes in photoperiod. *Plant Physiol.* **2019**, *180*, 185–197. [CrossRef] [PubMed]

33. Subki, A.; Abidin, A.; Balia Yusof, Z.N. The role of thiamine in plants and current perspectives in crop improvement. In *B Group Vitamins—Current Uses and Perspectives*; IntechOpen: London, UK, 2018; pp. 33–44.

34. Lorence, A.; Chevone, B.I.; Mendes, P.; Nessler, C.L. Myo-inositol oxygenase offers a possible entry point into plant ascorbate biosynthesis. *Plant Physiol.* **2004**, *134*, 1200. [CrossRef] [PubMed]

35. Wolucka, B.A.; Van Montagu, M. Gdp-mannose 3′,5′-epimerase forms gdp-l-gulose, a putative intermediate for the de novo biosynthesis of vitamin c in plants. *J. Biol. Chem.* **2003**, *278*, 47483–47490. [CrossRef] [PubMed]

36. Agius, F.; Gonzalez-Lamothe, R.; Caballero, J.L.; Munoz-Blanco, J.; Botella, M.A.; Valpuesta, V. Engineering increased vitamin c levels in plants by overexpression of a d-galacturonic acid reductase. *Nat. Biotechnol.* **2003**, *21*, 177–181. [CrossRef] [PubMed]

37. Simkin, A.J. Genetic engineering for global food security: Photosynthesis and biofortification. *Plants (Basel)* **2019**, *8*, E586. [CrossRef]

38. Gallie, D.R. The role of l-ascorbic acid recycling in responding to environmental stress and in promoting plant growth. *J. Exp. Bot.* **2013**, *64*, 433–443. [CrossRef]

39. Szarka, A.; Bánhegyi, G.; Asard, H. The inter-relationship of ascorbate transport, metabolism and mitochondrial, plastidic respiration. *Antioxid. Redox Signal.* **2013**, *19*, 1036–1044. [CrossRef]

40. Fernie, A.R.; Tóth, S.Z. Identification of the elusive chloroplast ascorbate transporter extends the substrate specificity of the pht family. *Mol. Plant* **2015**, *8*, 674–676. [CrossRef]

41. Miyaji, T.; Kuromori, T.; Takeuchi, Y.; Yamaji, N.; Yokosho, K.; Shimazawa, A.; Sugimoto, E.; Omote, H.; Ma, J.F.; Shinozaki, K.; et al. Atpht4; 4 is a chloroplast-localized ascorbate transporter in arabidopsis. *Nat. Commun.* **2015**, *6*, 5928. [CrossRef]

42. Dietz, K.-J. Efficient high light acclimation involves rapid processes at multiple mechanistic levels. *J. Exp. Bot.* **2015**, *66*, 2401–2414. [CrossRef] [PubMed]

43. Alsharafa, K.; Vogel, M.O.; Oelze, M.-L.; Moore, M.; Stingl, N.; König, K.; Friedman, H.; Mueller, M.J.; Dietz, K.-J. Kinetics of retrograde signalling initiation in the high light response of arabidopsis thaliana. *Philos. Trans. R. Soc. Lond. Ser. B Biol. Sci.* **2014**, *369*, 20130424. [CrossRef] [PubMed]

44. Mullineaux, P.M.; Exposito-Rodriguez, M.; Laissue, P.P.; Smirnoff, N. Ros-dependent signalling pathways in plants and algae exposed to high light: Comparisons with other eukaryotes. *Free Radic. Biol. Med.* **2018**, *122*, 52–64. [CrossRef]

45. Müller, P.; Li, X.-P.; Niyogi, K.K. Non-photochemical quenching. A response to excess light energy. *Plant Physiol.* **2001**, *125*, 1558. [CrossRef] [PubMed]

46. Müller-Moulé, P.; Havaux, M.; Niyogi, K.K. Zeaxanthin deficiency enhances the high light sensitivity of an ascorbate-deficient mutant of arabidopsis. *Plant Physiol.* **2003**, *133*, 748–760. [CrossRef]

47. Müller-Moulé, P.; Conklin, P.L.; Niyogi, K.K. Ascorbate deficiency can limit violaxanthin de-epoxidase activity in vivo. *Plant Physiol.* **2002**, *128*, 970–977. [CrossRef]

48. Giacomelli, L.; Rudella, A.; van Wijk, K.J. High light response of the thylakoid proteome in arabidopsis wild type and the ascorbate-deficient mutant vtc2-2. A comparative proteomics study. *Plant Physiol.* **2006**, *141*, 685–701. [CrossRef]

49. Berwal, M.K.; Ram, C. Superoxide dismutase: A stable biochemical marker for abiotic stress tolerance in higher plants. In *Abiotic and Biotic Stress in Plants*; IntechOpen: London, UK, 2018; p. 10.

50. Schroda, M.; Vallon, O.; Wollman, F.A.; Beck, C.F. A chloroplast-targeted heat shock protein 70 (hsp70) contributes to the photoprotection and repair of photosystem ii during and after photoinhibition. *Plant Cell* **1999**, *11*, 1165–1178. [CrossRef]

51. Li, X.P.; Bjorkman, O.; Shih, C.; Grossman, A.R.; Rosenquist, M.; Jansson, S.; Niyogi, K.K. A pigment-binding protein essential for regulation of photosynthetic light harvesting. *Nature* **2000**, *403*, 391–395. [CrossRef]

52. Nicol, L.; Nawrocki, W.J.; Croce, R. Disentangling the sites of non-photochemical quenching in vascular plants. *Nat. Plants* **2019**, *5*, 1177–1183. [CrossRef]

53. Giacomelli, L.; Masi, A.; Ripoll, D.R.; Lee, M.J.; Van Wijk, K.J. Arabidopsis thaliana deficient in two chloroplast ascorbate peroxidases shows accelerated light-induced necrosis when levels of cellular ascorbate are low. *Plant Mol. Biol.* **2007**, *65*, 627–644. [CrossRef] [PubMed]

54. Page, M.; Sultana, N.; Paszkiewicz, K.; Florance, H.; Smirnoff, N. The influence of ascorbate on anthocyanin accumulation during high light acclimation in arabidopsis thaliana: Further evidence for redox control of anthocyanin synthesis. *Plant Cell Environ.* **2012**, *35*, 388–404. [CrossRef] [PubMed]

55. Vanderauwera, S.; Zimmermann, P.; Rombauts, S.; Vandenabeele, S.; Langebartels, C.; Gruissem, W.; Inzé, D.; Van Breusegem, F. Genome-wide analysis of hydrogen peroxide-regulated gene expression in arabidopsis reveals a high light-induced transcriptional cluster involved in anthocyanin biosynthesis. *Plant Physiol.* **2005**, *139*, 806–821. [CrossRef] [PubMed]

56. Dowdle, J.; Ishikawa, T.; Gatzek, S.; Rolinski, S.; Smirnoff, N. Two genes in arabidopsis thaliana encoding gdp-l-galactose phosphorylase are required for ascorbate biosynthesis and seedling viability. *Plant J.* **2007**, *52*, 673–689. [CrossRef]

57. Laing, W.; Norling, C.; Brewster, D.; Wright, M.; Bulley, S. Ascorbate concentration in arabidopsis thaliana and expression of ascorbate related genes using rnaseq in response to light and the diurnal cycle. *bioRxiv* **2017**. [CrossRef]

58. Müller-Moulé, P. An expression analysis of the ascorbate biosynthesis enzyme vtc2. *Plant Mol. Biol.* **2008**, *68*, 31–41. [CrossRef]

59. Yabuta, Y.; Mieda, T.; Rapolu, M.; Nakamura, A.; Motoki, T.; Maruta, T.; Yoshimura, K.; Ishikawa, T.; Shigeoka, S. Light regulation of ascorbate biosynthesis is dependent on the photosynthetic electron transport chain but independent of sugars in arabidopsis. *J. Exp. Bot.* **2007**, *58*, 2661–2671. [CrossRef]

60. Bulley, S.M.; Rassam, M.; Hoser, D.; Otto, W.; Schünemann, N.; Wright, M.; MacRae, E.; Gleave, A.; Laing, W. Gene expression studies in kiwifruit and gene over-expression in arabidopsis indicates that gdp-l-galactose guanyltransferase is a major control point of vitamin c biosynthesis. *J. Exp. Bot.* **2009**, *60*, 765–778. [CrossRef]

61. Laing, W.A.; Wright, M.A.; Cooney, J.; Bulley, S.M. The missing step of the l-galactose pathway of ascorbate biosynthesis in plants, an l-galactose guanyltransferase, increases leaf ascorbate content. *Proc. Natl. Acad. Sci. USA* **2007**, *104*, 9534–9539. [CrossRef]

62. Mellidou, I.; Chagné, D.; Laing, W.A.; Keulemans, J.; Davey, M.W. Allelic variation in paralogs of gdp-l-galactose phosphorylase is a major determinant of vitamin c concentrations in apple fruit. *Plant Physiol.* **2012**, *160*, 1613. [CrossRef]

63. Li, J.; Liang, D.; Li, M.; Ma, F. Light and abiotic stresses regulate the expression of gdp-l-galactose phosphorylase and levels of ascorbic acid in two kiwifruit genotypes via light-responsive and stress-inducible cis-elements in their promoters. *Planta* **2013**, *238*, 535–547. [CrossRef] [PubMed]

64. Massot, C.; Stevens, R.; Génard, M.; Longuenesse, J.-J.; Gautier, H. Light affects ascorbate content and ascorbate-related gene expression in tomato leaves more than in fruits. *Planta* **2012**, *235*, 153–163. [CrossRef]

65. Urzica, E.I.; Adler, L.N.; Page, M.D.; Linster, C.L.; Arbing, M.A.; Casero, D.; Pellegrini, M.; Merchant, S.S.; Clarke, S.G. Impact of oxidative stress on ascorbate biosynthesis in chlamydomonas via regulation of the vtc2 gene encoding a gdp-l-galactose phosphorylase. *J. Biol. Chem.* **2012**, *287*, 14234–14245. [CrossRef] [PubMed]

66. Vidal-Meireles, A.; Neupert, J.; Zsigmond, L.; Rosado-Souza, L.; Kovács, L.; Nagy, V.; Galambos, A.; Fernie, A.R.; Bock, R.; Tóth, S.Z. Regulation of ascorbate biosynthesis in green algae has evolved to enable rapid stress-induced response via the vtc2 gene encoding gdp-l-galactose phosphorylase. *New Phytol.* **2017**, *214*, 668–681. [CrossRef] [PubMed]

67. Fenech, M.; Amaya, I.; Valpuesta, V.; Botella, M.A. Vitamin c content in fruits: Biosynthesis and regulation. *Front. Plant Sci.* **2018**, *9*, 2006. [CrossRef]

68. Smirnoff, N.; Wheeler, G.L. Ascorbic acid in plants: Biosynthesis and function. *Crit. Rev. Biochem. Mol. Biol.* **2000**, *35*, 291–314. [CrossRef]

69. Bartoli, C.G.; Yu, J.; Gomez, F.; Fernandez, L.; McIntosh, L.; Foyer, C.H. Inter-relationships between light and respiration in the control of ascorbic acid synthesis and accumulation in arabidopsis thaliana leaves. *J. Exp. Bot.* **2006**, *57*, 1621–1631. [CrossRef]

70. Leferink, N.G.; van Duijn, E.; Barendregt, A.; Heck, A.J.; van Berkel, W.J. Galactonolactone dehydrogenase requires a redox-sensitive thiol for optimal production of vitamin c. *Plant Physiol.* **2009**, *150*, 596–605. [CrossRef]

71. de Pinto, M.C.; Locato, V.; Paradiso, A.; De Gara, L. Role of redox homeostasis in thermo-tolerance under a climate change scenario. *Ann. Bot.* **2015**, *116*, 487–496. [CrossRef]

72. Zheng, X.T.; Zhang, X.H.; Wang, Y.Z.; Cai, M.L.; Li, M.; Zhang, T.-J.; Peng, C. Identification of a gldh-overexpressing arabidopsis mutant and its responses to high-light stress. *Photosynthetica* **2018**, *57*. [CrossRef]

73. Ntagkas, N.; Woltering, E.J.; Marcelis, L.F.M. Light regulates ascorbate in plants: An integrated view on physiology and biochemistry. *Environ. Exp. Bot.* **2018**, *147*, 271–280. [CrossRef]

74. Fukunaga, K.; Fujikawa, Y.; Esaka, M. Light regulation of ascorbic acid biosynthesis in rice via light responsive cis-elements in genes encoding ascorbic acid biosynthetic enzymes. *Biosci. Biotechnol. Biochem.* **2010**, *74*, 888–891. [CrossRef] [PubMed]

75. Gao, Y.; Badejo, A.A.; Shibata, H.; Sawa, Y.; Maruta, T.; Shigeoka, S.; Page, M.; Smirnoff, N.; Ishikawa, T. Expression analysis of the vtc2 and vtc5 genes encoding gdp-l-galactose phosphorylase, an enzyme involved in ascorbate biosynthesis, in arabidopsis thaliana. *Biosci. Biotech. Biochem.* **2011**, *75*, 1783–1788. [CrossRef] [PubMed]

76. Zhang, W.; Lorence, A.; Gruszewski, H.A.; Chevone, B.I.; Nessler, C.L. Amr1, an arabidopsis gene that coordinately and negatively regulates the mannose/l-galactose ascorbic acid biosynthetic pathway. *Plant Physiol.* **2009**, *150*, 942–950. [CrossRef]

77. Zhang, Z.; Wang, J.; Zhang, R.; Huang, R. The ethylene response factor aterf98 enhances tolerance to salt through the transcriptional activation of ascorbic acid synthesis in arabidopsis. *Plant J.* **2012**, *71*, 273–287. [CrossRef]

78. Ye, J.; Li, W.; Ai, G.; Li, C.; Liu, G.; Chen, W.; Wang, B.; Wang, W.; Lu, Y.; Zhang, J.; et al. Genome-wide association analysis identifies a natural variation in basic helix-loop-helix transcription factor regulating ascorbate biosynthesis via d-mannose/l-galactose pathway in tomato. *PLoS Genet.* **2019**, *15*, e1008149. [CrossRef]

79. Bulley, S.; Laing, W. The regulation of ascorbate biosynthesis. *Curr. Opin. Plant Biol.* **2016**, *33*, 15–22. [CrossRef]

80. Mellidou, I.; Kanellis, A.K. Genetic control of ascorbic acid biosynthesis and recycling in horticultural crops. *Front. Chem.* **2017**, *5*, 50. [CrossRef]

81. Wang, J.; Yu, Y.; Zhang, Z.; Quan, R.; Zhang, H.; Ma, L.; Deng, X.W.; Huang, R. Arabidopsis csn5b interacts with vtc1 and modulates ascorbic acid synthesis. *Plant Cell* **2013**, *25*, 625–636. [CrossRef]

82. Laing, W.A.; Martinez-Sanchez, M.; Wright, M.A.; Bulley, S.M.; Brewster, D.; Dare, A.P.; Rassam, M.; Wang, D.; Storey, R.; Macknight, R.C.; et al. An upstream open reading frame is essential for feedback regulation of ascorbate biosynthesis in arabidopsis. *Plant Cell* **2015**, *27*, 772–786. [CrossRef]

83. Conklin, P.L.; De Paolo, D.; Wintle, B.; Schatz, C.; Buckenmeyer, G. Identification of arabidopsis vtc3 as a putative and unique dual function protein kinase: Protein phosphatase involved in the regulation of the ascorbic acid pool in plants. *J. Exp. Bot.* **2013**, *64*, 2793–2804. [CrossRef] [PubMed]

84. Ntagkas, N.; Woltering, E.; Bouras, S.; de Vos, R.C.H.; Dieleman, J.A.; Nicole, C.C.S.; Labrie, C.; Marcelis, L.F.M. Light-induced vitamin c accumulation in tomato fruits is independent of carbohydrate availability. *Plants (Basel)* **2019**, *8*, E86. [CrossRef] [PubMed]

85. Massot, C.; Génard, M.; Stevens, R.; Gautier, H. Fluctuations in sugar content are not determinant in explaining variations in vitamin c in tomato fruit. *Plant Physiol. Biochem.* **2010**, *48*, 751–757. [CrossRef]

86. Ntagkas, N.; Woltering, E.; Nicole, C.; Labrie, C.; Marcelis, L.F.M. Light regulation of vitamin c in tomato fruit is mediated through photosynthesis. *Environ. Exp. Bot.* **2019**, *158*, 180–188. [CrossRef]

87. Bartoli, C.G.; Pastori, G.M.; Foyer, C.H. Ascorbate biosynthesis in mitochondria is linked to the electron transport chain between complexes iii and iv. *Plant Physiol.* **2000**, *123*, 335–343. [CrossRef]

88. Schimmeyer, J.; Bock, R.; Meyer, E.H. L-galactono-1,4-lactone dehydrogenase is an assembly factor of the membrane arm of mitochondrial complex i in arabidopsis. *Plant Mol. Biol.* **2016**, *90*, 117–126. [CrossRef] [PubMed]

89. Del-Saz, N.F.; Ribas-Carbo, M.; McDonald, A.E.; Lambers, H.; Fernie, A.R.; Florez-Sarasa, I. *An In Vivo Perspective of the Role(s) of the Alternative Oxidase Pathway*; Elsevier Ltd.: Amsterdam, The Netherlands, 2018; Volume 23, pp. 206–219.

90. Chew, O.; Whelan, J.; Millar, A.H. Molecular definition of the ascorbate-glutathione cycle in arabidopsis mitochondria reveals dual targeting of antioxidant defenses in plants. *J. Biol. Chem.* **2003**, *278*, 46869–46877. [CrossRef] [PubMed]

91. Karpinska, B.; Zhang, K.; Rasool, B.; Pastok, D.; Morris, J.; Verrall, S.R.; Hedley, P.E.; Hancock, R.D.; Foyer, C.H. The redox state of the apoplast influences the acclimation of photosynthesis and leaf metabolism to changing irradiance. *Plant Cell Environ.* **2018**, *41*, 1083–1097. [CrossRef]

92. Müller-Moulé, P.; Golan, T.; Niyogi, K.K. Ascorbate-deficient mutants of arabidopsis grow in high light despite chronic photooxidative stress. *Plant Physiol.* **2004**, *134*, 1163. [CrossRef]

93. Asada, K. The water-water cycle as alternative photon and electron sinks. *Philos. Trans. R. Soc. Lond. Ser. B Biol. Sci.* **2000**, *355*, 1419–1431. [CrossRef]

94. Foyer, C.H. The role of ascorbate in plants, interactions with photosynthesis, and regulatory significance. *Curr. Top. Plant Physiol.* **1991**, *6*, 131–144.

95. Heber, U. Irrungen, wirrungen? The mehler reaction in relation to cyclic electron transport in c3 plants. In *Discoveries in Photosynthesis*; Govindjee, G., Beatty, J.T., Gest, H., Allen, J.F., Eds.; Springer: Dordrecht, The Netherlands, 2005; pp. 551–559.

96. Mano, J.I.; Hideg, E.; Asada, K. Ascorbate in thylakoid lumen functions as an alternative electron donor to photosystem ii and photosystem i. *Arch. Biochem. Biophys.* **2004**, *429*, 71–80. [CrossRef] [PubMed]

97. Tóth, S.Z.; Puthur, J.T.; Nagy, V.; Garab, G. Experimental evidence for ascorbate-dependent electron transport in leaves with inactive oxygen-evolving complexes. *Plant Physiol.* **2009**, *149*, 1568. [CrossRef] [PubMed]

98. Kiddle, G.; Pastori, G.M.; Bernard, S.; Pignocchi, C.; Antoniw, J.; Verrier, P.J.; Foyer, C.H. Effects of leaf ascorbate content on defense and photosynthesis gene expression in arabidopsis thaliana. *Antioxid. Redox Signal.* **2003**, *5*, 23–32. [CrossRef]

99. Plumb, W.; Townsend, A.J.; Rasool, B.; Alomrani, S.; Razak, N.; Karpinska, B.; Ruban, A.V.; Foyer, C.H. Ascorbate-mediated regulation of growth, photoprotection, and photoinhibition in arabidopsis thaliana. *J. Exp. Bot.* **2018**, *69*, 2823–2835. [CrossRef]

100. Nunes-Nesi, A.; Carrari, F.; Lytovchenko, A.; Smith, A.M.; Loureiro, M.E.; Ratcliffe, R.G.; Sweetlove, L.J.; Fernie, A.R. Enhanced photosynthetic performance and growth as a consequence of decreasing mitochondrial malate dehydrogenase activity in transgenic tomato plants. *Plant Physiol.* **2005**, *137*, 611–622. [CrossRef]

101. Alhagdow, M.; Mounet, F.; Gilbert, L.; Nunes-Nesi, A.; Garcia, V.; Just, D.; Petit, J.; Beauvoit, B.; Fernie, A.R.; Rothan, C.; et al. Silencing of the mitochondrial ascorbate synthesizing enzyme l-galactono-1,4-lactone dehydrogenase affects plant and fruit development in tomato. *Plant Physiol.* **2007**, *145*, 1408–1422. [CrossRef]

102. Tomaz, T.; Bagard, M.; Pracharoenwattana, I.; Linden, P.; Lee, C.P.; Carroll, A.J.; Stroher, E.; Smith, S.M.; Gardestrom, P.; Millar, A.H. Mitochondrial malate dehydrogenase lowers leaf respiration and alters photorespiration and plant growth in arabidopsis. *Plant Physiol.* **2010**, *154*, 1143–1157. [CrossRef]

103. Bocobza, S.; Adato, A.; Mandel, T.; Shapira, M.; Nudler, E.; Aharoni, A. Riboswitch-dependent gene regulation and its evolution in the plant kingdom. *Genes Dev.* **2007**, *21*, 2874–2879. [CrossRef]

104. Raschke, M.; Burkle, L.; Muller, N.; Nunes-Nesi, A.; Fernie, A.R.; Arigoni, D.; Amrhein, N.; Fitzpatrick, T.B. Vitamin b1 biosynthesis in plants requires the essential iron sulfur cluster protein, thic. *Proc. Natl. Acad. Sci. USA* **2007**, *104*, 19637–19642. [CrossRef]

105. Kong, D.; Zhu, Y.; Wu, H.; Cheng, X.; Liang, H.; Ling, H.Q. Atthic, a gene involved in thiamine biosynthesis in *arabidopsis thaliana*. *Cell Res.* **2008**, *18*, 566–576. [CrossRef] [PubMed]

106. Machado, C.R.; de Oliveira, R.L.; Boiteux, S.; Praekelt, U.M.; Meacock, P.A.; Menck, C.F. Thi1, a thiamine biosynthetic gene in *arabidopsis thaliana*, complements bacterial defects in DNA repair. *Plant Mol. Biol.* **1996**, *31*, 585–593. [CrossRef] [PubMed]

107. Chatterjee, A.; Abeydeera, N.D.; Bale, S.; Pai, P.J.; Dorrestein, P.C.; Russell, D.H.; Ealick, S.E.; Begley, T.P. Saccharomyces cerevisiae thi4p is a suicide thiamine thiazole synthase. *Nature* **2011**, *478*, 542–546. [CrossRef] [PubMed]

108. Chatterjee, A.; Jurgenson, C.T.; Schroeder, F.C.; Ealick, S.E.; Begley, T.P. Biosynthesis of thiamin thiazole in eukaryotes: Conversion of nad to an advanced intermediate. *J. Am. Chem. Soc.* **2007**, *129*, 2914–2922. [CrossRef] [PubMed]

109. Chatterjee, A.; Schroeder, F.C.; Jurgenson, C.T.; Ealick, S.E.; Begley, T.P. Biosynthesis of the thiamin-thiazole in eukaryotes: Identification of a thiazole tautomer intermediate. *J. Am. Chem. Soc.* **2008**, *130*, 11394–11398. [CrossRef]

110. Ajjawi, I.; Tsegaye, Y.; Shintani, D. Determination of the genetic, molecular, and biochemical basis of the *arabidopsis thaliana* thiamin auxotroph th1. *Arch. Biochem. Biophys.* **2007**, *459*, 107–114. [CrossRef]

111. Hsieh, W.-Y.; Liao, J.-C.; Wang, H.-T.; Hung, T.-H.; Tseng, C.-C.; Chung, T.-Y.; Hsieh, M.-H. The arabidopsis thiamin-deficient mutant pale green1 lacks thiamin monophosphate phosphatase of the vitamin b1 biosynthesis pathway. *Plant J.* **2017**, *91*, 145–157. [CrossRef]

112. Dong, W.; Stockwell, V.O.; Goyer, A. Enhancement of thiamin content in arabidopsis thaliana by metabolic engineering. *Plant Cell Physiol.* **2015**, *56*, 2285–2296. [CrossRef]

113. Colinas, M.; Fitzpatrick, T.B. Natures balancing act: Examining biosynthesis de novo, recycling and processing damaged vitamin b metabolites. *Curr. Opin. Plant Biol.* **2015**, *25*, 98–106. [CrossRef]

114. Goyer, A. Thiamin biofortification of crops. *Curr. Opin. Biotechnol.* **2016**, *44*, 1–7. [CrossRef]

115. Colinas, M.; Eisenhut, M.; Tohge, T.; Pesquera, M.; Fernie, A.R.; Weber, A.P.M.; Fitzpatrick, T.B. Balancing of b6 vitamers is essential for plant development and metabolism in arabidopsis. *Plant Cell* **2016**, *28*, 439–453. [CrossRef] [PubMed]

116. Goyer, A. Thiamine in plants: Aspects of its metabolism and functions. *Phytochemistry* **2010**, *71*, 1615–1624. [CrossRef] [PubMed]

117. Wachter, A.; Tunc-Ozdemir, M.; Grove, B.C.; Green, P.J.; Shintani, D.K.; Breaker, R.R. Riboswitch control of gene expression in plants by splicing and alternative 3′ end processing of mrnas. *Plant Cell* **2007**, *19*, 3437–3450. [CrossRef] [PubMed]

118. Fenwick, M.K.; Mehta, A.P.; Zhang, Y.; Abdelwahed, S.H.; Begley, T.P.; Ealick, S.E. Non-canonical active site architecture of the radical sam thiamin pyrimidine synthase. *Nat. Commun.* **2015**, *6*, 6480. [CrossRef]

119. Rapala-Kozik, M.; Olczak, M.; Ostrowska, K.; Starosta, A.; Kozik, A. Molecular characterization of the thi3 gene involved in thiamine biosynthesis in zea mays: Cdna sequence and enzymatic and structural properties of the recombinant bifunctional protein with 4-amino-5-hydroxymethyl-2-methylpyrimidine (phosphate) kinase and thiamine monophosphate synthase activities. *Biochem. J.* **2007**, *408*, 149–159.

120. Li, C.-L.; Wang, M.; Wu, X.-M.; Chen, D.-H.; Lv, H.-J.; Shen, J.-L.; Qiao, Z.; Zhang, W. Thi1, a thiamine thiazole synthase, interacts with ca(2+)-dependent protein kinase cpk33 and modulates the s-type anion channels and stomatal closure in arabidopsis. *Plant Physiol.* **2016**, *170*, 1090–1104. [CrossRef]

121. Fitzpatrick, T.B.; Thore, S. Complex behavior: From cannibalism to suicide in the vitamin b1 biosynthesis world. *Curr. Opin. Struct. Biol.* **2014**, *29*, 34–43. [CrossRef]

122. Taiz, L.; Zeiger, E. Chapter 8—Photosynthesis: Carbon reactions. In *Plant Physiol*, 4th ed.; Sinauer Associates, Inc.: Sunderland, MA, USA, 2006; pp. 145–170.

123. Julliard, J.H.; Douce, R. Biosynthesis of the thiazole moiety of thiamin (vitamin b1) in higher plant chloroplasts. *Proc. Natl. Acad. Sci. USA* **1991**, *88*, 2042–2045. [CrossRef]

124. Binder, S. Branched-chain amino acid metabolism in arabidopsis thaliana. *Arab Book* **2010**, *8*, e0137. [CrossRef]

125. Du, Q.; Wang, H.; Xie, J. Thiamin (vitamin b_1) biosynthesis and regulation: A rich source of antimicrobial drug targets? *Int. J. Biol. Sci.* **2011**, *7*, 41–52. [CrossRef]

126. Bunik, V.I.; Fernie, A.R. Metabolic control exerted by the 2-oxoglutarate dehydrogenase reaction: A cross-kingdom comparison of the crossroad between energy production and nitrogen assimilation. *Biochem. J.* **2009**, *422*, 405–421. [CrossRef] [PubMed]

127. Araújo, W.L.; Martins, A.O.; Fernie, A.R.; Tohge, T. 2-oxoglutarate: Linking tca cycle function with amino acid, glucosinolate, flavonoid, alkaloid, and gibberellin biosynthesis. *Front. Plant Sci.* **2014**, *5*, 552. [CrossRef] [PubMed]

128. Araújo, W.L.; Nunes-Nesi, A.; Nikoloski, Z.; Sweetlove, L.J.; Fernie, A.R. Metabolic control and regulation of the tricarboxylic acid cycle in photosynthetic and heterotrophic plant tissues. *Plant Cell Environ.* **2012**, *35*, 1–21. [CrossRef] [PubMed]

129. Bunik, V. *Vitamin-Dependent Multienzyme Complexes of 2-Oxo Acid Dehydrogenases: Structure, Function, Regulation and Medical Implications*; Nova Science Publisher: Hauppauge, NY, USA, 2017.

130. Peng, C.; Uygun, S.; Shiu, S.H.; Last, R.L. The impact of the branched-chain ketoacid dehydrogenase complex on amino acid homeostasis in arabidopsis. *Plant Physiol.* **2015**, *169*, 1807–1820. [CrossRef] [PubMed]

131. Gass, N.; Glagotskaia, T.; Mellema, S.; Stuurman, J.; Barone, M.; Mandel, T.; Roessner-Tunali, U.; Kuhlemeier, C. Pyruvate decarboxylase provides growing pollen tubes with a competitive advantage in petunia. *Plant Cell* **2005**, *17*, 2355–2368. [CrossRef] [PubMed]

132. Bunik, V.I.; Tylicki, A.; Lukashev, N.V. Thiamin diphosphate-dependent enzymes: From enzymology to metabolic regulation, drug design and disease models. *FEBS J.* **2013**, *280*, 6412–6442. [CrossRef] [PubMed]

133. Martinis, J.; Gas-Pascual, E.; Szydlowski, N.; Crevecoeur, M.; Gisler, A.; Burkle, L.; Fitzpatrick, T.B. Long-distance transport of thiamine (vitamin b1) is concomitant with that of polyamines. *Plant Physiol.* **2016**, *171*, 542–553. [CrossRef]

134. Zhang, J.; Li, B.; Yang, Y.; Mu, P.; Qian, W.; Dong, L.; Zhang, K.; Liu, X.; Qin, H.; Ling, H.; et al. A novel allele of l-galactono-1,4-lactone dehydrogenase is associated with enhanced drought tolerance through affecting stomatal aperture in common wheat. *Sci. Rep.* **2016**, *6*, 30177. [CrossRef]

135. Ahn, I.P.; Kim, S.; Lee, Y.H. Vitamin b_1 functions as an activator of plant disease resistance. *Plant Physiol.* **2005**, *138*, 1505–1515. [CrossRef]

136. Ribeiro, D.T.; Farias, L.P.; de Almeida, J.D.; Kashiwabara, P.M.; Ribeiro, A.F.; Silva-Filho, M.C.; Menck, C.F.; Van Sluys, M.A. Functional characterization of the thi1 promoter region from arabidopsis thaliana. *J. Exp. Bot.* **2005**, *56*, 1797–1804. [CrossRef]

137. Ferreira, S.; Hjerno, K.; Larsen, M.; Wingsle, G.; Larsen, P.; Fey, S.; Roepstorff, P.; Salome Pais, M. Proteome profiling of populus euphratica oliv. Upon heat stress. *Ann. Bot.* **2006**, *98*, 361–377. [CrossRef] [PubMed]

138. Wong, C.E.; Li, Y.; Labbe, A.; Guevara, D.; Nuin, P.; Whitty, B.; Diaz, C.; Golding, G.B.; Gray, G.R.; Weretilnyk, E.A.; et al. Transcriptional profiling implicates novel interactions between abiotic stress and hormonal responses in thellungiella, a close relative of arabidopsis. *Plant Physiol.* **2006**, *140*, 1437–1450. [CrossRef] [PubMed]

139. Ahn, I.P.; Kim, S.; Lee, Y.H.; Suh, S.C. Vitamin b_1-induced priming is dependent on hydrogen peroxide and the npr1 gene in arabidopsis. *Plant Physiol.* **2007**, *143*, 838–848. [CrossRef] [PubMed]

140. Chi, W.; Sun, X.; Zhang, L. Intracellular signaling from plastid to nucleus. *Annu. Rev. Plant Biol.* **2013**, *64*, 559–582. [CrossRef]

141. Pogson, B.J.; Woo, N.S.; Forster, B.; Small, I.D. Plastid signalling to the nucleus and beyond. *Trends Plant Sci.* **2008**, *13*, 602–609. [CrossRef]

Perspective

Flexibility in the Energy Balancing Network of Photosynthesis Enables Safe Operation under Changing Environmental Conditions

Berkley J. Walker [1,2,*], **David M. Kramer** [1,3], **Nicholas Fisher** [1] **and Xinyu Fu** [1]

1   Plant Research Laboratory, Michigan State University, East Lansing, MI 48823, USA;
    kramerd8@msu.edu (D.M.K.); nefisher@msu.edu (N.F.); fuxinyu2@msu.edu (X.F.)
2   Department of Plant Biology, Michigan State University, East Lansing, MI 48823, USA
3   Department of Biochemistry and Molecular Biology, Michigan State University, East Lansing, MI 48823, USA
*   Correspondence: berkley@msu.edu; Tel.: +517-355-3928

Received: 17 January 2020; Accepted: 15 February 2020; Published: 1 March 2020

Abstract: Given their ability to harness chemical energy from the sun and generate the organic compounds necessary for life, photosynthetic organisms have the unique capacity to act simultaneously as their own power and manufacturing plant. This dual capacity presents many unique challenges, chiefly that energy supply must be perfectly balanced with energy demand to prevent photodamage and allow for optimal growth. From this perspective, we discuss the energy balancing network using recent studies and a quantitative framework for calculating metabolic ATP and NAD(P)H demand using measured leaf gas exchange and assumptions of metabolic demand. We focus on exploring how the energy balancing network itself is structured to allow safe and flexible energy supply. We discuss when the energy balancing network appears to operate optimally and when it favors high capacity instead. We also present the hypothesis that the energy balancing network itself can adapt over longer time scales to a given metabolic demand and how metabolism itself may participate in this energy balancing.

Keywords: energy balancing; cyclic electron flux; malate valve; photorespiration; photosynthesis; C3 cycle

---

1. Introduction

Photosynthetic organisms must match energy supply from the light reactions with metabolic demands to enable safe, flexible and efficient photosynthesis. Because of the interdependency between energy supply and metabolic demand, it is valuable to consider this linked energy network and not as a series of separate metabolic processes. This energy balancing network integrates ATP and reductant supply and metabolic demand to allow plants to efficiently and safely harvest energy from the sun under dynamic conditions (Figure 1). From this perspective, we will first discuss the basic mechanisms of energy balancing before presenting the demand for energy balancing under a variety of conditions. We will then discuss how the structure and efficiencies of the energy balancing network are poised to provide and turnover ATP and reducing power under a variety of conditions before exploring how the energy balancing network responds to long-term changes in metabolic demand.

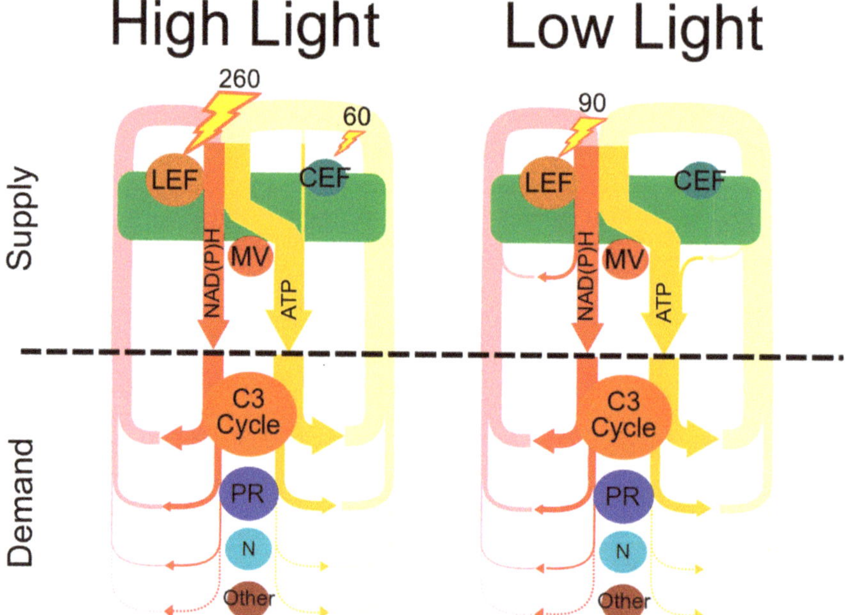

Figure 1. The energy balancing network configured for "high-light and low-efficiency" and "low-light and high-efficiency" conditions. Shown for each light intensity are the energy-producing supply processes of linear electron flux (LEF), cyclic electron flux around photosystem I (CEF) and the malate valve (MV), which produce reducing power (referred to generically as NAD(P)H, in red) and ATP (yellow). Metabolic demand comprises the primary ATP and NADPH consuming processes in C3 plants, the C3 cycle, photorespiration (PR), nitrate assimilation (N) and the remaining metabolic sinks for ATP and NAD(P)H (other). Numbers represent the amount of light energy absorbed by either LEF or CEF (in μmol photons m^{-2} s^{-1}) needed to supply and balance the needs of metabolism. The thickness of all lines is proportional to the fluxes modeled as part of Table 1 calculated using data from Miyake et al. 2005 [1].

2. Energy Balancing is Essential for Safe and Optimal Photosynthetic Systems

The light reactions of photosynthesis provide the chemical energy needed for plant metabolism. The core reactions of oxygenic photosynthesis involve a process called "linear electron flow" (LEF), in which light energy is used to extract electrons from water and transfer them to NADP$^+$ while generating ATP from ADP and P$_i$ [1,2], as detailed in Figure 2. These core processes store energy in two forms; ATP and NADPH. Extracting electrons from water and transferring them to NADP$^+$, energy is stored in the two redox half reactions 4H$^+$ + O$_2$/H$_2$O and NADP$^+$ + H$^+$/NADPH. In addition, the transfer of electrons results in the formation of the proton motive force (*pmf*), an electrochemical gradient of protons across the thylakoid membrane, which is dissipated by the ATP synthase to fuel the formation of ATP from ADP and P$_i$. The pmf is the sum of two energetic components; an electric field component ($\Delta\psi$) and the free energy stored in a chemical gradient of protons (ΔpH). Vectorial electron transfer from the lumenal to the stromal face of the thylakoid membrane, within photosystem II (PSII) and the cytochrome b$_6$f complex by the Q-cycle mechanism (reviewed in [3]) and photosystem I (PSI) results in the formation of $\Delta\psi$. Both $\Delta\psi$ and ΔpH are energetically equivalent drivers of the ATP synthase [4,5], but have very different impacts on photophysiological processes, as discussed below. One important feature of LEF is that it produces ATP and NADPH in a fixed stoichiometry, likely 2.6 ATP to 2 NADPH, or 1.28 ATP/NADPH [6].

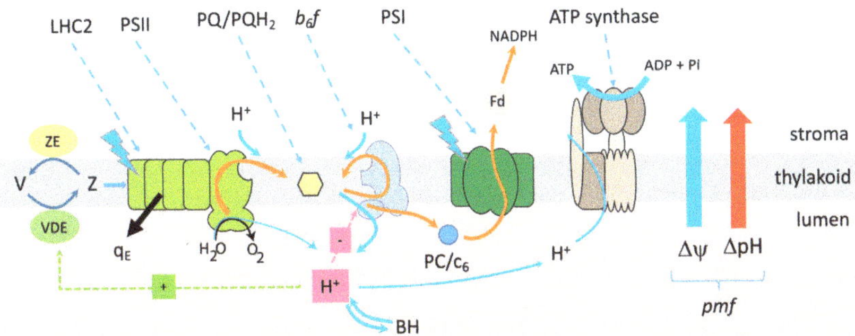

Figure 2. Basic Z-scheme model for the electronic and protonic circuits of the light reactions of photosynthesis, and the *pmf* paradigm for regulation of the light reactions. Scheme of linear electron flow (LEF) in oxygenic photosynthesis, in which light energy is captured by light harvesting complexes associated with photosystem II (PSII) and photosystem I (PSI), which initiates electron flow (orange arrows) from PSII, through the cytochrome b_6f complex, plastocyanin (PC), to PSI and ferredoxin (Fd) and finally to NADPH. Also shown is the formation of the LEF is coupled to proton flow (blue arrows) at PSII and the cytochrome b_6f complex, storing energy in the thylakoid proton motive force (*pmf*). Transfer of electrons from the lumenal to the stromal side of the thylakoid forms a transmembrane electric field (Δψ, blue arrow), while proton uptake from the stroma and deposition in the lumen lead to the formation of a transthylakoid pH gradient (ΔpH, red arrow), which together drive the synthesis of ATP from ADP + P_i at the thylakoid ATP synthase, storing energy in ΔGATP. The acidification of the lumen (indicated by the H⁺ in the red box) activates violaxanthin deepoxidase (VDE) which converts violaxanthin (V) to zeaxanthin (Z) and protonates the PsbS protein, which triggers the photoprotective dissipation of light energy by the qₑ (black arrow). Lumen pH also regulates electron flow (red box with '-') to PSI by slowing the rate of PQH_2 oxidation at the cytochrome b_6f complex.

The chloroplast must also balance the output of energy into the ATP and NADPH pools to perfectly match metabolic demands. The pool sizes of ATP and NADPH are small relative to the high fluxes of energy from the light reactions. Thus, any imbalance in the production and consumption of ATP or NADPH can rapidly lead to "metabolic congestion," depletion or buildup of metabolic intermediates, leading to the accumulation of high energy intermediates of the light reactions within seconds [7–10]. On the other hand, if too little ATP and NADPH are produced metabolic demand is energy limited, meaning that central metabolism is sub-optimal. The "correct" output of ATP and NADPH is a moving target since metabolic demand for ATP and NADPH changes dynamically based on environmental and physiological contexts (See below and [11]). The supply of ATP and NADPH must be matched with demand both in total capacity and stoichiometrically, and therefore plants have evolved mechanisms for regulating total energy output and fine tuning ATP/NADPH production ratios.

To regulate total energy production, chloroplasts partition light energy between photochemical processes which generate ATP and NADPH (LEF) and the energy dissipating process of "non-photochemical quenching" (NPQ) [12–17]. When metabolic demand for energy is less than current supply, the major form of NPQ, termed qₑ (for 'energy dependent' quenching), is triggered by acidification of the lumen (i.e., by the ΔpH component of *pmf*), through activation of violaxanthin deepoxidase, which catalyzes the conversion of violaxanthin to antheraxanthin and zeaxanthin [18], and through protonation of the antenna protein PsbS [19,20]. The ΔpH component of *pmf* also down-regulates electron flow by slowing plastoquinol (PQH_2) oxidation by the cytochrome b_6f complex, preventing accumulation of electrons on highly reducing components of PSI, a process called "photosynthetic control" (reviewed in [21,22]) and subsequent PSI photodamage. Lumen acidification is, in turn, modulated by several processes that respond to the physiological state of the cell [1]. When metabolic demand is low, the activity of the ATP synthase is also down-regulated to slow proton efflux, increasing *pmf* and down-regulation of the light reactions [1,23–27]. The fraction of *pmf* stored in the

ΔpH or $\Delta\psi$ is modulated to adjust its regulatory impact of a particular *pmf* [8,26,28]. The responses of q_E to lumen pH may also be modulated by altering the expression of q_E-related components [29–31]. Quantitatively, the dynamic range of NPQ is large, able to effectively partition from <5% to >80% of absorbed light energy to or away from energy production within tens of minutes. Importantly, even though increased light induces NPQ and decreases photochemical efficiency, the increase in total absorbed photons often more than compensates for this reduction and total LEF increases to safely produce sufficient NADPH to meet metabolic demand. Note that NPQ can only modulate total NADPH production from LEF with no change to the production stoichiometry of 1.28 ATP/NADPH.

3. The Structure of the Energy Network Simplifies ATP and NADPH Balancing

Once the total demand for NADPH is satisfied via the interplay between LEF and NPQ, other processes fine tune ATP/NADPH production ratios to match metabolic demand precisely. Downstream metabolism of an illuminated leaf (discussed in more detail below) requires ATP/NADPH ratios above 1.5, meaning that extra ATP is needed to achieve energy balancing. By poising metabolic demand at a higher ATP/NADPH ratio than that produced by LEF, the system can first produce the necessary NADPH, with coupled baseline production of ATP, before then producing the supplemental ATP needed for the specific metabolic context. This greatly simplifies the requirements of energy balancing since ATP and NADPH production ratios do not need to be independently re-adjusted following changes in total demand and total production capacity can be adjusted first based on a single factor (NADPH demand) before supplemental processes overcome the ATP deficit. In higher plants, three mechanisms are proposed to supply the additional ATP: (1) cyclic electron flux around PSI (CEF), (2) the malate valve and (3) the Mehler reaction. All three of these mechanisms have received extensive coverage in past reviews [6,32–34], and so we will focus on their basic mechanisms and relevance to the particular focus of this perspective.

4. Introduction to Supply-Side Mechanisms for Energy Balancing

4.1. Cyclic Electron Flux around Photosystem I

CEF contributes to the transthylakoid *pmf* without net production of NADPH by cycling electrons from photoexcited PSI via ferredoxin (Fd) back into the thylakoid plastoquinone (PQ) pool via the activity of Fd:PQ reductases (PQR) and the cytochrome b_6f complex [35]. Aside from ATP generation, the proton gradient generated by CEF may also serve a photoprotective function by triggering q_E ('energy dependent') NPQ, although CEF in itself is not essential for this process [8,36]. Many of the details of the electron transport pathways of CEF remain obscure. At least three PQR pathways have been postulated to function in CEF, which may operate in an organism-specific manner; (i) the antimycin A-sensitive Fd:PQ reductase (FQR), which has been proposed to be associated with the PGR5 and/or PGRL1 proteins [37–39], (but see [8,40–43] for additional viewpoints); (ii) the respiratory Complex I-like NADPH/Fd:PQ dehydrogenase (NDH) [44–47] and iii) direct reduction of b_6f-bound PQ through Q_i-associated FNR/Fd via b_6 hemes b_H/c_i [41,48–50]. Of these CEF pathways, those utilizing the proton motive NDH complex is likely to be the most energetically efficient, with a net $H^+/2e^-$ ratio of 8 [46], with the PGR5/PGRL1 and b_6f Q_i pathways yielding an $H^+/2e^-$ ratio of 4 by virtue of the (b_6f-associated) Q-cycle alone [35].

The NDH pathway is, for the most part, associated with plant (and cyanobacterial) CEF, as this enzyme is absent from the majority of algal genera, although it should be noted that it is also absent from certain orchids, cacti and gymnosperms [51]. In general, the electron flux through CEF during steady-state photosynthesis in healthy, non-stressed C3 plants is considered to be small compared to LEF (i.e., $\lesssim$15%) [8,9,50], although it is likely to be (significantly) up-regulated during environmental stress like drought or during the induction of photosynthesis in dark-adapted plants, conditions under which increased ATP demand may be expected [48,52,53]. Nevertheless, this small flux is of vital importance for balancing the ATP and NADPH demands of metabolic supply and demand.

Furthermore, CEF is likely to be of particular importance to C4 photosynthetic species and aquatic algae to generate ATP and proton/ion gradients necessary for the carbon-concentrating mechanisms of these organisms [46,54–56]

4.2. The Malate Valve

The malate valve operates to adjust cellular ATP/NADPH supply by shuttling reducing power from the chloroplast to other organelles like the mitochondria via malate/oxaloacetate shuttles [32,57,58]. In the chloroplast, NADPH reduces oxaloacetate to malate via chloroplastic malate dehydrogenase (MDH). This malate is then exported from the chloroplast where it can be oxidized to form NADH in the cytosol, peroxisome or mitochondria via organelle-specific MDH enzymes. Reducing power shuttled to the mitochondria can fuel mitochondrial electron transport following transfer through the full complement of the electron transport complex proteins, generating additional *pmf* and ATP, or through only a portion of the electron transport complex proteins by dissipation of electrons via the alternative oxidase (AOX) or alternative mitochondrial electron carrier proteins. In all cases, the net effect is to increase ATP/NADPH supply either by decreasing NADPH or by simultaneously decreasing NADPH and increasing ATP. Chloroplastic NADP-MDH operates under tight light regulation via the Fd-thioredoxin (Fd-Trx) system, suggesting a role in photosynthetic energy balancing [59,60]. Importantly, the malate valve offers a way to "trade" NADPH for ATP by diverting reducing equivalence directly into mitochondrial electron transport.

4.3. The Mehler Peroxidase Reaction (Water–Water Cycle)

The term water–water cycle was coined by Asada (1999) [33] to indicate a process wherein electrons from LEF are extracted from water at the oxygen evolving complex of PSII, through the intermediate electron transfer chain, to PSI and to O_2, reforming H_2O. In plants, most of the O_2 reduction occurs the by the Mehler peroxidase reaction (sometimes referred to as the water–water cycle (WWC)), electrons are transferred from low-potential donors (most probably $F_{(X/A/B)}$) within PSI to molecular oxygen, bypassing terminal $NADP^+$ reduction, and producing superoxide. The resulting reactive oxygen species are rapidly detoxified by the activities of superoxide dismutase and the plastid ascorbate peroxidases [33]. This results in H^+ translocation through water oxidation and the Q-cycle, without parallel NADPH production, increasing ATP/NADPH supply. Note, however, that PSI-involvement is a not a strict requirement of a WWC (i.e., the 'traditional' Mehler peroxidase reaction), and the activity of the plastid terminal (plastoquinol) oxidase, also serves to bypass NADPH production, albeit at low capacity [61]. Alternative forms of the WWC are also found in moss, algae and cyanobacteria, where flavodiiron proteins function as NADPH-dependent oxygen reductases [62]. While the WWC may be important under certain stress conditions, current evidence suggests that it does not operate at significant rates under a variety of conditions (i.e., <5% of LEF in tobacco when the C3 was inhibited [63]), and so it will not be further considered in the context of energy balancing [64,65].

5. Metabolic Demand for ATP and NADPH

While plant metabolism employs ATP and NADPH in a myriad of biochemical reactions, the vast majority of ATP and NADPH flux in an illuminated leaf enters metabolic networks at relatively few nodes of central metabolism, most notably CO_2 assimilation and related processes, making it possible to reasonably estimate total ATP/NADPH demand [66]. Some reactions require reductive energy from alternative redox carriers (i.e., Fd or NADH) but for convenience in calculation and discussion, we will refer to reductive demand in terms of NADPH (2 e^-) equivalents. The fixation of each CO_2, and subsequent regeneration of the C3 cycle intermediates requires 3 ATP and 2 NADPH for a total demand of 1.5 ATP/NADPH [67]. In C3 plants growing under ambient conditions, the next largest demand for ATP and NADPH is photorespiration, which results from the molecular fixation of O_2 by the first enzyme of the C3 cycle (rubisco, [66,68]). Photorespiration requires 3.5 ATP and 2 NADPH for complete operation, meaning that as photorespiration increases relative to CO_2 fixation, ATP/NADPH demand increases as well.

Altering the relative rates of photorespiration and carbon fixation will alter the relative demands for ATP and NADPH. Rates of rubisco carboxylation (V_c) and oxygenation (V_o) determine downstream rates and energy requirements for carbon fixation and photorespiration respectively. Since V_c and V_o in C3 plants are constrained by rubisco kinetics, rates of each can be estimated for a given rate of net CO_2 exchange (A) and CO_2 and O_2 concentration to calculate subsequent ATP and NADPH demand [68–72]. While these calculations have been presented in part across many publications, we compile them all herein to make their use more convenient for the non-specialist to use measured gas exchange data to calculate V_c, V_o, ATP and NADPH demand, and extra ATP needed above that provided from LEF (see Appendix A). This quantitative framework requires several simplifying assumptions, but these estimates are close enough to show the magnitude of fluxes and relative impact between conditions.

While carbon fixation and photorespiration comprise the largest portion of central metabolic demand, other metabolic processes such as nitrate assimilation requires a significant contribution. Nitrogen assimilation in leaves involves nitrate reduction into nitrite by nitrate reductase (NR) in the cytosol, translocation of nitrite to chloroplast where it is reduced to ammonium by nitrite reductase (NiR), followed by ammonium assimilation into amino nitrogen via the glutamine synthetase (GS)-glutamine-2-oxoglutarate aminotransferase (GOGAT) pathway in the chloroplasts [73]. Nitrate assimilation to glutamine requires 5 NAD(P)H and 1 ATP. Specifically, reduction of one molecule of nitrate (oxidation state +5) to ammonium (oxidation state −3) requires eight electrons (equivalent to four NADPH), whereas the production of a glutamate via the GS-GOGAT pathway requires an additional two electrons (equivalent to 1 NADPH) and 1 ATP [66]. The reducing power required by the plastidic NiR and GS-GOGAT is supplied from photosynthetic electron transport via the reduced Fd. Higher rates of nitrate assimilation in the light than in the dark [74] reflects the tight connection between photosynthetic metabolism and nitrate assimilation. The reducing power needed for nitrate reduction via the cytosolic NR could be provided by the plastidic NAD-driven malate valve [75]. The NADPH demand for nitrate assimilation is estimated to range from ~ 0.35 to 3 μmol m^{-2} s^{-1} on an area basis, based on the nitrate assimilation rate measured by prior studies [76,77]. These rates of nitrate reduction would require ~2.5%–23% of total LEF in the sample dataset examined in Table 1, making nitrate assimilation a significant electron sink in terms of total electron flux.

Lipid biosynthesis represents another sink for NADPH and ATP consumption in plants, but quantitative estimates of its magnitude have not been reported. Lipids, being an important structural component of membranes, constitute approximately 5% to 10% of the dry weight of vegetative cells of plants [78]. The major constituents of lipids are fatty acids, which can represent up to 10% of the chemical energy of leaves on a biomass basis [79]. The synthesis and breakdown of fatty acids occur constitutively during leaf development. As much as 4% of total fatty acid content in leaves is degraded per day [80]. The turnover of fatty acids is exceeded by the rate of de novo fatty acid synthesis in non-senescent leaves. The net fatty acid accumulation generally increases during leaf expansion, with a rate ranging from 0.16 to 8 μmol carbon atoms mg^{-1} chlorophyll h^{-1} [80–83]. Plant de novo fatty acid synthesis is an energy-demanding process occurred in plastids. ATP drives the first committed step of fatty acid synthesis, the formation of malonyl-CoA from acetyl-CoA catalyzed by acetyl-CoA carboxylase. Reducing power in the form of NADPH and NADH is required for the two reductases involved in each round of fatty acid synthesis [78]. The predominant carbon source of plastidic acetyl-CoA is pyruvate, which is generated from photosynthetically fixed 3-phospho-D-glycerate (3-PGA) via the intermediate phosphoenolpyruvate. For every molecule of palmitic acid (16:0) produced, eight molecules of acetyl-CoA (generation of each acetyl-CoA from 3-PGA regenerates one ATP and one NADH), seven molecules of ATP, and 14 molecules of NAD(P)H are needed, resulting in the consumption of six molecules of NAD(P)H and surplus of one ATP collectively. Based on the total fatty acid content measured in Arabidopsis and *Brachypodium distachyon* (40 μg cm^{-2} leaf area, [83]), we estimate that the NADPH demand to maintain the 4% turnover rate of fatty acids is ~0.5

μmol m^{-2} s^{-1}, which represents approximately 2% and 0.5% of the total NADPH demand under low light and high light, respectively (Table 1). Due to the small pool size of fatty acids in young leaves, the NADPH demand for fatty acid synthesis would be even smaller in the developing leaves. Although the NADPH demand for fatty acid synthesis is relatively small, this process is highly dependent on light and subject to redox regulation [84]. Nevertheless, while up to 2% of total NADPH demand has potential implications to some situations, this is insufficient to significantly affect calculations for total leaf energy balancing.

Table 1. Requirements for energy production for the supply and demand of the energy balancing network under low and high light in *Nicotiana tabacum*. For metabolic demand, shown are rates of CO_2 assimilation (A), intercellular and chloroplastic CO_2 concentration, rates of rubisco carboxylation (v_c) and oxygenation (v_o), rates of nitrate reduction (V_n), rates of lipid production (V_l) and total ATP and NADPH demand. For energy supply shown are photosynthetically active radiation (PAR), measured rates of electron transport through PSII (LEF) and PSI (J_{PSI}), rates of linear electron flux needed to provide sufficient NADPH for metabolic demand (LEF$_{pred}$), ATP produced from LEF$_{pred}$ (ATP$_{LEF}$) and the ATP deficit. For energy balancing, shown are the electron and photon demands for the ATP deficit to be provided by CEF via the NDH, FQR or *b6f* pathways or the malate valve. Details for these calculations found in the text. Values taken from Miyake et al. 2005 [1] indicated with a star (*), with remaining values calculated or assumed herein. For these calculations R$_l$, Γ* and g$_m$ were assumed to be 1.5 μmol CO_2 m^{-2} s^{-1}, 4.7 Pa and 6 μmol CO_2 Pa CO_2^{-1} m^{-2} s^{-1}.

	High Light (1100 PAR)	Low Light (150 PAR)
Metabolic demand		
A (μmol CO_2 m^{-2} s^{-1})	21.3 *	5.7 *
Intercellular CO_2 (Pa)	23.0 *	25.0 *
Chloroplastic CO_2 (Pa)	19.5	24.1
v_c (μmol CO_2 m^{-2} s^{-1})	30.1	8.9
v_o (μmol O_2 m^{-2} s^{-1})	14.5	3.5
v_n (μmol N m^{-2} s^{-1})	1.5	0.5
v_l (μmol N m^{-2} s^{-1})	0.3	0.1
Total ATP demand (μmol ATP m^{-2} s^{-1})	143	40
Total NADPH demand (μmol NADPH m^{-2} s^{-1})	97	27
Total ATP/NADPH ratio	1.47	1.45
Energy supply		
LEF (μmol m^{-2} s^{-1})	132.0 *	45.6 *
J_{PSI} (μmol m^{-2} s^{-1})	192.0 *	56.4 *
LEF$_{pred}$ (μmol m^{-2} s^{-1})	193.9	54.4
ATP$_{LEF}$ (μmol m^{-2} s^{-1})	124.1	34.8
ATP deficit (μmol m^{-2} s^{-1})	18.6	4.7
Energy balancing requirements via CEF		
NDH e$^-$ (μmol m^{-2} s^{-1})	43.3	11.1
NDH photons (μmol m^{-2} s^{-1})	43.3	11.1
FQR/*b6f* e$^-$ (μmol m^{-2} s^{-1})	86.6	22.2
FQR/*b6f* photons (μmol m^{-2} s^{-1})	86.6	22.2
Energy balancing requirements via malate valve		
e$^-$ (μmol m^{-2} s^{-1})	9.3	2.4
Photons (μmol m^{-2} s^{-1})	18.5	4.7

6. Determining the Efficiency of Energy Balancing Mechanisms

As an autotrophic organism, the energy that fuels metabolism in plants is derived ultimately from absorbed photons, providing a metric by which to gauge the efficiency of an energy balancing mechanism. Photon use efficiency has thus provided a logical objective function for approaches that assume photoautotrophs use light energy optimally (i.e., [85]), but given the massive amount of absorbed energy that is dissipated as NPQ under high light, it is not clear that light energy is always limiting to growth. Additionally, given the dynamic fluctuations in energy demand and light supply many plants face under growing conditions, it is likely that the capacity for a given energy balancing mechanism may become more important than its efficiency when light energy supply is adequate. In this section we outline the photon costs of various energy balancing mechanism and incorporate them into a quantitative framework. We then use this framework to examine past work and hypothesize that the energy balancing network operates in a high or low-efficiency mode based on light availability. To examine the energy requirements for energy balancing under various light conditions, the ATP and NADPH demand for the C3 cycle and photorespiration has been determined from past work which paired concurrent gas exchange with measurements of electron flux through PSII and PSI (Figure 1, Table 1 and [86]).

Different pathways of CEF have different costs for energy balancing, depending on how many H^+ are pumped per electron excited by an absorbed photon. As outlined above, the highest efficiency CEF pathway proceeds through NDH, which pumps 4 H^+/2 e^- (Table 2). The FQR and b_6f pathways have identical yields of 2 H^+/2 e^-. Since 14 H^+ are required to generate 3 ATP in the chloroplast, CEF has an ATP/photon or e^- stoichiometry of 0.43 via NDH and 0.21 via FQR or b_6f. Additionally, the e- and photon demand for energy balancing can be calculated for the low and high-light conditions presented in Table 1 and data from Miyake et al., 2005. Under low light, 11 or 22 µmol photons m^{-2} s^{-1} are needed to produce the ATP needed to balance energy supply via NDH or FQR/b_6f pathways, respectively, or between 7% and 15% of the total incident light (Table 1). Under high light, this requirement drops to 4%–8%. To gain a more complete picture of the relative energy cost of these mechanisms under their respective light conditions, the photon demand can be expressed in terms of actual absorbed photon energy that enters photochemistry by adding the rates of flux through PSII and PSI. Interestingly, this recalculation reveals that as light level increases, a greater percentage of photon energy absorbed and passed through the photosystems would need to be partitioned to CEF processes for energy balancing, specifically 11%–22% under low light and 13%–27% under high light. Under high light, however, energy from more photons is dissipated as NPQ compared to low light. If energy to NPQ is considered as excess, this means that there is more excess energy under high light that can be used to drive CEF. Specifically, based on the data from Miyake et al. 2005 [86], energy from only 48 µmol photons m^{-2} s^{-1} was dissipated via NPQ under low-light conditions, but energy from 776 µmol photons m^{-2} s^{-1} was dissipated via NPQ under high-light conditions. These numbers reveal that while photon energy could be limiting to drive CEF under low-light conditions, there appears to be enough surplus photon energy available under high light to drive CEF. This is expected because if NPQ limits excitation of PSII, it should also limit flux to PSI through LEF, but it will not necessarily limit PSI electron flow of electrons through CEF.

Table 2. Energy requirements and efficiencies of CEF pathways and the malate valve to produce supplemental ATP. Shown are the number of absorbed photons used for the calculation of each pathway. Further details and assumptions for calculations are found in the text.

	CEF Pathways			
	NDH	**FQR**	b_6f	**Malate Valve**
Chloroplast				
e-/photons	1	1	1	0.5
H^+/e^-	2	1	1	3
ATP/H^+	0.21	0.21	0.21	0.21
ATP/photon in chloroplast	0.43	0.21	0.21	0.32
Mitochondria				
H^+/e^-	-	-	-	5
ATP/H^+	-	-	-	0.27
ATP/photon in mitochondria	-	-	-	0.68
Total ATP/photon	0.43	0.21	0.21	1.00

The energetics of the malate valve are more difficult to assess given the added complexity of transport and flexibility of mitochondrial electron transport. The initial energetics and efficiency of the malate valve are tied to LEF; eight photons produce two NADPH and 12 H$^+$, resulting in 2.57 ATP. The energetics following the transport of the reducing power of 2 NADPH into mitochondrial electron transport and ATP generation depend on the e$^-$/H$^+$ and, more generally, the e$^-$/ATP efficiency of the mitochondria. For our theoretical evaluation of malate valve energetics, we will first assume mitochondrial electron transport operates optimally and each electron contributes maximally to the proton gradient, passing through Complex I, III and IV to pump 10H$^+$/2e$^-$. To produce ATP, these protons pass through a ring of c-subunits of ATPase, with each full rotation producing three ATP and the number of H$^+$ per rotation depending on the number of c-subunits present in the ring [87–91]. We assume the number of c-subunits is the same as found in animal cells since there is no available data on plant mitochondrial c-subunit number, requiring eight H$^+$/3 ATP, although in yeast there are 10 c-subunits [92]. Since each molecule of ATP synthesized requires the (electroneutral) transport of one P$_i$ with the associated (electrogenic) ADP^{3-}/ATP^{4-} exchange activity of the mitochondrial adenine nucleotide translocase (equivalent to the uptake of an additional proton per molecule of ATP synthesized) [93,94], the final stoichiometry is 11 H$^+$/3 ATP, making a theoretically maximum ATP/oxygen ratio of 2.7 [95]. This stoichiometry is closely matched in experimental measurements of the ratio of 2.6 ADP/oxygen consumed in intact mitochondria in potato [96], suggesting that these stoichiometries reasonably approximate mitochondrial respiration in plants despite the highly-branched potential of mitochondrial electron transport. Therefore, for every two NADPH (4 e$^-$) that are processed via the malate valve, 5.45 ATP are produced in the mitochondria. The above discussion focuses specifically on the ATP produced via mitochondrial respiration fueled by electrons provided from the light reactions, we recognize that some ATP may be produced in the light from pyruvate produced during "dark-type" glycolysis. Exact rates of glycolysis-supplied mitochondrial respiration in the light are not available, but estimates from CO$_2$ gas exchange indicate these rates are rather small compared to net assimilation and lower than rates measured in the dark [97–99], suggesting that the bulk of ATP generated in the mitochondria may come from other sources (such as the Mehler valve), but more information is needed to explore this in more detail.

To integrate the production stoichiometries into a complete malate valve cycle, the costs of transporting ATP from the mitochondria back into the chloroplast where it is primarily needed for the phosphorylation of C3 and photorespiratory cycle intermediates must also be considered. Transport of ATP from the mitochondria proceeds via the ADP/ATP translocase [100,101] and into the chloroplast via the plastidic ADP/ATP transporter [102,103].

The above energetics determine that the malate valve is a highly efficient ATP producer on a photon basis. For every eight photons of light energy, 2.57 ATP are produced in the chloroplast and 5.45 are produced in the mitochondria for a total ATP/photon ratio of 1, much higher than the 0.21–0.43 determined for CEF (Table 2). This high ATP/photon ratio means that much less absorbed light energy is needed for energy balancing assuming low and high-light conditions (Table 1). Specifically, only 4.7 and 18.5 μmol photons m^{-2} s^{-1} were needed under low and high light, respectively (Table 1). This comprises only 3.2$ and 1.7% of incident irradiance for the low and high light intensities.

7. With an Efficient Malate Valve, Why is CEF Important?

Given the theoretically greater efficiency of the malate valve, how can we explain the commonly observed participation of CEF in energy balancing? We propose that the real energy cost of CEF will depend on the light intensity and other factors. At low light, when energy capture is strongly limited by the number of photons captured by the photosystems, activating CEF will require diverting energy from LEF, limiting the overall efficiency of energy capture. However, as light input nears saturation imposed by downstream reactions, PSII efficiency drops substantially, either by decreased efficiency of antenna (increased NPQ), or by increases in the fraction of closed PSII centers. In this case, activating CEF may have little effect on the efficiency of LEF (because it is already light saturated) but will increase total energy capture, albeit with a higher fraction stored in ATP/NADPH. Indeed, experimentally, CEF appears to play a larger role in energy balancing under high, but not low irradiances when there is limiting energy available from absorbed photons [72,86].

For example, when high light intensities drive light-saturated photosynthetic rates, CEF shifts proportionally in response to changing CO_2 to cover ATP deficient predicted from gas exchange [72]. These measurements were made across CO_2 concentrations and reveal the capacity for CEF to respond to changes in leaf energy demand (Figure 3). We refer to the measured change of CEF in response to changing energy balancing requirements as the "dynamic range of CEF". By comparing the measured dynamic range of CEF to the change predicted from gas exchange modeling, the degree by which CEF participates in energy balancing can be determined. The dynamic range of CEF can run into a self-regulating upper limit when the high rates of CEF increase the ΔpH and initiate q_E-dependent NPQ, as occurs in shade leaves exposed to increasing light [104]. This could serve as a protective mechanism, when CEF increases ΔpH to sufficient levels, light harvesting is down-regulated and further energy mismatch is avoided when the capacity for energy balancing by CEF is reached. Interestingly, the dynamic range of CEF was minimal in response to changing energy demanding conditions when measured under low light, indicating that processes other than CEF (e.g., the malate valve) accomplish energy balancing under these low flux conditions [72].

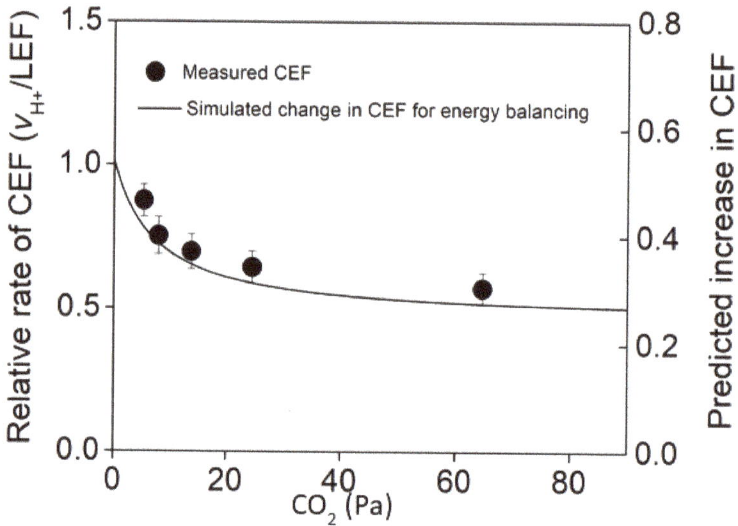

Figure 3. Comparison of the measured relative rate of cyclic electron flux (CEF: circle symbols) to the predicted change in CEF required to match ATP/NADPH supply with demand across CO_2 concentration (line). Shown are n = 3–4 ± SE. This data is a replotted subset of measurements from Walker et al. (2014) [72].

A role for "low-light, high-efficiency" and "high-light, low-efficiency" energy balancing networks is further supported by flux balance analysis of photosynthetic systems and in mutant lines deficient in CEF. Flux balance analysis of photosynthetic systems that are optimized for energy production per photon of absorbed light predict the malate valve to be the optimal mechanism of energy balancing unless the additional costs of enzymatic interconversions are introduced into the model [105]. This is also supported in work using a modified flux balance analysis approach, which weights flux solutions based on pathway complexity [106]. As light intensity increases, absorbed light energy is actively released as NPQ, indicating that under high light, the system is no longer light limited and the energy balancing network could trade the more light-optimal malate valve for CEF. This position is supported by work showing that CEF is critical for plant growth under high, but not low-light conditions [107].

It is not clear what the advantages of CEF might be over the malate valve, but the speed and flexibility of CEF may provide an explanation. The operation of the complete malate valve requires tight coordination of enzymatic and transport activity between the chloroplast and mitochondria, limiting the dynamic range of its energy balancing capacity over short time scales. The malate valve also requires the careful coordination of two electron transport chains in separate organelles, further complicating the upregulation of this pathway under greater energy balancing demand. CEF occurs only in the chloroplast, simplifying the signaling network required to up- or down-regulate ATP production. By contrast, CEF is likely regulated by stromal ATP levels as well as stromal redox state [108] and thus may also be more rapidly responsive to alterations in energy demands, e.g., during induction or rapid changes in light, CO_2, etc. whereas the malate valve appears to require (potentially slower) redox activation (see below). Thus, having at two routes of ATP/NADPH balancing, provides photosynthetic systems with greater flexibility to balance diverse metabolic imbalances as well as providing optimal efficiency under low light (malate valve) or more rapid/responsive responses (CEF) pathways. We further hypothesize that the baseline requirements for energy balance are achieved by the more light-optimal malate valve. This baseline activity satisfies the needs for energy balancing until greater capacity is needed, such as occurs under higher light regimes. Under high light, CEF acts as a highly flexible stop-gap to allow energy balancing to occur under dynamic conditions.

8. Demand-Side Energy Balancing Processes

While there is much focus on how supply-side reactions mediate energy balancing, there is less focus on how metabolic demand itself changes. For example, under increased ATP/NADPH demand, metabolism could either increase the supply of ATP via supply-side mechanisms like CEF or the malate valve or reprogram metabolism itself to use the supply more optimally. A simple example of this is in the redox regulation of the C3 cycle, where multiple redox post-translational modifications may tune activity to available reducing power availability [109].

There is additional evidence for this demand-side reprogramming to achieve energy balancing in the unique link between nitrate assimilation and photorespiration. C3 plants have lower nitrate assimilation rates when photorespiration is reduced through altered atmospheres [76,77,110–112]. This link could be explained if under ambient conditions, the increased demand for ATP/NADPH imposed by photorespiration is offset by increased rates of nitrate assimilation, which has a much lower ATP/NADPH demand. This would achieve ATP/NADPH balance not exclusively by increased ATP supply, but by repartitioning demand-side processes themselves. Interestingly, expression of nitrate assimilation genes increase in NADP-MDH mutants, suggesting that nitrate assimilation could be a compensatory response to achieve energy balancing when the malate valve is disrupted [75]. NADP-MDH mutants also show improved growth on nitrate-rich media [75,113]. These experiments demonstrate that nitrate assimilation and the malate valve may cooperate to maintain a baseline level of energy balancing, increasing the complexity of the energy balancing network.

9. The Acclimation of Energy Balancing Networks to Long-Term Change in Energy Demand

As mentioned above, plants must cope in the long term to changing ATP and NADPH demand to achieve energy balance. It is unclear whether the same mechanisms balance energy mismatches under long time scales as occur under shorter time scales. Furthermore, it is unknown to what degree energy balancing networks poise to a given demand and how this poise acclimates to changing demands. The acclimation of the energy balancing network can be investigated experimentally either via transition experiments or by examining mutants with an altered network capacity that forces flux through alternative facets of the network. We will first discuss the potential for acclimation of supply-side processes to changing energy demand before outlining how metabolic demand itself may acclimate to changes in energy balancing requirements.

According to our model of the two-component supply-side energy balancing system, malate valve activity should scale with excess ATP demand over long-term transitions to optimally use absorbed light. The largest driver of excess ATP demand for any condition is increased light, and so this model predicts that malate valve activity should increase with light. Indeed there appears to be a light and dark malate valve cycle, with the dark cycle relying on plastidic NAD-MDH and the light shuttle using plastidic NADP-MDH [58]. The switch to the NADP-MDH cycle is mediated through the light-dependent Fd-Trx system [114,115]. This activation (at least for NADP-MDH in isolated Pea chloroplasts) occurs within 10–20 min, and so activation of this component is likely integrated in our short-term measurements [116]. Activation occurs even more rapidly during a high-light transition [117]. This is an effective regulatory strategy at short time scales, since it activates malate valve activity when there is a surplus of NADPH, as when the C3 cycle is not consuming it fast enough and reduces it when there is too much $NADP^+$ [58]. It is also likely that as the malate valve is reaching full capacity, CEF plays a role in vivo during very short time scales, at least in C3 plants [118].

Other factors increase malate valve capacity over longer time scales. For example, after transfer to sustained high light, NADP-MDH expression and protein levels increase, suggesting increased capacity of the malate valve following hours of exposure to the new condition [119]. Interestingly, the same response is not observed when photoperiod increases, suggesting that a photosynthetic steady-state solution must be found and that the effect is not cumulative.

Interestingly, chloroplastic *nadh-mdh* mutants show no phenotype, even under stress conditions, potentially due to additional compensatory redox strategies [75,113]. It is important to note that the malate valve shuttles reducing power not only between the chloroplast and mitochondria, but also the peroxisome during photorespiration [120]. In contrast to chloroplastic *nadh-mdh* mutants, mitochondrial *nadh-mdh* mutants lacking both MDH isoforms (*mmdh1mmdh2*) show lowered photosynthetic rates and growth rates [121]. These decreases were likely due to impaired shuttling of reducing power from the mitochondrion to the peroxisome via the malate valve to provide the reducing power for hydroxypyruvate reduction in photorespiration, a viewpoint supported using ^{13}C flux analysis of *mmdh1* [122]. Indeed *mmdh1mmdh2* show reduced photorespiratory capacity, but the reduced growth and photosynthesis is not explained strictly by decreased availability of reductant to photorespiration since mutants lacking the peroxisomal MDH isoforms show an even more subtle phenotype than *mmdh1mmdh2* [123,124]. This work with mitochondrial MDH indicates that the malate valve is not strictly required for energy balancing, but it is important for optimal photosynthesis and long-term growth.

The capacity for CEF itself may also increase over longer time scales to allow for increased energy balancing demand. Notably, NDH and FQR content change under different growth conditions [51]. The ratio of PSI and PSII re-proportions when plants are grown under light regimes with outputs that favor PSI or PSII. After hours or days, this results in changes to the actual stoichiometry of PSI and PSII photosystems, in green algae [125–127] and plants [128]. This re-proportioning also occurs days following transition between different light qualities, which can increase the capacity for CEF [129]. At short time scales, repartitioning of light energy between PSI and PSII occurs when reduced PQ builds up and triggers the phosphorylation of the PSII light harvesting complex. These then migrate to PSI to balance out energy capture [130]. While these state transitions occur in response to long-term differences in energy supply (changes in light regimes), it is not clear if this happens in response to long-term changes in energy demand. Such a change would predict that as conditions decrease in the ratio of photorespiration, there should be a decrease in demand for CEF and, therefore, a decrease in the PSI/PSII ratio.

Measurements of PSI/PSII from plants grown under conditions of different ATP/NADPH demand did not indicate that the capacity for CEF change with energy demand via changes in photosystem stoichiometry. Specifically, there was no difference in PSI/PSII in aspen trees exposed to elevated CO_2 (560 PPM) over a single season following 5 years of elevated high CO_2 treatment [131]. However, this increase in CO_2 is not expected to change the demand for CEF by all this much (~1% of LEF change). Additionally, micro-array work in soybeans exposed to 550 PPM also show no difference in photosystem expression, but interestingly show an increased expression of a mitochondrial ATP/ADP antiporter [132]. Overall, these findings do not point clearly to the acclimation of the capacity of CEF in response to changing energy demand, but the treatments resulted in relatively modest changes in energy demand and CEF was not evaluated specifically. There is clearly room for more work examining this question specifically.

10. Conclusions

The energy-balancing network comprises a flexible set of possibilities that enable partitioning through pathways with different ATP and NADPH production stoichiometries that require different amounts of light energy. We hypothesize that the network partitions flux through high-efficiency pathways (e.g., the Malate valve) when light is limiting and high-efficiency pathways (e.g., CEF) when light is abundant. Furthermore, the energy balancing network may adapt to long-term energy demand through enzyme expression.

Author Contributions: This article was prepared with the following contributions: Conception, B.J.W. and D.M.K.; Formal Analysis, B.J.W.; Writing – original draft, B.J.W., D.M.K., N.F., and X.F.; writing—review and editing, B.J.W., D.M.K., N.F., and X.F.; Funding acquisition, B.J.W. and D.M.K. All authors have read and agreed to the published version of the manuscript.

Funding: This work was supported by the Division of Chemical Sciences, Geosciences and Biosciences, Office of Basic Energy Sciences of the United States Department of Energy [Grant DE-FG02-91ER20021].

Acknowledgments: We would like to thank Yair Shachar-Hill for discussions on flux balance analysis work examining the malate valve and cyclic electron flux.

Conflicts of Interest: The authors declare no conflict of interest.

Abbreviations

Abbreviation	Full Name
ΔpH	The gradient potential component of the proton motive force
$\Delta\psi$	Electric force component of the proton motive force
A	Net CO_2 assimilation
AOX	Alternative oxidase
CEF	Cyclic electron flux around photosystem I
Fd	Ferredoxin
Fd-Trx	Ferredoxin-thioredoxin system
FQR	Ferredoxin:plastoquinone reductase
GS	glutamine synthetase
GOGAT	glutamine-2-oxoglutarate aminotransferase
LEF	Linear electron flow
NDH	NADPH/Ferredoxin:plastoquinone dehydrogenase (NDH)
pmf	Proton motive force
MDH	Malate dehydrogenase
NiR	Nitrite reductase
NPQ	Non-photochemical quenching
NR	Nitrate reductase
PQ	Plastoquinone
PQH_2	Plastoquinol
PQR	Ferredoxin:plastoquinone reductases
PSI	Photosystem I
PSII	Photosystem II
q_E	Energy-dependent quenching
R_l	Respiration in the light
V_c	Rate of rubisco carboxylation
V_o	Rate of rubisco oxygenation

Appendix A

Calculating ATP and NADPH Ratios from Gas Exchange Data

While the theory behind calculating ATP and NADPH demand ratios for rubisco carboxylation (V_c) and oxygenation (V_o) from gas exchange data has been published on previously [6,72], the underlying derivation and final equations have been admittedly lacking. Here, we present a complete derivation, complete with underlying assumptions, appropriate to allow the non-specialist to apply these calculations to their gas exchange data. We will not attempt a full derivation of the underlying biochemical model of leaf photosynthesis, but will instead refer to the relevant equations directly as presented completely previously [133,134].

The cornerstone equation for modelling net CO_2 assimilation (A) is the mass balance subtracting from rates of carbon fixation via V_c and rates of CO_2 loss from photorespiration and respiration in the light (R_l). Rates of photorespiratory CO_2 loss are calculated by multiplying CO_2 loss per rubisco oxygenation (usually assumed to be 0.5) by V_o and total A is represented by Equation (2.1) in von Caemmerer 2000 [133]:

$$A = V_c - 0.5V_o - R_l \tag{A1}$$

Since our goal is to use measured rates of A to estimate V_c and V_o, and subsequent ATP and NADPH demand, it becomes convenient to express Equation (A1) in terms of one unknown variable (V_c) and then solve for V_o. Equation (A1) is expressed in terms of V_c in principle by combining rubisco specificity for CO_2 relative to O_2 ($S_{c/o}$) with measured gas concentrations to determine what catalytic rates of V_c and V_o would produce the measured A. This is accomplished based on the following relationships (Equations (2.16) and (2.18) from von Caemmerer 2000 [133])

$$\phi = \frac{V_o}{V_c} \tag{A2}$$

$$\phi = \frac{2\Gamma^*}{C_c} \tag{A3}$$

where C_c is the partial pressure of CO_2 at the site of rubisco catalysis and Γ^* is the CO_2 compensation point in the absence of R_l defined as

$$\Gamma^* = \frac{0.5O}{S_{c/o}} \tag{A4}$$

where O is the oxygenation partial pressure. Note that since O is part of the definition of Γ^*, it must be scaled according to the measurement concentration if an altered oxygen background is used during the experiment. In using Equation (A1), R_l is assumed or independently measured under the experimental conditions using a variety of gas exchange approaches and treated as a constant [98,135,136]. With Equations (A1)–(A3) we are able to represent the relationship between A and V_c with no other unknown variables

$$V_c = \frac{A + R_l}{1 - \frac{\Gamma^*}{C_c}} \tag{A5}$$

The solution for V_c can then be used with Equation (A1) to solve for V_o.

$$V_o = \frac{V_c - A - R_l}{0.5} \tag{A6}$$

With V_c and V_o determined from the above, the rate of demand for ATP (V_{ATP}) and NAD(P)H (V_{NADPH}) can then be determined based on the requirements for the C3 cycle (3 ATP and 2 NADPH) and photorespiration (3.5 ATP and 2 NAD(P)H, [66,67]) according to

$$V_{ATP} = 3V_c + 3.5V_o \tag{A7}$$

And

$$V_{NADPH} = 2V_c + 2V_o \tag{A8}$$

Of course additional energy demanding processes can be added to Equations (A6) and (A7) to determine total leaf energy demand [72], but we have limited these calculations to those most directly measured using gas exchange.

It should be noted that several of the constants assumed above require additional interpretation depending on the species and conditions they are measured under. These calculations depend on C_c to account for the chloroplastic supply of CO_2, but standard gas exchange measurements can only practically resolve the concentration of CO_2 in the intercellular airspace (C_i). C_i can be converted to C_c assuming a simple linear conductance using Fick's law as

$$C_c = C_i - \frac{A}{g_m} \tag{A9}$$

where g_m is the mesophyll conductance to CO_2 diffusion. Selecting an appropriate g_m to use experimentally is complicated since it varies by species, temperature and the underlying theory used for it estimation [137–145]. Fortunately, under most conditions, V_{ATP} and V_{NADPH} are not extremely sensitive to small errors in g_m assumptions, but a sensitivity analysis can be performed to confirm that the findings of a study are robust. Note that stomatal conductance has a similar impact on changing C_i for a given photosynthetic rate. Since stomata close during drought, this means that the ratio V_o/V_c increases under these water-limiting conditions, increasing metabolic demand for ATP/NADPH. Additionally, the temperature response of Γ^* should be accounted for in addition to its linear dependence on O [138].

References

1. Miyake, C.; Miyata, M.; Shinzaki, Y.; Tomizawa, K.-I. CO_2 Response of Cyclic Electron Flow around PSI (CEF-PSI) in Tobacco Leaves-Relative Electron fluxes through PSI and PSII Determine the Magnitude of Non-photochemical Quenching (NPQ) of Chl Fluorescence. *Plant Cell Physiol.* **2005**, *46*, 629–637. [CrossRef]
2. Kramer, D.M.; Avenson, T.J.; Edwards, G.E. Dynamic flexibility in the light reactions of photosynthesis governed by both electron and proton transfer reactions. *Trends Plant Sci.* **2004**, *9*, 349–357. [CrossRef]
3. Ort, D.R.; Yocum, C.F. Light reactions of oxygenic photosynthesis. In *Oxygenic Photosynthesis: The Light Reactions*; Ort, D.R., Yocum, C.F., Eds.; Kluwer Academic Publishers: Dordrecht, The Netherlands, 1996; pp. 1–9.
4. Cape, J.L.; Bowman, M.K.; Kramer, D.M. Understanding the cytochrome *bc* complexes by what they don't do. The Q-cycle at 30. *Trends Plant Sci.* **2006**, *11*, 46–55. [CrossRef]
5. Fischer, S.; Graber, P. Comparison of DpH- and Dy-driven ATP synthesis catalyzed by the H(+)-ATPases from Escherichia coli or chloroplasts reconstituted into liposomes. *FEBS Lett.* **1999**, *457*, 327–332. [CrossRef]
6. Hangarter, R.P.; Good, N.D. Energy thresholds for ATP synthesis in chloroplasts. *Biochim. Biophys. Acta* **1982**, *681*, 396–404. [CrossRef]
7. Kramer, D.M.; Evans, J.R. The importance of energy balance in improving photosynthetic productivity. *Plant Phys.* **2011**, *155*, 70–78. [CrossRef] [PubMed]
8. Avenson, T.J.; Kanazawa, A.; Cruz, J.A.; Takizawa, K.; Ettinger, W.E.; Kramer, D.M. Integrating the proton circuit into photosynthesis: Progress and challenges. *Plant Cell Environ.* **2005**, *28*, 97–109. [CrossRef]
9. Avenson, T.J.; Cruz, J.A.; Kanazawa, A.; Kramer, D.M. Regulating the proton budget of higher plant photosynthesis. *Proc. Natl. Acad. Sci. USA* **2005**, *102*, 9709–9713. [CrossRef] [PubMed]
10. Cruz, J.A.; Avenson, T.J.; Kanazawa, A.; Takizawa, K.; Edwards, G.E.; Kramer, D.M. Plasticity in light reactions of photosynthesis for energy production and photoprotection. *J. Exp. Bol.* **2005**, *56*, 395–406. [CrossRef] [PubMed]
11. Amthor, J.S. From sunlight to phytomass: On the potential efficiency of converting solar radiation to phyto-energy. *New Phytol.* **2010**, *188*, 939–959. [CrossRef]

12. Blankenship, R.E.; Tiede, D.M.; Barber, J.; Brudvig, G.W.; Fleming, G.; Ghirardi, M.; Gunner, M.R.; Junge, W.; Kramer, D.M.; Melis, A.; et al. Comparing Photosynthetic and Photovoltaic Efficiencies and Recognizing the Potential for Improvement. *Science* **2011**, *332*, 805–809. [CrossRef] [PubMed]

13. Müller, P.; Li, X.-P.; Niyogi, K.K. Non-photochemical quenching. A response to excess light energy. *Plant Physiol.* **2001**, *125*, 1558–1566. [CrossRef] [PubMed]

14. Anderson, B.; Barber, J. Mechanisms of photodamage and protein degradation during photoinhibition of photosystem II. In *Photosynthesis and the Environment*; Baker, N.R., Ed.; Kluwer Academic Publishers: Dordrecht, The Netherlands, 1996; pp. 101–121.

15. Styring, S.; Virgin, I.; Ehrenberg, A.; Andersson, B. Strong light photoinhibition of electron transport in photosystem II. Impairment of the function of the first quinone acceptor, Q_A. *Biochim. Biophys. Acta* **1990**, *1015*, 269. [CrossRef]

16. Aro, E.; Virgin, I.; Andersson, B. Photoinhibition of photosystem II. Inactivation, protein damage and turnover. *Biochim. Biophys. Acta* **1993**, *1143*, 113–134. [CrossRef]

17. Baker, N.R.; Bowyer, J.R. Photoinhibition of photosynthesis from molecular mechanisms to the field. In *Environmental Plant Biology Series*; Davies, W.J., Ed.; Bios Scientific Publishers: Institute of Environmental and Biological Sciences; Bios Scientific Publishers: Institute of Environmental and Biological Sciences, Division of Biological Sciences, University of Lancaster: Oxford, UK, 1994; pp. 1–471.

18. Demmig-Adams, B.; Adams-III, W.W. The role of xanthophyll cycle carotenoids in the protection of photosynthesis. *Trends Plant Sci.* **1996**, *1*, 21–26. [CrossRef]

19. Eskling, M.; Emanuelsson, A.; Akerlund, H.-E. Enzymes and mechanisms for violaxanthin-zeaxanthin conversion. In *Regulation of Photosynthesis*; Aro, E.-M., Anderson, B., Eds.; Kluwer Academic Publishers: Dordrecht, The Netherlands, 2001; Volume 100, pp. 806–816.

20. Li, X.P.; Muller-Moule, P.; Gilmore, A.M.; Niyogi, K.K. PsbS-dependent enhancement of feedback de-excitation protects photosystem II from photoinhibition. *Proc. Natl. Acad. Sci. USA* **2002**, *99*, 15222–15227. [CrossRef]

21. Niyogi, K.K.; Li, X.-P.; Rosenberg, V.; Jung, H.-S. Is PsbS the site of non-photochemical quenching in photosynthesis? *J. Exp. Bot.* **2004**, *56*, 375–382. [CrossRef]

22. Takizawa, K.; Kanazawa, A.; Cruz, J.A.; Kramer, D.M. In vivo thylakoid proton motive force. Quantitative non-invasive probes show the relative lumen pH-induced regulatory responses of antenna and electron transfer. *Biochim. Biophys. Acta* **2007**, *1767*, 1233–1244. [CrossRef]

23. Tikhonov, A.N. The cytochrome b6f complex at the crossroad of photosynthetic electron transport pathways. *Plant Physiol. Biochem.* **2014**, *81*, 163–183. [CrossRef]

24. Cruz, J.A.; Avenson, T.J.; Takizawa, K.; Kramer, D.M. The contribution of cyclic electron flux (CEF1) to formation of proton motive force (pmf). In Proceedings of the 13th International Congress of Photosynthesis: Fundamental Aspects to Global Perspectives, Lawrence, KA, USA, 29 August–3 September 2005; pp. 1033–1035.

25. Kanazawa, A.; Kramer, D.M. In vivo modulation of nonphotochemical exciton quenching (NPQ) by regulation of the chloroplast ATP synthase. *Proc. Natl. Acad. Sci. USA* **2002**, *99*, 12789–12794. [CrossRef]

26. Avenson, T.; Cruz, J.; Kramer, D. Modulation of CF_O-CF_1 ATP synthase conductivity and proton motive force (pmf) partitioning regulate light capture. In Proceedings of the 13th International Congress of Photosynthesis: Fundamental Aspects to Global Perspectives, Lawrence, KA, USA, 29 August–3 September 2005; pp. 575–577.

27. Avenson, T.; Cruz, J.A.; Kramer, D. Modulation of energy dependent quenching of excitons (q_E) in antenna of higher plants. *Proc. Natl. Acad. Sci. USA* **2004**, *101*, 5530–5535. [CrossRef] [PubMed]

28. Cruz, J.A.; Sacksteder, C.A.; Kanazawa, A.; Kramer, D.M. Contribution of Electric Field ($\Delta\psi$) to Steady-State Transthylakoid Proton Motive Force (pmf) in Vitro and in Vivo. Control of pmf Parsing into $\Delta\psi$ and ΔpH by Ionic Strength. *Biochemistry* **2001**, *40*, 1226–1237. [CrossRef] [PubMed]

29. Kramer, D.M.; Cruz, J.A.; Kanazawa, A. Balancing the central roles of the thylakoid proton gradient. *Trends Plant Sci.* **2003**, *8*, 27–32. [CrossRef]

30. Zhang, R.; Cruz, J.A.; Kramer, D.M.; Magallanes-Lundback, M.; DellaPenna, D.; Sharkey, T.D. Heat stress reduces the pH component of the transthylakoid proton motive force in light-adapted intact tobacco leaves. *Plant Cell Environ.* **2009**, *32*, 1538–1547. [CrossRef] [PubMed]

31. Murchie, E.H.; Niyogi, K.K. Manipulation of photoprotection to improve plant photosynthesis. *Plant Physiol.* **2011**, *155*, 86–92. [CrossRef]

32. Li, X.P.; Gilmore, A.M.; Caffarri, S.; Bassi, R.; Golan, T.; Kramer, D.; Niyogi, K.K. Regulation of photosynthetic light harvesting involves intrathylakoid lumen pH sensing by the PsbS protein. *J. Biol. Chem.* **2004**, *279*, 22866–22874. [CrossRef]

33. Scheibe, R. Malate valves to balance cellular energy supply. *Physiol. Plant* **2004**, *120*, 21–26. [CrossRef]

34. Asada, K. The water-water cycle in chloroplasts: Scavenging of active oxygens and dissipation of excess photons. *Annu. Rev. Plant Biol.* **1999**, *50*, 601–639. [CrossRef]

35. Alric, J.; Johnson, X. Alternative electron transport pathways in photosynthesis: A confluence of regulation. *Curr. Opin. Plant Biol.* **2017**, *37*, 78–86. [CrossRef]

36. Strand, D.D.; Fisher, N.; Kramer, D.M. *Distinct Energetics and Regulatory Functions of the Two Major Cyclic Electron Flow Pathways in Chloroplasts*; Caister Academic Press: Poole, UK, 2016.

37. Ishikawa, N.; Endo, T.; Sato, F. Electron transport activities of Arabidopsis thaliana mutants with impaired chloroplastic NAD(P)H dehydrogenase. *J. Plant Res.* **2008**, *121*, 521–526. [CrossRef]

38. Munekage, Y.; Hojo, M.; Meurer, J.; Endo, T.; Tasaka, M.; Shikanai, T. *PGR5* Is Involved in Cyclic Electron Flow around Photosystem I and Is Essential for Photoprotection in *Arabidopsis*. *Cell* **2002**, *110*, 361–371. [CrossRef]

39. DalCorso, G.; Pesaresi, P.; Masiero, S.; Aseeva, E.; Schünemann, D.; Finazzi, G.; Joliot, P.; Barbato, R.; Leister, D. A Complex Containing PGRL1 and PGR5 Is Involved in the Switch between Linear and Cyclic Electron Flow in Arabidopsis. *Cell* **2008**, *132*, 273–285. [CrossRef] [PubMed]

40. Hertle, A.P.; Blunder, T.; Wunder, T.; Pesaresi, P.; Pribil, M.; Armbruster, U.; Leister, D. PGRL1 Is the Elusive Ferredoxin-Plastoquinone Reductase in Photosynthetic Cyclic Electron Flow. *Mol. Cell* **2013**, *49*, 511–523. [CrossRef] [PubMed]

41. Nandha, B.; Finazzi, G.; Joliot, P.; Hald, S.; Johnson, G.N. The role of PGR5 in the redox poising of photosynthetic electron transport. *Biochim. Biophys. Acta Bioenerg.* **2007**, *1767*, 1252–1259. [CrossRef] [PubMed]

42. Joliot, P.; Johnson, G.N. Regulation of cyclic and linear electron flow in higher plants. *Proc. Natl. Acad. Sci. USA* **2011**, *108*, 13317–13322. [CrossRef] [PubMed]

43. Fisher, N.; Kramer, D.M. Non-photochemical reduction of thylakoid photosynthetic redox carriers in vitro: Relevance to cyclic electron flow around photosystem I? *Biochim. Biophys. Acta Bioenerg.* **2014**, *1837*, 1944–1954. [CrossRef]

44. Mosebach, L.; Heilmann, C.; Mutoh, R.; Gäbelein, P.; Steinbeck, J.; Happe, T.; Ikegami, T.; Hanke, G.; Kurisu, G.; Hippler, M. Association of Ferredoxin:NADP+ oxidoreductase with the photosynthetic apparatus modulates electron transfer in Chlamydomonas reinhardtii. *Photosynth. Res.* **2017**, *134*, 291–306. [CrossRef]

45. Burrows, P.A.; Sazanov, L.A.; Svab, Z.; Maliga, P.; Nixon, P.J. Identification of a functional respiratory complex in chloroplasts through analysis of tobacco mutants containing disrupted plastid ndh genes. *EMBO J.* **1998**, *17*, 868–876. [CrossRef] [PubMed]

46. Shikanai, T. Chloroplast NDH: A different enzyme with a structure similar to that of respiratory NADH dehydrogenase. *Biochim. Biophys. Acta Bioenerg.* **2016**, *1857*, 1015–1022. [CrossRef]

47. Strand, D.D.; Fisher, N.; Kramer, D.M. The higher plant plastid NAD(P)H dehydrogenase-like complex (NDH) is a high efficiency proton pump that increases ATP production by cyclic electron flow. *J. Biol. Chem.* **2017**, *292*, 11850–11860. [CrossRef]

48. Laughlin, T.G.; Bayne, A.N.; Trempe, J.-F.; Savage, D.F.; Davies, K.M. Structure of the complex I-like molecule NDH of oxygenic photosynthesis. *Nature* **2019**, *566*, 411–414. [CrossRef] [PubMed]

49. Joliot, P.; Joliot, A. Cyclic electron transfer in plant leaf. *Proc. Natl. Acad. Sci. USA* **2002**, *99*, 10209–10214. [CrossRef] [PubMed]

50. Stroebel, D.; Choquet, Y.; Popot, J.-L.; Picot, D. An atypical haem in the cytochrome b6f complex. *Nature* **2003**, *426*, 413–418. [CrossRef] [PubMed]

51. Alric, J.; Pierre, Y.; Picot, D.; Lavergne, J.; Rappaport, F. Spectral and redox characterization of the heme c_i of the cytochrome b_6f complex. *Proc. Natl. Acad. Sci. USA* **2005**, *102*, 15860–15865. [CrossRef]

52. Strand, D.D.; D'Andrea, L.; Bock, R. The plastid NAD(P)H dehydrogenase-like complex: Structure, function and evolutionary dynamics. *Biochem. J.* **2019**, *476*, 2743–2756. [CrossRef]

53. Joliot, P.; Béal, D.; Joliot, A. Cyclic electron flow under saturating excitation of dark-adapted Arabidopsis leaves. *Biochim. Biophys. Acta Bioenerg.* **2004**, *1656*, 166–176. [CrossRef]

54. Kohzuma, K.; Cruz, J.A.; Akashi, K.; Hoshiyasu, S.; Munekage, Y.N.; Yokota, A.; Kramer, D.M. The long-term responses of the photosynthetic proton circuit to drought. *Plant Cell Environ.* **2009**, *32*, 209–219. [CrossRef]

55. Kubicki, A.; Funk, E.; Westhoff, P.; Steinmüller, K. Differential expression of plastome-encoded ndh genes in mesophyll and bundle-sheath chloroplasts of the C4 plant Sorghum bicolor indicates that the complex I-homologous NAD(P)H-plastoquinone oxidoreductase is involved in cyclic electron transport. *Planta* **1996**, *199*, 276–281. [CrossRef]

56. Takabayashi, A.; Kishine, M.; Asada, K.; Endo, T.; Sato, F. Differential use of two cyclic electron flows around photosystem I for driving CO_2-concentration mechanism in C_4 photosynthesis. *Proc. Natl. Acad. Sci. USA* **2005**, *102*, 16898–16903. [CrossRef]

57. Lucker, B.; Kramer, D.M. Regulation of cyclic electron flow in Chlamydomonas reinhardtii under fluctuating carbon availability. *Photosynth. Res.* **2013**, *117*, 449–459. [CrossRef]

58. Scheibe, R. Maintaining homeostasis by controlled alternatives for energy distribution in plant cells under changing conditions of supply and demand. *Photosynth. Res.* **2019**, *139*, 81–91. [CrossRef] [PubMed]

59. Selinski, J.; Scheibe, R. Malate valves: Old shuttles with new perspectives. *Plant Biol.* **2019**, *21*, 21–30. [CrossRef]

60. Fickenscher, K.; Scheibe, R. Limited proteolysis of inactive tetrameric chloroplast NADP-Malate dehydrogenase produces active dimers. *Arch. Biochem. Biophys.* **1988**, *260*, 771–779. [CrossRef]

61. Ocheretina, O.; Harnecker, J.; Rother, T.; Schmid, R.; Scheibe, R. Effects of N-terminal truncations upon chloroplast NADP-malate dehydrogenases from pea and spinach. *Biochim. Biophys. Acta BBA Protein Struct. Mol. Enzymol.* **1993**, *1163*, 10–16. [CrossRef]

62. Krieger-Liszkay, A.; Feilke, K. The Dual Role of the Plastid Terminal Oxidase PTOX: Between a Protective and a Pro-oxidant Function. *Front. Plant Sci.* **2016**, *6*, 1147. [CrossRef] [PubMed]

63. Allahverdiyeva, Y.; Suorsa, M.; Tikkanen, M.; Aro, E.-M. Photoprotection of photosystems in fluctuating light intensities. *J. Exp. Bot.* **2014**, *66*, 2427–2436. [CrossRef] [PubMed]

64. Ruuska, S.A.; Badger, M.R.; Andrews, T.J.; von Caemmerer, S. Photosynthetic electron sinks in transgenic tobacco with reduced amounts of Rubisco: Little evidence for significant Mehler reaction. *J. Exp. Bot.* **2000**, *51*, 357–368. [CrossRef]

65. Heber, U. Irrungen, Wirrungen? The Mehler reaction in relation to cyclic electron transport in C3 plants. *Photosynth. Res.* **2002**, *73*, 223–231. [CrossRef]

66. Driever, S.M.; Baker, N.R. The water–water cycle in leaves is not a major alternative electron sink for dissipation of excess excitation energy when CO_2 assimilation is restricted. *Plant Cell Environ.* **2011**, *34*, 837–846. [CrossRef]

67. Noctor, G.; Foyer, C. Review article. A re-evaluation of the ATP:NADPH budget during C_3 photosynthesis: A contribution from nitrate assimilation and its associated respiratory activity? *J. Exp. Bot.* **1998**, *49*, 1895–1908. [CrossRef]

68. Edwards, G.E.; Walker, D.A. *C_3, C_4: Mechanisms, and Cellular and Environmental Regulation, of Photosynthesis*; Blackwell Scientific Publications: Oxford, UK, 1983.

69. Walker, B.J.; VanLoocke, A.; Bernacchi, C.J.; Ort, D.R. The costs of photorespiration to food production now and in the future. *Annu. Rev. Plant Biol.* **2016**, *67*, 107–129. [CrossRef] [PubMed]

70. Farquhar, G.D.; von Caemmerer, S.; Berry, J.A. A biochemical model of photosynthetic CO_2 assimilation in leaves of C_3 species. *Planta* **1980**, *149*, 78–90. [CrossRef] [PubMed]

71. von Caemmerer, S.; Farquhar, G.D. Some relationships between the biochemistry of photosynthesis and the gas exchange of leaves. *Planta* **1981**, *153*, 376–387. [CrossRef] [PubMed]

72. Sharkey, T.D. Estimating the rate of photorespiration in leaves. *Physiol. Plant* **1988**, *73*, 147–152. [CrossRef]

73. Walker, B.J.; Strand, D.D.; Kramer, D.M.; Cousins, A.B. The response of cyclic electron flow around photosystem I to changes in photorespiration and nitrate assimilation. *Plant Phys.* **2014**, *165*, 453–462. [CrossRef]

74. Masclaux-Daubresse, C.; Daniel-Vedele, F.; Dechorgnat, J.; Chardon, F.; Gaufichon, L.; Suzuki, A. Nitrogen uptake, assimilation and remobilization in plants: Challenges for sustainable and productive agriculture. *Ann. Bot.* **2010**, *105*, 1141–1157. [CrossRef]

75. Reed, A.J.; Canvin, D.T. Light and Dark Controls of Nitrate Reduction in Wheat (*Triticum aestivum* L.) Protoplasts. *Plant Phys.* **1982**, *69*, 508–513. [CrossRef]

76. Selinski, J.; Scheibe, R. Lack of malate valve capacities lead to improved N-assimilation and growth in transgenic A. thaliana plants. *Plant Signal. Behav.* **2014**, *9*, e29057. [CrossRef]

77. Rachmilevitch, S.; Cousins, A.B.; Bloom, A.J. Nitrate assimilation in plant shoots depends on photorespiration. *Proc. Natl. Acad. Sci. USA* **2004**, *101*, 11506–11510. [CrossRef]

78. Bloom, A.J.; Burger, M.; Asensio, J.S.R.; Cousins, A.B. Carbon Dioxide Enrichment Inhibits Nitrate Assimilation in Wheat and Arabidopsis. *Science* **2010**, *328*, 899–903. [CrossRef]

79. Ohlrogge, J.; Browse, J. Lipid biosynthesis. *Plant Cell* **1995**, *7*, 957–970. [CrossRef] [PubMed]

80. Troncoso-Ponce, M.A.; Cao, X.; Yang, Z.; Ohlrogge, J.B. Lipid turnover during senescence. *Plant Sci.* **2013**, *205–206*, 13–19. [CrossRef] [PubMed]

81. Bao, X.; Focke, M.; Pollard, M.; Ohlrogge, J. Understanding in vivo carbon precursor supply for fatty acid synthesis in leaf tissue. *Plant J.* **2000**, *22*, 39–50. [CrossRef] [PubMed]

82. Pollard, M.; Ohlrogge, J. Testing Models of Fatty Acid Transfer and Lipid Synthesis in Spinach Leaf Using in Vivo Oxygen-18 Labeling. *Plant Phys.* **1999**, *121*, 1217–1226. [CrossRef]

83. Bonaventure, G.; Bao, X.; Ohlrogge, J.; Pollard, M. Metabolic responses to the reduction in palmitate caused by disruption of the FATB gene in Arabidopsis. *Plant Phys.* **2004**, *135*, 1269–1279. [CrossRef]

84. Yang, Z.; Ohlrogge, J.B. Turnover of Fatty Acids during Natural Senescence of Arabidopsis, *Brachypodium*, and Switchgrass and in Arabidopsis β-Oxidation Mutants. *Plant Phys.* **2009**, *150*, 1981–1989. [CrossRef]

85. Geigenberger, P.; Kolbe, A.; Tiessen, A. Redox regulation of carbon storage and partitioning in response to light and sugars. *J. Exp. Bot.* **2005**, *56*, 1469–1479. [CrossRef]

86. Shastri, A.A.; Morgan, J.A. Flux Balance Analysis of Photoautotrophic Metabolism. *Biotechnol. Prog.* **2005**, *21*, 1617–1626. [CrossRef]

87. Von Ballmoos, C.; Cook, G.M.; Dimroth, P. Unique Rotary ATP Synthase and Its Biological Diversity. *Annu. Rev. Biophys.* **2008**, *37*, 43–64. [CrossRef]

88. Von Ballmoos, C.; Wiedenmann, A.; Dimroth, P. Essentials for ATP Synthesis by F1F0 ATP Synthases. *Annu. Rev. Biochem.* **2009**, *78*, 649–672. [CrossRef]

89. Junge, W.; Sielaff, H.; Engelbrecht, S. Torque generation and elastic power transmission in the rotary FOF1-ATPase. *Nature* **2009**, *459*, 364–370. [CrossRef] [PubMed]

90. Ferguson, S.J. ATP synthase: From sequence to ring size to the P/O ratio. *Proc. Natl. Acad. Sci. USA* **2010**, *107*, 16755–16756. [CrossRef] [PubMed]

91. Silverstein, T. The mitochondrial phosphate-to-oxygen ratio is not an integer. *Biochem. Mol. Biol. Educ.* **2005**, *33*, 416–417. [CrossRef] [PubMed]

92. Watt, I.N.; Montgomery, M.G.; Runswick, M.J.; Leslie, A.G.W.; Walker, J.E. Bioenergetic cost of making an adenosine triphosphate molecule in animal mitochondria. *Proc. Natl. Acad. Sci. USA* **2010**, *107*, 16823–16827. [CrossRef] [PubMed]

93. Ferguson, S.J.; Sorgato, M.C. Proton Electrochemical Gradients and Energy-Transduction Processes. *Annu. Rev. Biochem.* **1982**, *51*, 185–217. [CrossRef]

94. Nicholls, D.G. Mitochondrial ion circuits. *Essays Biochem.* **2010**, *47*, 25–35.

95. Bhatla, S.C.; Lal, M.A. ATP Synthesis. In *Plant Physiology, Development and Metabolism*; Springer: Singapore, 2018; pp. 315–337.

96. Lambers, H.; Oliveira, R.S. Photosynthesis, Respiration, and Long-Distance Transport: Respiration. In *Plant Physiological Ecology*; Springer International Publishing: Cham, Switzerland, 2019; pp. 115–172.

97. Haupt-Herting, S.; Klug, K.; Fock, H.P. A new approach to measure gross CO_2 fluxes in leaves. Gross CO_2 sssimilation, photorespiration, and mitochondrial respiration in the light in tomato under drought stress. *Plant Phys.* **2001**, *126*, 388–396. [CrossRef]

98. Villar, R.; Held, A.A.; Merino, J. Comparison of methods to estimate dark respiration in the light in leaves of two woody species. *Plant Phys.* **1994**, *105*, 167–172. [CrossRef]

99. Kromer, S. Respiration During Photosynthesis. *Annu Rev. Plant Physiol Plant Mol. Biol* **1995**, *46*, 45–70. [CrossRef]

100. Kunji, E.R.S.; Aleksandrova, A.; King, M.S.; Majd, H.; Ashton, V.L.; Cerson, E.; Springett, R.; Kibalchenko, M.; Tavoulari, S.; Crichton, P.G.; et al. The transport mechanism of the mitochondrial ADP/ATP carrier. *Biochim. Biophys. Acta BBAMol. Cell Res.* **2016**, *1863*, 2379–2393. [CrossRef]

101. Palmieri, F.; Monné, M. Discoveries, metabolic roles and diseases of mitochondrial carriers: A review. *Biochim. Biophys. Acta BBAMol. Cell Res.* **2016**, *1863*, 2362–2378. [CrossRef]

102. Heldt, H.W. Adenine nucleotide translocation in spinach chloroplasts. *FEBS Lett.* **1969**, *5*, 11–14. [CrossRef]
103. Schunemann, D.; Borchert, S.; Flugge, U.I.; Heldt, H.W. ADP/ATP Translocator from Pea Root Plastids (Comparison with Translocators from Spinach Chloroplasts and Pea Leaf Mitochondria). *Plant Phys.* **1993**, *103*, 131–137. [CrossRef] [PubMed]
104. Huang, W.; Yang, Y.-J.; Hu, H.; Zhang, S.-B. Different roles of cyclic electron flow around photosystem I under sub-saturating and saturating light intensities in tobacco leaves. *Front. Plant Sci.* **2015**, *6*. [CrossRef] [PubMed]
105. Cheung, C.Y.M.; Ratcliffe, R.G.; Sweetlove, L.J. A Method of Accounting for Enzyme Costs in Flux Balance Analysis Reveals Alternative Pathways and Metabolite Stores in an Illuminated Arabidopsis Leaf. *Plant Phys.* **2015**, *169*, 1671–1682. [CrossRef] [PubMed]
106. Shameer, S.; Ratcliffe, R.G.; Sweetlove, L.J. Leaf Energy Balance Requires Mitochondrial Respiration and Export of Chloroplast NADPH in the Light. *Plant Phys.* **2019**, *180*, 1947–1961. [CrossRef] [PubMed]
107. Munekage, Y.; Hashimoto, M.; Miyake, C.; Tomizawa, K.-I.; Endo, T.; Tasaka, M.; Shikanai, T. Cyclic electron flow around photosystem I is essential for photosynthesis. *Nature* **2004**, *429*, 579–582. [CrossRef]
108. Strand, D.D.; Fisher, N.; Davis, G.A.; Kramer, D.M. Redox regulation of the antimycin A sensitive pathway of cyclic electron flow around photosystem I in higher plant thylakoids. *Biochim. Biophys. Acta Bioenerg.* **2016**, *1857*, 1–6. [CrossRef]
109. Michelet, L.; Zaffagnini, M.; Morisse, S.; Sparla, F.; Pérez-Pérez, M.E.; Francia, F.; Danon, A.; Marchand, C.; Fermani, S.; Trost, P.; et al. Redox regulation of the Calvin–Benson cycle: Something old, something new. *Front. Plant Sci.* **2013**, *4*. [CrossRef]
110. Bloom, A.J.; Caldwell, R.M.; Finazzo, J.; Warner, R.L.; Weissbart, J. Oxygen and Carbon Dioxide Fluxes from Barley Shoots Depend on Nitrate Assimilation. *Plant Physiol.* **1989**, *91*, 352–356. [CrossRef]
111. Bloom, A.J.; Smart, D.R.; Nguyen, D.T.; Searles, P.S. Nitrogen assimilation and growth of wheat under elevated carbon dioxide. *Proc. Natl. Acad. Sci. USA* **2002**, *99*, 1730–1735. [CrossRef]
112. Bloom, A.J.; Burger, M.; Kimball, B.A.; Pinter, P.J. Nitrate assimilation is inhibited by elevated CO_2 in field-grown wheat. *Nat. Clim. Chang.* **2014**, *4*, 477–480. [CrossRef]
113. Hebbelmann, I.; Selinski, J.; Wehmeyer, C.; Goss, T.; Voss, I.; Mulo, P.; Kangasjärvi, S.; Aro, E.-M.; Oelze, M.-L.; Dietz, K.-J.; et al. Multiple strategies to prevent oxidative stress in Arabidopsis plants lacking the malate valve enzyme NADP-malate dehydrogenase. *J. Exp. Bot.* **2011**, *63*, 1445–1459. [CrossRef] [PubMed]
114. Buchanan, B.B. The Ferredoxin/Thioredoxin System: A Key Element in the Regulatory Function of Light in Photosynthesis. *Bioscience* **1984**, *34*, 378–383. [CrossRef] [PubMed]
115. Buchanan, B.B.; Balmer, Y. REDOX REGULATION: A Broadening Horizon. *Annu. Rev. Plant Biol.* **2005**, *56*, 187–220. [CrossRef] [PubMed]
116. Scheibe, R. Thioredoxinm in pea chloroplasts: Concentration and redox state under light and dark conditions. *FEBS Lett.* **1981**, *133*, 301–304. [CrossRef]
117. Scheibe, R.; Stitt, M. Comparison of NADP-malate dehydrogenase activation, Q_A reduction and O_2 evolution in spinach leaves. *Plant Physiol. Biochem.* **1988**, *26*, 473–481.
118. Joliot, P.; Joliot, A. Cyclic electron flow in C3 plants. *Biochim. Biophys. Acta Bioenerg.* **2006**, *1757*, 362–368. [CrossRef]
119. Becker, B.; Holtgrefe, S.; Jung, S.; Wunrau, C.; Kandlbinder, A.; Baier, M.; Dietz, K.-J.; Backhausen, J.E.; Scheibe, R. Influence of the photoperiod on redox regulation and stress responses in Arabidopsis thaliana L. (Heynh.) plants under long- and short-day conditions. *Planta* **2006**, *224*, 380–393. [CrossRef]
120. Bauwe, H.; Hagemann, M.; Fernie, A.R. Photorespiration: Players, partners and origin. *Trends Plant Sci.* **2010**, *15*, 330–336. [CrossRef]
121. Tomaz, T.; Bagard, M.; Pracharoenwattana, I.; Linden, P.; Lee, C.P.; Carroll, A.J.; Stroher, E.; Smith, S.M.; Gardestrom, P.; Millar, A.H. Mitochondrial malate dehydrogenase lowers leaf respiration and alters photorespiration and plant growth in Arabidopsis. *Plant Phys.* **2010**, *154*, 1143–1157. [CrossRef] [PubMed]
122. Lindén, P.; Keech, O.; Stenlund, H.; Gardeström, P.; Moritz, T. Reduced mitochondrial malate dehydrogenase activity has a strong effect on photorespiratory metabolism as revealed by 13C labelling. *J. Exp. Bot.* **2016**, *67*, 3123–3135. [CrossRef] [PubMed]

123. Cousins, A.B.; Pracharoenwattana, I.; Zhou, W.; Smith, S.M.; Badger, M.R. Peroxisomal malate dehydrogenase is not essential for photorespiration in Arabidopsis but its absence causes an increase in the stoichiometry of photorespiratory CO_2 release. *Plant Phys.* **2008**, *148*, 786–795. [CrossRef]

124. Cousins, A.B.; Walker, B.J.; Pracharoenwattana, I.; Smith, S.M.; Badger, M.R. Peroxisomal hydroxypyruvate reductase is not essential for photorespiration in arabidopsis but its absence causes an increase in the stoichiometry of photorespiratory CO_2 release. *Photosynth. Res.* **2011**, *108*, 91–100. [CrossRef]

125. Anderson, J.M.; Chow, W.S.; Park, Y.-I. The grand design of photosynthesis: Acclimation of the photosynthetic apparatus to environmental cues. *Photosynth. Res.* **1995**, *46*, 129–139. [CrossRef] [PubMed]

126. Fujita, Y. A study on the dynamic features of photosystem stoichiometry: Accomplishments and problems for future studies. *Photosynth. Res.* **1997**, *53*, 83–93. [CrossRef]

127. Melis, A.; Murakami, A.; Nemson, J.A.; Aizawa, K.; Ohki, K.; Fujita, Y. Chromatic regulation inChlamydomonas reinhardtii alters photosystem stoichiometry and improves the quantum efficiency of photosynthesis. *Photosynth. Res.* **1996**, *47*, 253–265. [CrossRef] [PubMed]

128. Pfannschmidt, T.; Allen, J.F.; Oelmüller, R. Principles of redox control in photosynthesis gene expression. *Physiol. Plant* **2001**, *112*, 1–9. [CrossRef]

129. Dietzel, L.; Bräutigam, K.; Pfannschmidt, T. Photosynthetic acclimation: State transitions and adjustment of photosystem stoichiometry—Functional relationships between short-term and long-term light quality acclimation in plants. *FEBS J.* **2008**, *275*, 1080–1088. [CrossRef]

130. Allen, J.F. Protein phosphorylation in regulation of photosynthesis. *Biochim. Biophys. Acta Bioenerg.* **1992**, *1098*, 275–335. [CrossRef]

131. Gupta, P.; Duplessis, S.; White, H.; Karnosky, D.F.; Martin, F.; Podila, G.K. Gene expression patterns of trembling aspen trees following long-term exposure to interacting elevated CO_2 and tropospheric O_3. *New Phytol.* **2005**, *167*, 129–142. [CrossRef] [PubMed]

132. Ainsworth, E.A.; Rogers, A.; Vodkin, L.O.; Walter, A.; Schurr, U. The Effects of Elevated CO_2 Concentration on Soybean Gene Expression. An Analysis of Growing and Mature Leaves. *Plant Phys.* **2006**, *142*, 135–147. [CrossRef] [PubMed]

133. Von Caemmerer, S. *Biochemical Models of Leaf Photosynthesis*; CSIRO: Collingwood, Australia, 2000; Volume 2.

134. Von Caemmerer, S. Steady-state models of photosynthesis. *Plant Cell Environ.* **2013**, *36*, 1617–1630. [CrossRef] [PubMed]

135. Walker, B.J.; Ort, D.R. Improved method for measuring the apparent CO_2 photocompensation point resolves the impact of multiple internal conductances to CO_2 to net gas exchange. *Plant Cell Environ.* **2015**, *38*, 2462–2474. [CrossRef] [PubMed]

136. Laisk, A. Kinetics of photosynthesis and photorespiration in C_3 plants. *Nauka Mosc.* **1977**. (In Russian)

137. Laisk, A.; Loreto, F. Determining photosynthetic parameters from leaf CO_2 exchange and chlorophyll fluorescence. *Plant Physiol.* **1996**, *110*, 903–912. [CrossRef] [PubMed]

138. Bernacchi, C.J.; Portis, A.R.; Nakano, H.; von Caemmerer, S.; Long, S.P. Temperature response of mesophyll conductance. Implications for the determination of Rubisco enzyme kinetics and for limitations to photosynthesis in vivo. *Plant Phys.* **2002**, *130*, 1992–1998. [CrossRef]

139. Tazoe, Y.; von Caemmerer, S.; Estavillo, G.M.; Evans, J.R. Using tunable diode laser spectroscopy to measure carbon isotope discrimination and mesophyll conductance to CO_2 diffusion dynamically at different CO_2 concentrations. *Plant Cell Environ.* **2011**, *34*, 580–591. [CrossRef]

140. Gilbert, M.E.; Pou, A.; Zwieniecki, M.A.; Holbrook, N.M. On measuring the response of mesophyll conductance to carbon dioxide with the variable J method. *J. Exp. Bot.* **2012**, *63*, 413–425. [CrossRef]

141. Tholen, D.; Ethier, G.; Genty, B.; Pepin, S.; Zhu, X.-G. Variable mesophyll conductance revisited: Theoretical background and experimental implications. *Plant Cell Environ.* **2012**, *35*, 2087–2103. [CrossRef]

142. Evans, J.R.; von Caemmerer, S. Temperature response of carbon isotope discrimination and mesophyll conductance in tobacco. *Plant Cell Environ.* **2013**, *36*, 745–756. [CrossRef]

143. Jahan, E.; Amthor, J.S.; Farquhar, G.D.; Trethowan, R.; Barbour, M.M. Variation in mesophyll conductance among Australian wheat genotypes. *Funct. Plant Biol.* **2014**, *41*, 568–580. [CrossRef]

144. von Caemmerer, S.; Evans, J.R. Temperature responses of mesophyll conductance differ greatly between species. *Plant Cell Environ.* **2014**, *38*, 629–637. [CrossRef] [PubMed]
145. Loreto, F.; Harley, P.C.; Di Marco, G.; Sharkey, T.D. Estimation of mesophyll conductance to CO_2 flux by three different methods. *Plant Phys.* **1992**, *98*, 1437–1443. [CrossRef] [PubMed]

MDPI

St. Alban-Anlage 66

4052 Basel

Switzerland

Tel. +41 61 683 77 34

Fax +41 61 302 89 18

www.mdpi.com

Plants Editorial Office

E mail: plants@mdpi.com

www.mdpi.com/journal/plants

9 783036 511047